# Passage of High-Energy Particles through Matter

ranslation Series

# ?assage of Iigh-Energy Particles hrough Matter

A. N. Kalinovskii
N. V. Mokhov
Yu. P. Nikitin

Translated by
S. J. Amoretty

**Library of Congress Cataloging-in-Publication Data**

Kalinovskiĭ, A. N. (Aleksandr Nikolaevich)
[Prokhozhdenie chastits vysokikh energiĭ cherez veshchestvo. English]
Passage of high-energy particles through matter/A. N. Kalinovskii, N. V. Mokhov, Yu. P. Nikitin; translated by S. J. Amoretty
p. cm. --

Translation of: Prokhozhdenie chastits vysokikh energiĭ cherez veshchestvo.
Bibliography: p.
Includes index.
ISBN 978-0-88318-618-3
1. Transport theory. 2. Particles (Nuclear physics). 3. Cascade shower. 4. Electromagnetic interactions. 5. Hadron interactions. I. Mokhov, N. V. (Nikolaĭ Veniaminovich) II. Nikitin, IUriĭ Petrovich. III. Title. IV. Series: AIP translation series.
QC793.3.T7K3513 1989 530.1'38--dc20 89-6949
CIP

# Contents



## Chapter 1. Electromagnetic cascade



## Chapter 2. Electromagnetic interaction of particles with matter. Decay of unstable particles



## Chapter 3. Description of multiple production of hadrons at high energies

## Chapter 4. Electron–photon showers



## Chapter 5. Solution of equations for nucleon–meson cascades



## Chapter 6. Study of electromagnetic cascades by the Monte Carlo method

## Chapter 7. High-energy muons



## Chapter 8. Applications of the methods and results of experimental studies of electromagnetic cascades

# Preface

Although recent development of accelerator technology, experimental techniques, and theory has provided answers to many questions in high-energy physics and closely related high-energy radiation physics, many problems still require further study. This book deals with one of these problems: nuclear and atomic interactions associated with the passage of high-energy particles through matter and the solution of the corresponding boundary-value transport problem.

The interaction of fast particles with matter is a topic of current importance primarily in applied physics: simulation of accelerator experiments and cosmic-ray experiments, radiation shielding of accelerators and space vehicles, study of the effect of high-energy elementary-particle and nuclear beams on the environment, electronuclear method based on the use of breeder reactors, cancer therapy, and other applied fields. There are currently no monographs or reviews in the Russian language, dealing with calculation methods and the physical aspects of problems, which could be used as a base to write new computer programs or as a source of available programs.

In this book we focus attention principally on the interaction of hadrons at $E > 10$ GeV* and on the methods of calculating electromagnetic cascades in matter. In contrast with monographs dealing with classical theory of low-energy radiation transport, a considerable part of this book deals with the description of physical processes and relevant theory, since at high energies the only way in which a closed solution of the transport problem can now be obtained is by using models. The material contained in this book has been selected and presented in such a way that the reader can, where necessary, use an algorithm independently, without referring to additional literature.

The characteristic features of high-energy electromagnetic and hadron cascades that develop in condensed matter are analyzed in detail using the calculation methods and computer programs described in this book. Specific examples of the application of these methods in today's physics experiments are given.

Since some of the topics discussed in this book are still being developed, some subjectivity in the emphasis given to the topics was unavoidable, although considerable effort was made to minimize it. In selecting the contents of this book we have drawn upon the considerable experience gained from many years of research using the largest accelerators in the USSR and abroad. This book is intended for scientific workers and new researchers spe-

*The interaction of hadrons at lower energies has been analyzed in detail in a book by V. S. Barashenkov and V. D. Toneev, entitled *Interaction of High-Energy Particles and Atomic Nuclei with Nuclei* (Atomizdat, Moscow, 1972).

cializing in high-energy physics and radiation physics based on high-energy accelerators, as well as in cosmic-ray physics.

We extend our deep appreciation to V. S. Barashenkov for useful comments.

We welcome any comments or suggestions concerning this book.

# Chapter 1
# Electromagnetic cascade

## 1. General remarks

The distinguishing feature of the interaction of particles with nuclei at high energies is its multiple nature. In this process the reaction products contain nearly the entire spectrum of known elementary particles. Having sufficient energy, these particles in turn interact with the nuclei of the medium, producing new generations of hadrons, leptons, and fragments which give rise to an internuclear or electromagnetic cascade. In principle, a cascade can be described completely only if the information about all particles and their strong, electromagnetic, and weak interactions over a broad energy range is available.

Cascade calculations are important in the study of the passage of high-energy particles through matter, in the design and design modifications of shielding for accelerators and space vehicles, in determining the radiation environment of the active accelerators and the radiation background in experiments with particle beams, in calculating particle channels, extended targets, etc. The use of superconducting elements and hence strong magnetic fields in accelerators has made it necessary to calculate the radiation-induced heating of superconducting windings, the internuclear cascades in the presence of a magnetic field, and other effects. A valid accelerator-based physics experiment must start with a simulation of cascades in the detecting elements of the experimental apparatus.

The particle energy of accelerators increases by approximately an order of magnitude every ten years. The following accelerators are currently in operation. A 76-GeV proton synchrotron at Serpukhov, a proton colliding beam accelerator with an effective energy up to 2000 GeV at CERN, a 500-GeV proton accelerator at FNAL, and a 400-GeV proton supersynchrotron at CERN. An accelerator with colliding proton–antiproton beams with an energy of $2\times 270$ GeV was recently started up at CERN and the first superconducting 1000-GeV proton accelerator was built at FNAL. A construction of a 3000-GeV proton accelerator is under way at Serpukhov. The feasibility of constructing a proton colliding beam accelerator with an energy of $2\times 20\,000$ GeV is now under study.

An effective implementation of all these projects requires, in particular, the development of methods for calculating electromagnetic cascades produced as a result of passage through matter of hadrons with energies of order $10^8$–$10^{12}$ eV. The study of processes with increasingly smaller cross sections and the increasingly more complex designs of experimental apparatus and shielding have sharply increased the demands on the performance and accuracy of the mathematical methods of solving the problem of the transfer of radiation through matter. The collimators, targets, and shielding for accelerators and space vehicles can be upgraded only if effective algorithms are used in calculations over a broad energy range.

In the discussion below, the kinetic energy of particles is called low energy in the energy interval from 0.025 eV to 15 MeV, intermediate energy in the energy interval from 15 MeV to 1 GeV, and high energy in the energy interval above 1 GeV.

In a developing electromagnetic cascade the range of energy variation in an accelerator may reach fourteen orders of magnitude (from $10^{-2}$ to $10^{12}$ eV) in the case of neutrons. In this energy interval there is a large variety of interactions of particles with matter, and at high energies these interactions are far from being fully understood. The lack of complete understanding of these interactions applies particularly to the energy region of primary hadrons, $E_0 > 10$ GeV, to the description of which a substantial part of this book is devoted.

## 2. Interaction of fast particles with the medium

All known processes in nature are ultimately traceable to the interaction of elementary particles. In the present classification there are only two types of elementary particles: *leptons* and *quarks*. It is assumed that there are six leptons: $e$, $\mu$, $\tau$, $\nu_e$, $\nu_\mu$, and $\nu_\tau$ and six quarks: $u$, $d$, $s$, $c$, $b$, and $t$. The quarks and the corresponding antiquarks are the structural elements of *hadrons*—the actually observable particles.

There are four types of interactions between elementary particles. In increasing order of strengths, they are the *gravitational*, the *weak*, the *electromagnetic*, and the *strong interactions*. The gravitational and weak forces act between particles of all species, the electromagnetic forces act only between particles that carry an electric charge and photons, and the strong forces act only between hadrons and the quarks that constitute them.

The interaction occurs by means of the exchange of intermediate particles—the *quanta* of the corresponding fields. The quantum of gravitational interaction is the *graviton*, the quantum of weak interaction is the *intermediate boson* ($W^\pm$, $Z^0$), the quantum of electromagnetic interaction is the *photon*, and the quantum of strong interaction is the *gluon*. With the exception of intermediate bosons, which have a mass of about 80 GeV, the rest of the interaction quanta are massless.

The role of gravitational interaction in the phenomena considered in this book is negligible.

*Electromagnetic interaction* plays an important role in the passage of charged particles and photons through matter. Moving in a medium, a charged particle changes its direction of motion as a result of interaction with the electric fields of atomic nuclei. The direction of motion changes in most cases by only a very small angle. However, because the frequency of these collisions is high [they occur on the average about every $(v/c)^2 \times 10^{-5}\rho^{-1}$ cm, where $v$ is the particle velocity, $c$ is the velocity of light, and $\rho$ is the density of matter], the particle changes the direction of its motion markedly after passing through a thick layer of matter. This process is called *multiple Coulomb scattering*. During its random walk, the charged particle loses energy due to the ionization and excitation of atoms of the material, due to the bremsstrahlung, and due to the direct production of $e^+e^-$ pairs in the field of the nucleus and atomic electrons (the total cross section for the production of $\mu^+\mu^-$ pairs is four orders of magnitude smaller[1]).

The average energy losses due to radiation and direct production of $e^+e^-$ pairs per unit length increase linearly with increasing energy. The ionization losses of relativistic particles, however, depend on the energy only slightly. At electron energies $E \gtrsim 700/(Z+1)$ MeV, the radiation losses dominate over the ionization losses. The electron energy loss due to the direct production of $e^+e^-$ pairs is approximately $\alpha^{-1} = 137$ times lower than the energy loss due to the bremsstrahlung. For all particles other than the electron, the cross section for the direct production of an $e^+e^-$ pair is larger than the cross section for bremsstrahlung, which is inversely proportional to the square of the mass of the emitting particle. In the case of heavy particles, the energy losses due to these processes are detectable only at a very high energy. In the case of muons in iron, for example, the radiation and pair-production losses are comparable to the ionization loss at $E \sim 350$ GeV. If a heavy relativistic particle participates only in the electromagnetic interactions, it can penetrate deeply into the matter: A 1000-GeV muon can traverse, with a probability of 0.1, a distance of 320 m in iron.[2,3]

The interaction of photons with matter differs from the ionization caused by charged particles in that each time a photon interacts it is either absorbed or scattered at a large angle, whereas the ionization involves the loss of only a negligible part of the energy and virtually no scattering at significant angles. At low energy (below the ionization energy) the *Rayleigh scattering* is the most important photon–nucleus-interaction process. With further increase in the photon energy, the following processes become dominant sequentially: *photoelectric effect*—the absorption of a photon by an atom with a subsequent emission of an electron; *Compton effect*—incoherent scattering of a photon by atomic electrons; and the $e^+e^-$ *pair production* by a photon in the electromagnetic field of the nucleus and atomic electrons.

The energy of electrons produced in the last process (the energy of photons produced due to the electron bremsstrahlung) is comparable to the energy of the primary photon (electron). If the energy of the primary photon (electron) is large, the electrons (photons) arising in these processes can produce the next generation of photons (electrons). These processes occur alternately, and the particles continue to increase in number until their energy drops to a level at which they no longer can efficiently produce new particles. During this process, increasingly more electrons enter the energy region in which the radiation loss cannot compete with the ionization loss, until the total energy of the primary electron (photon) is completely expended on the excitation and ionization of atoms. This phenomenon is the *electron–photon shower*.

The photons have several more channels through which they can interact with matter: the photonuclear reactions with neutron, pion, muon, etc. yield. The maximum cross section for a photonuclear reaction with a neutron yield in lead is, however, 350 times smaller than the minimum of the sum of the cross section for the Compton effect and cross section for the production of $e^+e^-$ pairs.

*The strong hadron-nucleus* ($hA$) *interactions* can be divided into two groups: processes in which no secondary particles are formed (the elastic $hA$ interaction) and processes in which at least one secondary particle is formed (the inelastic $hA$ interaction). In the first case, the hadron does not change its quantum numbers, and the nucleus may become excited and break up or it may acquire a momentum, remaining in the ground state. During its passage through matter, the hadron participates in the inelastic $hA$ interactions. The particles produced as a result of this interaction may be arbitrarily divided into two groups: fast particles ($\beta = v/c > 0.7$), which are primarily newly produced particles, and slow particles ($\beta < 0.7$), which are primarily nucleons and nuclear fragments. The average number of fast particles produced in the inelastic $hA$ interaction increases approximately logarithmically with the energy and when the energy of the primary proton is 3 TeV, there are approximately eighteen of these particles in the $pFe$ interaction.

These particles and some of the slow nucleons in turn interact inelastically with the nuclei of the material, producing new generations of hadrons, leptons, and fragments. There are approximately 20 long-lived hadrons ($c\tau p/m \gtrsim \lambda_{\text{abs}}$, where $\tau$ is the lifetime of a hadron in its rest frame, $p$ and $m$ are its momentum and mass, and $\lambda_{\text{abs}}$ is its range before the inelastic $hA$ interaction). Of these hadrons the nucleons and charged $\pi$ and $K$ mesons constitute almost 100% of the total number of the produced long-lived particles. The short-lived ($c\tau p/m \ll \lambda_{\text{abs}}$) hadrons, not having enough time to interact, decay into long-lived hadrons, leptons, and photons. The cascade occurring in the material is therefore frequently called a nucleon–meson cascade in the literature.

The process of gradual multiplication begins to decay when the particle energy falls below a certain level. The energy of the individual parts of the cascade decreases because of the distribution of the energy of the initial particle among the newly produced particles, which gradually increase in number, and because of the energy loss due to the electromagnetic interaction with matter. In the case of slow nucleons, the role of elastic $hA$ scattering increases. The loss of energy due to ionization causes the protons to stop and upon capturing electrons they become hydrogen atoms or are absorbed, like neutrons and mesons, by the nuclei of the material. A decrease in energy causes the probability for a $\pi^-$ capture—absorption of a $\pi^-$ meson by a nucleus which subsequently decays—to increase. The nuclear fragments that are produced are generally low-energy fragments with a short range because of the large electric charge.

The electrons and photons which are produced as a result of inelastic $hA$ interactions initiate an electron–photon shower. Experimental studies of the photon yield in $pp$ and $\pi p$ interactions in the energy ranges 2–2000 GeV and 5–200 GeV, respectively, showed that the $\pi^0 \to \gamma\gamma$ decay is the principal source of the photons.[4] The experimentally measured ratio of the number of electrons to the number of charged $\pi$ mesons created in the inelastic $hA$ interaction is small $[(e^+ + e^-)/(\pi^+ + \pi^-) \sim (1\text{–}5) \times 10^{-4}$ in the $pp$ interaction at[5] $E_p = 1500$ GeV and $\langle n_\pi{}^0 \rangle \approx (\langle n_\pi{}^+ \rangle + \langle n_\pi{}^- \rangle)/2$, where $\langle n_\pi i \rangle$ is the average multiplicity of the pions of type $i$]. Simple estimates show that the contribution from electromagnetic bremsstrahlung and from direct production of $e^+e^-$ pairs by protons, pions, and kaons is negligible in comparison with the contribution from the $\pi^0 \to \gamma\gamma$ decay, which is the principal source of an electron–photon shower. The $\pi^0$ mesons have a short lifetime ($c\tau = 2.5 \times 10^{-6}$ cm) and most of them are produced directly in the interaction event and through the decay of short-lived hadrons (principally that of the $\eta$ meson). It can be assumed, therefore, that an electron–photon shower begins essentially at the point of the $hA$ interaction.

The neutrinos and muons are the most penetrating particles. The muons are divided into two groups: the so-called *direct* muons, which are produced by the decay of short-lived $\rho$, $\omega$, $\varphi$, $D$, and $J/\Psi$ mesons, and the *decay* muons, which are produced by the decay of long-lived $\pi^\pm$ and $K^\pm$ mesons. It had been established experimentally that the ratio of the number of direct muons to the number of $\pi^\pm$ mesons is small $(\mu^+ + \mu^-)/(\pi^+ + \pi^-) \sim 10^{-4}$ over a broad energy range (70–400 GeV in the $pp$ interaction) and a broad range of kinematic variables.[5] The probability for the decay of a particle along a path length $l$ is $P(l) = 1 - \exp(-l/\lambda_D)$, where $\lambda_D = c\tau p/m$. $P(l) \approx 1$ for the direct muons and $P(l) \ll 1$, with $l \sim \lambda_{\rm abs}$, for the decay muons. The data of Refs. 5 and 6 can be used to crudely estimate the ratio of the differential cross sections for the production of direct and decay muons:

$$\xi = (d^3\sigma/dp^3)_\mu^{\rm direct}/(d^3\sigma/dp^3)_\mu^{\rm decay} \sim 2 \times 10^{-3} E_0\, x/l$$

for $x > 0.5$, $E_0 \gtrsim 10$ GeV, and $l \ll 26E_0$, where $E_0$ is the energy (GeV) of the proton incident on matter, $x$ is the fraction of proton energy carried away by a muon, and $l$ is the length ($m$) of the decay interval. $\xi \approx 5$ for a 3000-GeV proton in the ground. For small decay intervals the direct muons may therefore play an important role and for large decay intervals they can be ignored.

Accordingly, when a high-energy hadron impinges on bulk matter, an electromagnetic cascade begins to develop. The $\pi^0$ mesons which initiate electron–photon showers may be produced at each point of the inelastic $hA$ interaction, and the decay of $\pi^\pm$ and $K^\pm$ mesons produces muons that travel long distances in matter. During the interaction, nucleons largely with a momentum $p \approx 1$ GeV/$c$ are knocked out of the nuclei, giving rise to purely nucleonic cascades. At a time long (on the nuclear scale) after the interaction, the slowest ($p > 0.2$ GeV/$c$) neutrons, which travel a considerable distance from the point of their production, protons, and heavier nuclear fragments, which are absorbed by matter near the interaction point, are emitted during the nuclear deexcitation. Such a process occurs when $\pi^-$ mesons are captured by nuclei.

The weak interaction, which is responsible for the decay of many elementary particles, accounts for various processes that occur during their collisions. The cross sections of these processes are on the order of $10^{-35}$–$10^{-40}$ cm$^2$, while the cross sections of strong interactions are on the order of $10^{-26}$–$10^{-27}$ cm$^2$. In certain problems the weak processes are nevertheless important. An example of such a process is the production of muons in neutrino–nucleus interactions in the case of a strong neutrino flux and in the case of a suppression or absence of muons from other sources.

These are briefly the principal processes that occur when high-energy radiation interacts with matter. The role of the various processes changes, depending on the class of problems considered. The particle fluxes behind the accelerator radiation shields are determined primarily by the nucleon–meson cascade, and the radiation heating of the accelerator systems and superconducting magnets is determined by the electron–photon showers from the $\pi^0 \to \gamma\gamma$ decays. The characteristic features of the biological action of high-energy hadrons are attributable to the nuclear fragments, $\pi^-$ capture, and possibly the collective effect of produced particles on matter in the neighborhood of the interaction point. Only the decay muons and muons from the neutrino–nucleus interactions can be taken into account in upgrading of the muon shield for accelerator neutrino channels and only the decay and direct muons can be taken into account in designing of various absorbers. The solution of a given problem can be facilitated considerably by choosing the relevant processes that occur in the interaction of high-energy particles with matter. This choice can be made if the general behavior of these interactions is understood and if the special features of the processes occurring in them are known.

## 3. Basic concepts of the physics of multiple-production processes

The reactions involving the interaction of elementary particles are studied in different *reference frames* for purely experimental reasons and because of the convenience of describing a particular phenomenon theoretically. We restrict the discussion here to the three most frequently used reference frames.

### Laboratory frame of reference

This reference frame corresponds to a standard arrangement of proton-accelerator experiments, in which a beam of particles of species $a$ strikes a fixed target consisting of particles of species $b$ (hydrogen protons, atomic nuclei, atomic electrons). In this reference frame, the 4-momenta of the interacting particles have the following components: $P_a = (E_a, \mathbf{p}_a)$ and $P_b = (m_b, 0)$, where $E_a$ and $\mathbf{p}_a$ are the energy and momentum vector of particle species $a$; $E_a = (p_a^2 + m_a^2)^{1/2}$, where $m_a$ and $m_b$ are the particle masses; and the velocity of particle $a$, $v_a = |\mathbf{p}_a|/E_a$ ($\hbar = c = 1$).

**Antilaboratory frame of reference**

In this reference frame the particle $a$ is at rest and the target particle $b$ strikes it at a velocity $v_b$. The 4-momenta have the components $P_a = (m_a, 0)$ and $P_b = (E'_b, \mathbf{p}'_b)$. It is clear that the antilaboratory frame of reference is moving at a velocity $v_a$ relative to the laboratory frame.

According to the relativistic transformation equations, we have

$$E'_b = m_b(1 - v_a^2)^{-1/2} = \frac{m_b}{m_a} E_a, \; \mathbf{p}'_b = -\frac{m_b \mathbf{v}_a}{\sqrt{1 - v_a^2}} = -\frac{m_b}{m_a} \mathbf{p}_a.$$

The zero-mass particles ($m_a = 0$ for photons and neutrinos) have no antilaboratory frame of reference. For $m_a \neq 0$ the antilaboratory frame of reference is physically completely equivalent to the laboratory frame of reference, especially when particle $a$ plays the role of a stable target.

**The center-of-mass (c.m.) frame of the reaction**

This reference frame is determined by the relation $\mathbf{p}_a^* + \mathbf{p}_b^* = 0$. In the center-of-mass frame the colliding particles have equal but oppositely directed momenta. The velocity of the c.m. frame relative to the laboratory frame of reference is directed opposite the momentum $\mathbf{p}_a$ and is given by

$$\mathbf{v}_c = -\mathbf{p}_a/(E_a + m_b). \tag{1.1}$$

The velocity of the c.m. frame relative to the antilaboratory frame of reference is directed opposite the momentum $\mathbf{p}_b$ and is given by

$$v'_c = -\frac{\mathbf{p}'_b}{E'_b + m_a} = \frac{\mathbf{p}_a}{E_a + m_a^2/m_b}.$$

**Kinematic variables in different reference frames**

In the c.m. frame the particle energies are expressed in terms of the relativistic invariant $S$ in the form

$$\left.\begin{aligned} &S = (P_a + P_b)^2 = (E_a + E_b)^2 - (\mathbf{p}_a + \mathbf{p}_b)^2;\\ &E_a^* = (S + m_a^2 - m_b^2)/2\sqrt{S}; \; E_b^* = (S + m_b^2 - m_a^2)/2\sqrt{S};\\ &\sqrt{S} = E_a^* + E_b^*. \end{aligned}\right\} \tag{1.2}$$

In the c.m. frame the momentum-vector moduli can be calculated from the equations

$$|\mathbf{p}_a^*| = |\mathbf{p}_b^*| = \lambda(S, m_a, m_b)/2\sqrt{S},$$

where

$$\lambda(S, m_a, m_b) = [S - (m_a + m_b)^2]^{1/2}[S - (m_a - m_b)^2]^{1/2}.$$

The invariant $S$ is expressed in terms of the particle mass and energy in the laboratory or antilaboratory frame of reference,

$$S = m_a^2 + m_b^2 + 2m_b E_a = m_a^2 + m_b^2 + 2m_a E_b'.$$

Hence we can write

$$E_a = (S - m_a^2 - m_b^2)/2m_b;\ E_b' = (S - m_a^2 - m_b^2)/2m_a;$$
$$p_a = \lambda(S, m_a, m_b)/2m_b;\ p_b' = \lambda(S, m_a, m_b)/2m_a.$$

Since $\mathbf{v}_a$, $\mathbf{v}_c$, and $\mathbf{v}_c'$ are known, we can convert any kinematic characteristics of the primary and secondary particles that participate in the reaction from one reference frame into another by means of the well-known relativistic transformation equations.[7]

In the relativistic transformations along the collision axis of the primary particles the transverse components of the momenta $\mathbf{p}$ of the secondary particles remain constant:

$$p_\perp = p_\perp^* = p_\perp'.$$

The energies of the secondary particles and their longitudinal momenta in different reference frames can easily be expressed in terms of the scalar products of the 4-vectors of the problem*

$$E = (PP_b)/m_b;\quad E^* = (P, P_a + P_b)/\sqrt{S};\quad E' = (PP_a)/m_a;$$
$$p_\parallel = \frac{(PP_b)(P_aP_b) - m_b^2(PP_a)}{m_b\sqrt{(P_aP_b)^2 - m_a^2 m_b^2}};$$
$$p_\parallel^* = \frac{(PP_b)[m_a^2 + (P_aP_b)] - (PP_a)[m_b^2 + (P_aP_b)]}{\sqrt{S}\sqrt{(P_aP_b)^2 - m_a^2 m_b^2}};$$
$$p_\parallel' = \frac{(PP_a)(P_aP_b) - m_a^2(PP_b)}{m_a\sqrt{(P_aP_b)^2 - m_a^2 m_b^2}}.$$

### Light-cone variables

It is sometimes convenient to use the light-cone variables $p_+ = E + p_\parallel$ and $p_- = E - p_\parallel$, instead of the variables $(E, p_\parallel)$, which characterize the kinematic state of a particle. These variables satisfy the relation $p_+p_- = m^2 + p_\perp^2 = m_\perp^2$. The quantity $m_\perp$ is called the *transverse mass*.

The dimensionless ratio of the cone variables of different particles, $p/k$, does not change under relativistic transformations along the collision axis: $p_+/k_+ = p'_+/k'_+$, $p_-/k_- = p'_-/k'_-$.

*The scalar product of the 4-vectors $A$ and $B$ is defined as $(AB) = A_0B_0 - \mathbf{AB}$, where $A_0$ and $B_0$ are the time-dependent components.

### The Feynman dimensionless variable

At high energies the average transverse momentum of secondary particles during multiple production of hadrons is virtually independent of energy (for secondary pions $\langle p_\perp \rangle \simeq 0.35$ GeV/$c$). At the same time, the average longitudinal momenta of the secondary particles increase with energy almost linearly. It is therefore convenient to use as an independent variable the dimensionless variable

$$x_F = p^*_\parallel / p^*_{\parallel\,\max}, \tag{1.3}$$

which is called the *Feynman variable*, $x_F$. This variable changes at a fixed value of $p_\perp$ in the range

$$-1 \leqslant x_F \leqslant 1. \tag{1.4}$$

If we are dealing with a single secondary particle with a 4-momentum $P$, with the other secondary particles having arbitrary momenta which are consistent with the energy-momentum conservation laws, then the maximum energy of this isolated particle in the c.m. system can be determined from the equation

$$E^*_{\max} = (S + m^2 - \omega^2)/2\sqrt{S}. \tag{1.5}$$

where $\omega$ is the minimum allowable value of the effective mass of the particles that accompany this particle. Using (1.5), we find

$$p^*_{\parallel\,\max} = [E^{*2}_{\max} - m^2 - p^2_\perp]^{1/2}. \tag{1.6}$$

For arbitrary values of $p_\perp$ we have $p^*_{\parallel\,\max} = |\mathbf{p}^*|_{\max}$. At a reasonably high energy ($S \gg m^2, \omega^2$) and $|p^*| \gg \langle p_\perp \rangle$ we have

$$x_F \simeq 2p^*_\parallel/\sqrt{S}. \tag{1.7}$$

Expression (1.7) is frequently used to determine $x_F$, although the limits in (1.4) are ony approximate and actually unattainable in the case $|p^*_\parallel| \sim p_\perp$.

The variable $x_F$ can easily be used to distinguish the events occurring in the front cone from those in the back cone in the c.m. system, in which both interacting primary particles are on an equal footing. The c.m. system is used to describe the secondary-particle spectra for values of $x_F$ that are not too close to zero. It can also be used for $(1 - |x_F|) \ll 1$, i.e., near the kinematic boundaries of the spectrum.

### The rapidity variable

In the region of small values of $x_F$ in the modulus the *rapidity variable* is a useful $p^*_\parallel$-dependent dimensionless variable,

$$y^* = \frac{1}{2}\ln\frac{E^* + p^*_\parallel}{E^* - p^*_\parallel} = \frac{1}{2}\ln\left(\frac{p^*_+}{p^*_-}\right) = \ln\frac{p^*_+}{m_\perp}.$$

In the relativistic transformations along the collision axis the rapidity variable changes additively. In the laboratory frame of reference, for example, $y = y^* + \frac{1}{2}\ln[(1 + v_c)/(1 - v_c)]$, where $v_c = |\mathbf{v}_c|$ and is defined by Eq. (1.1).

In the transformation from one reference frame to another moving along the collision axis at a velocity **v**, the distribution of events in the rapidity $y$ is therefore shifted to the right or to the left parallel to the $y$ axis by the amount $|\Delta y|_v = \frac{1}{2}\ln[(1+v)/(1-v)]$ relative to the new reference point.

In any reference frame the variables $E$ and $p_\parallel$ are given by the equations $E = m_\perp \cosh y$ and $p_\parallel = m_\perp \sinh y$. For $|p_\parallel^*| \ll m_\perp (|x_F| \to 0)$ the rapidity variable $y^* \approx p_\parallel^*/m_\perp$. Clearly, the range of variation of $y^*$ is

$$-\ln \frac{E^*_{\max} + p^*_{\parallel\max}}{m_\perp} \leqslant y^* \leqslant \ln \frac{E^*_{\max} + p^*_{\parallel\max}}{m_\perp}.$$

In other reference frames $y_v = y^* + \frac{1}{2}\ln[(1+v)/(1-v)]$, so that the variation range can also be easily determined ($v > 0$ or $v < 0$, depending on the direction of motion of the reference frame relative to the c.m. system). In the rapidity variables the region of small but finite values of $x_F$, such that $(2m_\perp/\sqrt{S}) \ll |x_F| \lesssim x_0 \ll 1$, is extended and it increases logarithmically with increasing energy:

$$\Delta y^* = 2\ln\left(\frac{\sqrt{S}}{m_\perp} x_0\right).$$

The rapidity variable can therefore be used for small values of $x_F$.

### The reaction cross section

At high energies, aside from the elastic scattering of the type $a + b \to a + b$, there can be a wide variety of transformations of the primary particles $a$ and $b$ into secondary particles:

$$a + b \to c + d + \cdots + f. \tag{1.8}$$

With increase in energy, the number of secondary particles increases, on the average, in the reactions involving the hadron–nucleon, photon–nucleon, and deep inelastic lepton–nucleon interactions of the type in (1.8). An increase in energy also leads to an increase in the number of different types of reactions (reaction channels) and to the production of particles of various species, even those that were not among the primary colliding particles. Here all the conservation laws characterizing a given type of interaction (strong, electromagnetic, weak) hold: (1) energy-momentum conservation, (2) electric-charge conservation, (3) the baryon and lepton quantum-number conservation, (4) conservation of the isotopic spin and strangeness in strong interactions, etc.

A quantitative parameter of reaction (1.8) is the *differential cross section* $d\sigma_i$ which is defined as follows ($i$ is an index which denotes the type of reaction). Let us assume that a beam of particle species $a$, with a velocity $v_a$, strikes a typical target which contains particle species $b$ and which has a volume $dV$. The number of reactions of a given type, $dN_i$, in a time $dt$ in the volume $dV$ will then be proportional to the number of particles $b$ in the volume $dV$ and to the number of beam particles that traverse the cross section $dS$ of the volume $dV$ in a time $dt$:

$$dN_i = d\sigma_i(n_b^0 dV)(n_a v_a dt).$$

Here $n_b^0$ and $n_a$ are the densities of the target particles and beam particles in a coordinate system with the target at rest, $n_b^0 dV$ is the number of particles $b$ in the volume $dV$, $n_a v_a dt$ is the number of particles $a$ that traverse the cross section of the target in a time $dt$, and $n_a v_a$ is the flux density of the incident particles. The differential cross section $d\sigma_i$, by definition, characterizes *the number of reactions of type i occurring in a unit volume per unit time for a unit flux density of incident particles and unit density of the target.* The differential cross section is measured in square centimeters.

The differential nature of $d\sigma_i$ can be found from the observation of reaction (1.8). If the reaction products (particles $c$, $d$,...) are detected in certain intervals of momenta from $\mathbf{p}_c$ to $\mathbf{p}_c + d\mathbf{p}_c$, from $\mathbf{p}_d$ to $\mathbf{p}_d + d\mathbf{p}_d$, etc., then $dN_i$, and hence $d\sigma_i$, are measured for the given range of variation of the momenta $\mathbf{p}_c$, $\mathbf{p}_d$,... . The *total cross section* $\sigma_i$ of a given reaction can be found by integrating (summing) over all possible momenta of the secondary particles in reaction (1.8).

The quantities $d\sigma_i$ and $\sigma_i$ remain constant in the relativistic transformations in the same (opposite) direction as the velocity $\mathbf{v}_a$, i.e., these quantities are relativistic invariants. This conclusion follows from the relativistic invariance of the number of reactions, $dN_i$, and the volume of the four-dimensional space–time, $dVdt$, in which the reactions occur. To prove this assertion, we need only to consider the properties of $n_b^0 n_a \mathbf{v}_a$ in the relativistic transformation. We shall see that it can be represented in the form $n_b^0 n_a v_a = n_b n_a |\mathbf{v}_a - \mathbf{v}_b|$, where the particle densities $n_{a,b}$ and the particle velocities $\mathbf{v}_{a,b}$ are given in the coordinate in which the target particle $b$ is moving with a velocity $\mathbf{v}_b$ in the same (opposite) direction as the particle $a$. Since $n_b^0 n_a = (j_b j_a)^*$ and $v_a = [(P_a P_b)^2 - m_a^2 m_b^2]^{1/2}/(P_a P_b)$, the relation which we are seeking can easily be proved [ $j_{a,b}$ are the 4-vectors of the flux densities of particle species $a(b)$, and $m_{a,b}$ are the masses of these particles].

Accordingly, if in the determination of $d\sigma_i$ the quantity $\mathbf{v}_a$ is taken to mean the difference in velocities of the particles that collide in the same direction, and $n_a$ and $n_b$ are the densities of the beams of these particles, then $d\sigma_i$ will not depend on the reference frame. If, on the other hand, the directions of the particle velocities are not the same, then one should use $\{(\mathbf{v}_a - \mathbf{v}_b)^2 - [\mathbf{v}_a v_b]^2\}^{1/2}$, instead of $|\mathbf{v}_a - \mathbf{v}_b|$. The quantity $d\sigma_i$, which is measured in an arbitrary reference frame, will then be the same as that measured in the laboratory frame of reference, c.m. frame, or the antilaboratory frame of reference.[7]

*Here $(j_a j_b)$ is the scalar product of the 4-vectors of the current densities which are defined as $j = (n^0/\sqrt{1-v^2}, n^0\mathbf{v}/\sqrt{1-v^2}) = (n, n\mathbf{v})$.

The differential cross section can be written in the form

$$d\sigma_i = \varphi(\{p_j\}; S) \prod_{j=1}^{N} \frac{d^3p_j}{E_j} \delta^{(4)}\left(\sum_{j=1}^{N} P_j - P_a - P_b\right). \tag{1.9}$$

where $\delta^{(4)}$-function represents the action of the energy-momentum conservation law in reaction (1.8); $N$ is the number of final-state particles; $\varphi$ is the relativistically invariant function which depends on the interaction dynamics; $S = (P_a + P_b)^2$ is the invariant in (1.2), which is equal to the square of the total energy in the c.m. frame of reaction (1.8); $\{p_j\}$ is the set of linearly independent invariant variables of the type $(P_a - P_c)^2$, $(P_a - P_d)^2$, etc., on which the function $\varphi$ depends; and $d^3p_j/E_j$ is an element of the relativistically invariant phase space of the particle $j$, which is proportional to the standard phase space* divided by the particle energy. It follows from (1.9) that

$$d\sigma_i = F(\{p_j\}; S) \prod_{j=1}^{N-1} d^3p_j/E_j. \tag{1.10}$$

The total cross section $\sigma_i$ of the reaction under consideration [reaction (1.8)] can be obtained by integrating (1.10) over all permissible kinematic momenta. The sum of the cross sections of all possible processes that occur upon the collision of particles $a$ and $b$ is called the *total interaction cross section*,

$$\sigma_{\text{tot}} = \sum_i \sigma_i. \tag{1.11}$$

The *relative probability* for the occurrence of a given reaction channel or the number of reactions of a given type per interaction is $w_i = \sigma_i/\sigma_{\text{tot}}$. Accordingly, the relative *differential probability* for the occurrence of a given reaction channel is $dw_i = d\sigma_i/\sigma_{\text{tot}}$.

The total number of reactions of type $i$ can be obtained by multiplying $w_i$ by the total number of interactions $N_0$: $N_i = N_0\sigma_i/\sigma_{\text{tot}}$.

### Inclusive method of studying the multiple production of hadrons

At high energies, when the average multiplicity of secondary particles is large and the number of various channels like (1.8) is also large, it is basically impossible to study each channel and each particle in a given channel. Important information on the multiple-production mechanisms and on the kinematic characteristics of various types of hadrons can nevertheless be obtained by studying not the individual *exclusive processes* [reaction (1.8)], but rather the *inclusive reactions* of the type

$$a + b \rightarrow c + X, \tag{1.12}$$

*The number of quantum states in the volume element of the phase space $d^3p\,dV$ is determined in quantum mechanics from the equation $d^3p\,dV/(2\pi)^3$ and is called the phase-space element. The factors of the type $(2\pi)^3$ will be included in the definition of $\varphi$ and $F$.

where $X$ is an arbitrary system of undetectable particles which is formed along with the detectable particle $c$. In the study of reaction (1.12) the detector is adjusted so that it can detect only particles of species $c$ and measure their momenta $\mathbf{p}_c$. If several particles of species $c$ are produced in reaction (1.12), then any one of these particles may be detected.

Reaction (1.12) is characterized by an *invariant single-particle differential cross section* of the inclusive reaction

$$E_c d^3\sigma_c / d^3p_c = f(\mathbf{p}_c; S), \tag{1.13}$$

where the function $f(\mathbf{p}_c; S)$ is the result of the integration of (1.10) over the momenta of all particles except one (of species $c$), the summation over all reactions in which at least one particle of species $c$ is produced, and the summation over all particles of species $c$ which are produced in a given reaction.

The definition of the function $f(\mathbf{p}_c; S)$ implies the normalization relation (the sum rule)

$$\int f(\mathbf{p}_c; S) \frac{d^3p_c}{E_c} = \sum_{n_c} n_c \sigma_c^{n_c}(S) = \langle n_c \rangle \sigma_{in}(S), \tag{1.14}$$

where $\langle n_c \rangle$ is the average multiplicity of particles of species $c$ in the $ab$ collision reaction, $\sigma_{\text{in}}(S)$ is the total cross section of the inelastic $ab$ collisions, $\sigma_c(S) = \sum_{n_c} \sigma_c^n c(S)$; $\sigma_c^{n_c}$ is the cross section for the production of $n_c$ particles of species $c$ in the $ab$ collision [reaction (1.12)], and $\sigma_c(S)$ is the cross section for the production of particles of species $c$ (at least one particle) in reaction (1.12). Equation (1.14) contains the quantity $\sigma_{\text{in}}(S)$, instead of the total cross section $\sigma_{\text{tot}}(S)$ of the $ab$ interaction, because the elastic $ab$ scattering events usually are disregarded in the determination of $\langle n_c \rangle$.

**The sum rules for the inclusive single-particle distributions**

Aside from the sum rule (1.14), there are other obvious relations. If $\sigma_c(S)$ is summed over all particle species, we shall have

$$\sum_c \sigma_c(S) = \sum_c \sum_{n_c} \sigma_c^{n_c}(S) = \sigma_{\text{tot}}(S).$$

In exactly the same manner we find $\sum_c \langle n_c \rangle = \langle N \rangle$, where $\langle N \rangle$ is the average multiplicity of secondary particles in the reactions for the interaction of particles $a$ and $b$. The integral of the 4-momentum of the inclusive particle, $P_\mu$, is

$$\int (P_c)_\mu f(\mathbf{p}_c; S)\, d^3p_c / E_c = \langle P_\mu^c \rangle \sigma_{\text{in}}(S),$$

where $\langle P_\mu^c \rangle$ is the total average 4-momentum transferred by all particles of species $c$ produced in the $ab$ collision. The following relations, which are the corollary of the law of conservation of the electric charge and total 4-momentum, can also be easily written out:

$$\sum_c Q_c\langle n_c\rangle = Q_a + Q_b; \quad \sum_c \langle P^c_\mu\rangle = (P_a + P_b)_\mu.$$

Here $Q_{a,b,c}$ are the particle charges, and $P_a$ and $P_b$ are the 4-momenta of the initial particles. Similar methods can be used to obtain various sum rules for the conservation laws of other discrete quantum numbers (strangeness, baryon number, etc.).

If the *single-particle density function* for the inclusive particles of species $c$ is defined as

$$\rho(\mathbf{p}_c; S) = f(\mathbf{p}_c; S)/\sigma_{\text{in}}(S), \tag{1.15}$$

then the function (1.15) is normalized to allow for the average number of particles of species $c$ and $\rho(\mathbf{p}_c; S)d^3p_c/E_c$ is normalized to allow for the average number of particles of species $c$ which enter the relativistic phase space $d^3p_c/E_c$. The function $\rho(\mathbf{p}_c; S)$ is used in the calculations of the secondary-particle spectra.

### Inclusive two-particle density functions and the correlation function

Let us examine the inclusive production of two particles in the reaction

$$a + b \rightarrow c + d + X, \tag{1.16}$$

where $c$ and $d$ are the detectable particles, and $X$ are the accompanying undetectable hadrons. The *two-particle density function* is

$$\rho(\mathbf{p}_c, \mathbf{p}_d; S) = E_c E_D d^6\sigma_{cd}/\sigma_{\text{in}}(S)\, d^3p_c d^3p_d, \tag{1.17}$$

where the invariant differential cross section of reaction (1.16) is defined by the detector which detects only the particles of species $c$ and $d$ and which measures their momenta $\mathbf{p}_c$ and $\mathbf{p}_d$. Since any particle of species $c$ or $d$ can be detected, Eq. (1.17) automatically contains the sum over all particles of species $c$ and $d$, which are produced in the reaction involving the $ab$ collision. The function (1.17) is therefore normalized as follows:

$$\int \rho(\mathbf{p}_c, \mathbf{p}_d; S)\, d^3p_c d^3p_d/E_c E_d = \langle n_c n_d\rangle, \tag{1.18}$$

if $c \neq d$. If two particles, $c_1$ and $c_2$, of the same species are detected, the normalization has the form

$$\int \rho(\mathbf{p}_{c1}, \mathbf{p}_{c2}; S)\frac{d^3p_{c1}}{E_{c1}}\frac{d^3p_{c2}}{E_{c2}} = \langle n_c(n_c - 1)\rangle, \tag{1.19}$$

since the second particle of species $c$ is detected after the detection of the first particle of the same species.

By analogy one can determine the $n$-particle density functions of the inclusive particles. In principle, a consideration of all possibilities leads in the limit to a comprehensive description of the $ab$ interaction processes with different final states.

If particles $c$ and $d$ are produced independently in reaction (1.16) (there are always trivial corrections associated with the laws of conservation of the energy-momentum and discrete quantum numbers), we have

$$\rho(\mathbf{p}_c, \mathbf{p}_d; S) = \rho(\mathbf{p}_c; S)\,\rho(\mathbf{p}_d; S). \qquad (1.20)$$

Accordingly, if this condition is satisfied, we have $\langle n_c n_d \rangle = \langle n_c \rangle \langle n_d \rangle$. If, on the other hand, relation (1.20) does not hold, then the production of inclusive particles $c$ and $d$ is called *correlated production*. This correlation is described by the *correction function*

$$C(\mathbf{p}_c, \mathbf{p}_d; S) = \rho(\mathbf{p}_c, \mathbf{p}_d; S) - \rho(\mathbf{p}_c; S)\,\rho(\mathbf{p}_d; S).$$

The production of two inclusive particles can be described simultaneously by experimentally measuring this function and the single-particle distribution densities.

**General properties of single-particle density functions**

In the analysis of the experimental data, the region of permissible values of the variables $x_F$ (or $y$) is customarily divided into three intervals:

(1) The region where the effect (correlation) of the primary particle, which is the incident particle in the laboratory frame, on the newly produced particles is still a factor: $x_0 \lesssim x_F \lesssim 1$, where $x_0 = \text{const}\,(x_0 \ll 1)$, or $(2m_\perp/\sqrt{S}) \lesssim x_F \leqslant 1$. This region is called the *incident-particle fragmentation region.*

(2) The region where the effect (correlation) of the target particle (in the laboratory frame) on the newly produced particles is a factor: $-1 \leqslant x_F \lesssim -x_0$ or $-1 \leqslant x_F \leqslant -2m_\perp/\sqrt{S}$. This region is called the *target-particle fragmentation region.*

(3) The intermediate or *central region* where the primary particles $a$ and $b$ have no effect on the secondary particles: $-x_0 \leqslant x_f \lesssim x_0$ or $-2m_\perp/\sqrt{S} \leqslant x_F \leqslant 2m_\perp/\sqrt{S}$.

In all three cases two of the most common types of dynamic hypotheses are considered. In the first type of hypothesis $(2m_\perp/\sqrt{S} \lesssim |x_0| \ll 1)$ the central region in $x_F$ remains constant at high energies ($\sqrt{S} \geqslant 2m_\perp/x_0$, where $x_0$ is a fixed quantity in the experiments or in the theoretical models). In the second type of hypothesis the central region in $x_F$ vanishes with increasing energy (for $\sqrt{S} \gg 2m_\perp$) and the regions of fragmentation of the particles $a$ and $b$ merge as $|x_F| \to 0$ (Table 1.1).

Let us consider how the dynamics of the process, which is hypothesized on the basis of intuitive physical assumptions, affects the functions of the single-particle density function. If the particle $c$ is a spin-zero particle or if the primary particles are not polarized, the function $\rho(\mathbf{p}_c; S)$ will depend exclusively on the transverse-momentum modulus of the particle $c$: $\rho(\mathbf{p}_c; S) = \rho(p_{c\parallel}, p_{c\perp}; S)$.

The variable which depends on the longitudinal momentum can be written in dimensionless form in terms of $x_F$ or $y^*$. In general, the rapidity variable is more useful, since it can easily be represented in covariant form:

**Table 1.1. Boundaries of the regions in the variables $x_F$ and $y^*$ (rapidity in the c.m. frame) for two dynamic hypotheses.**

| | $x_F$ | $y^*$ |
|---|---|---|
| Fragmentations of particle $a$ | $x_0 \lesssim x_F \leqslant 1$ | $\ln \frac{\sqrt{S} x_0}{m_\perp} \lesssim y^* \lesssim \ln \frac{\sqrt{S}}{m_\perp}$ |
| | $\frac{2m_\perp}{\sqrt{S}} \lesssim x_F \lesssim 1$ | $l \lesssim y^* \lesssim \ln \frac{\sqrt{S}}{m_\perp}$ |
| Central | $-x_0 \lesssim x_F \lesssim x_0$ | $-\ln \frac{\sqrt{S} x_0}{m_\perp} \lesssim y^* \lesssim \ln \frac{\sqrt{S} x_0}{m_\perp}$ |
| | $-\frac{2m_\perp}{\sqrt{S}} \lesssim x_F \lesssim \frac{2m_\perp}{\sqrt{S}}$ | $-l \lesssim y^* \lesssim l$ |
| Fragmentations of particle $b$ | $-1 \leqslant x_F \lesssim -x_0$ | $-\ln \frac{\sqrt{S}}{m_\perp} \lesssim y^* \lesssim -\ln \frac{\sqrt{S} x_0}{m_\perp}$ |
| | $-1 \leqslant x_F \lesssim -\frac{2m_\perp}{\sqrt{S}}$ | $-\ln \frac{\sqrt{S}}{m_\perp} \lesssim y^* \lesssim -l$ |

Note. $l$ is on the order of 1.

$y^*_{max} - y^*$ or $y^* - y^*_{min}$. In principle, the function $\rho(\mathbf{p}_c; S)$ may depend on each of these variables separately:

$$\rho(\mathbf{p}_c; S) = \rho(y^*_{max} - y^*, y^* - y^*_{min}, p_{c\perp}; S). \tag{1.21}$$

In the region of fragmentation of the particle $a$, function (1.21) must depend solely on $y^*_{max} - y^*$, since a dependence on $y^* - y^*_{min}$ would correspond to the correlation with the target particle $b$. In this region we therefore have

$$\rho(\mathbf{p}_c; S) = \rho_a(y^*_{max} - y^*, p_{c\perp}; S). \tag{1.22}$$

Analogously, in the region of fragmentation of the particle $b$ we have

$$\rho(\mathbf{p}_c; S) = \rho_b(y^* - y^*_{min}, p_{c\perp}; S). \tag{1.23}$$

For unlike primary particles ($a \neq b$) the dependence of (1.22) on the rapidity is different from that of (1.23).

In the central region, there should be, as expected, no correlation with the particle $a$ or with the particle $b$, implying that the single-particle density function in this case should be independent of $y$:

$$\rho(\mathbf{p}_c; S) = \rho(p_{c\perp}; S). \tag{1.24}$$

It follows from (1.22) and (1.23) that at high energies ($S \gg m_\perp^2$) in the fragmentation region we have

$$\rho(\mathbf{p}_c; S) = \rho_{a,b}(x_F; p_{c\perp}; S), \tag{1.25}$$

and in the central region there is no $x_F$ dependence. In Eqs. (1.22)–(1.25) the dependence on the primary energy is retained, since a slight $S$ dependence was found to be present experimentally over the energy interval that has thus far been studied.

### Limiting fragmentation, scaling

Let us now discuss the principal hypotheses on the behavior of $\rho(\mathbf{p}_c; S)$ with increasing energy, which have been advanced on the basis of semi-intuitive physical considerations (see, e.g., Refs. 7 and 8). Chou and Yang advanced a hypothesis on the *limiting fragmentation* which states that at high energies the multiple production of hadrons becomes such that the longitudinal momenta of the target fragments stop increasing in the laboratory frame and remain bounded as the energy is raised. In fact, this means that the spectrum of the hadron-target fragments is no longer dependent on the kinematic characteristics of the incident particle and on its quantum numbers. This hypothesis can be written in the form

$$\rho(\mathbf{p}_c;S) = \rho_b(p_{c\parallel}, p_{c\perp}).$$

We express $p_{c\parallel}$ in terms of the variable $x_F$ for $S \gg m_a^2$, $m_b^2$, and $m_{c\perp}^2$. Using the law describing the relativistic transformation of longitudinal momenta, we find

$$p_{c\parallel} = \frac{E_c^* + v_c p_{c\parallel}^*}{\sqrt{1-v_c^2}} = \frac{E_c^*(E_a + m_b) + p_a p_{c\parallel}^*}{\sqrt{S}}.$$

In the target fragmentation region, $p_{c\parallel}^* < 0$, assuming $x_F = 2^*_{c\parallel}/\sqrt{S}$, we therefore find

$$p_{c\parallel} \approx \frac{m_{c\perp}^2 - m_b^2 x_F^2}{m_b(\sqrt{x_F^2 + 4m_{c\perp}^2/S} - x_F)}.$$

The value $p_{c\parallel}$ is bounded if $|x_F| \geqslant 2m_{c\perp}/\sqrt{S}$ $(x_F < 0)$. Here we have $p_{c\parallel} \simeq (m_b x_F - m_{c\perp}^2/m_b x_F)/2$.

Accordingly, $p_{c\parallel}$ depends exclusively on $x_F$, and the limiting-fragmentation hypothesis states that the single-particle density function in the fragmentation region of the target is solely a function of the dimensionless variables $x_F$ and $p_{c\perp}$, which does not depend on the primary energy,

$$\rho_b(p_{c\parallel}, p_{c\perp}) = \rho_b(x_F, p_{c\perp}).$$

This property is called *scaling* or *scaling invariance.*

Making use of the analogy between the laboratory frame and the antilaboratory frame, we can show that the boundedness of $p'_{c\parallel}$ in the system where particle $a$ is at rest leads to a scaling in the fragmentation region of the incident particle in the c.m. frame:

$$\rho_a(p'_{c\parallel}, p_{c\perp}) = \rho_a(x_F, p_{c\perp}).$$

These properties are the consequence of the limiting-fragmentation hypothesis.[7] This hypothesis, however, does not deal in any way with the central region and most likely corresponds to the complete disappearance of this region in the limit $S \to \infty$, since this region in this case is limited by the inequalities $-2m_\perp/\sqrt{S} \lesssim x_F \lesssim 2m_\perp/\sqrt{S}$.

In the limiting-fragmentation hypothesis, the multiplicity of secondary hadrons increases as $S \to \infty$ because of the logarithmic increase of the fragmentation regions in the rapidity space:

$$\langle N \rangle = \langle N_a \rangle + \langle N_b \rangle;$$

$$\langle N_{a,b} \rangle = \int_0^{y^*_{\max}} \rho_{a,b}(|y^*_{\max} - y^*|, p_{c\perp})\, dy^* dp_{c\perp}$$

$$= \int_0^{\ln(\sqrt{S}/m)} \rho_{a,b}(|y^*_{\max} - y^*|)\, dy^* = \int_0^{\ln(\sqrt{S}/m)} \rho_{a,b}(Y)\, dY$$

$$= \rho_{a,b}(\overline{Y}) \ln(\sqrt{S}/m),$$

where $\rho_{a,b}(\overline{Y})$ is the distribution density at a certain point in the integration interval.

If there is a central region, whose width in the rapidity space increases logarithmically with increasing energy, then the scaling may manifest itself in the entire region of $x_F$. Since the distribution density is generally independent of $x_F$ in the central region, and since it has the shape of a plateau in the variable $y$, the principal multiplicity of the secondary particles is governed by the contribution from this region,

$$\langle N \rangle \simeq \int_{-\ln(\sqrt{S}/m_\perp)}^{\ln(\sqrt{S}/m_\perp)} \rho(\mathbf{p}_c)\, dy^* dp_{c\perp} \sim \ln(\sqrt{S}/m),$$

while the fragmentation regions in this case give a finite contribution to $\langle N \rangle$, which does not depend on $S$ in the limit $S \to \infty$. The hypothesis which states that $\rho(\mathbf{p}_c; S) = \rho(x_F, p_{c\perp})$ and that it does not depend on the primary energy in the entire range of values of $x_F$ is called the *Feynman scaling*.[7–9] This hypothesis is supported by the multiperipheral and simple Regge models for the multiple production of hadrons. Even in these models, however, the Feynman scaling can be broken (see the discussion below). The concept of scaling is nonetheless very attractive, since this simple mode of behavior of the single-particle distribution densities at high energies would indicate that in the theory there are no quantities with the dimensionality of energy other than the primary- and secondary-particle energies. If scaling holds, the data obtained at the presently attainable energies can be extrapolated to the region of superhigh energies which will be studied with the new generation of accelerators. The shield design and calculations for new proton accelerators and the planning of new fundamental experiments can now be carried out only on the basis of the available experimental data and their present-day interpretation in terms of various models.

# 4. Basic concepts and transport theory equations

The nuclear electromagnetic cascade is a process involving multiple stochastic interactions of particles with nuclei and atoms of a substance. The calculation of this process is based on the formalism in the section of statistical physics called *transport theory*. If the particle energy is so high that the quantum-mechanical effects due to the interference of waves from various scatterers are negligible,* we can derive linear integrodifferential equations which describe the particle balance in the volume element of the phase space $(\mathbf{r}, E, \mathbf{\Omega})$. These equations are called radiation *transport equations* or Boltzmann kinetic equations.†

The linearity of the transport equations stems from the independence of the sequential collision events, since the number of propagating particles in any volume is assumed to be much smaller than the number of nuclei in the same volume, and since there is no self-interaction of the beam particles. Furthermore, it is assumed that the effects associated with the ordering of atoms in the crystal lattice and with the polarization of particles due to the spin-orbit coupling can be ignored.

The derivation of transport equations takes into account the balance of particles of species $j$: $dN_j = \Phi_j(\mathbf{r}, E, \mathbf{\Omega})\, d\mathbf{r}\, dE\, d\mathbf{\Omega}$ in the volume $d\mathbf{r}$ in the neighborhood of the point $\mathbf{r}$. The energy of these particles is in the range from $E$ to $E + dE$ and the direction of the particle velocities is confined within the solid angle $d\mathbf{\Omega}$ near $\mathbf{\Omega}$. Here $\Phi_j(\mathbf{r}, E, \mathbf{\Omega})$ is the *differential flux density* of the particles of species $j$ at the point $\mathbf{r}$, which is sometimes called the angular flux density.

The number of particles in the phase space may change because of (1) the escape of particles from the volume element $d\mathbf{r}$ of the phase space, with $E$ and $\mathbf{\Omega}$ remaining constant; (2) the departure of particles as a result of interaction with the atomic nuclei; (3) the decay of unstable particles; (4) the continuous loss of energy of the charged particles as a result of electromagnetic interaction with a small energy transfer (the *continuous slowing down* of particles); (5) the increase in the number of particles due to the elastic or inelastic scattering at which a particle of species $i$ with the coordinates $E'$ and $\mathbf{\Omega}'$ in the phase space near the point $\mathbf{r}$ produces a particle of species $j$ in the energy interval $dE$ near $E$ and in the interval of directions $d\mathbf{\Omega}$ near $\mathbf{\Omega}$ (in the special case, $i = j$); and (6) the source which emits the particles of species $j$ directly into the relevant volume of the phase space.

Establishing a balance of particles of species $j$ in the volume element of the phase space, $d\mathbf{r}\, dE\, d\mathbf{\Omega}$, we find the *system of steady-state transport equations* which we are seeking

*The wavelength of the particle must be much shorter than the distance between the atoms of the substance, while inside the nucleus the wavelength of the particle must be smaller than the internucleon distances.

† In 1872, L. Boltzmann was first to derive these equations in the gaskinetic theory.

$$\hat{L}_j \Phi_j = Q_j + G_j \tag{1.26}$$

with the boundary conditions at the convex surface $S$

$$\Phi_j(\mathbf{r}, E, \boldsymbol{\Omega})|_S = \Phi_0 j(E, \boldsymbol{\Omega}), \boldsymbol{\Omega}\cdot\mathbf{n} < 0, \tag{1.27}$$

where $\mathbf{n}$ is the normal to the surface $S$.

In the system of equations (1.26) $j$ specifies all particle species that are considered in the problem. Furthermore, we use the following notation:

$$\hat{L}_j = \boldsymbol{\Omega}\nabla + \Sigma_j(\mathbf{r}, E) + \Sigma_{jD}(\mathbf{r}, E) - \frac{\partial}{\partial E}\beta_j(\mathbf{r}, E); \tag{1.28}$$

$$Q_j = \sum_i \int d\boldsymbol{\Omega} \int dE' \, \Sigma^s_{ij}(\mathbf{r}, E' \to E, \boldsymbol{\Omega}' \to \boldsymbol{\Omega})\Phi_i(\mathbf{r}, E', \boldsymbol{\Omega}'); \tag{1.29}$$

$$G_j = G_j(\mathbf{r}, E, \boldsymbol{\Omega}). \tag{1.30}$$

Here $G_j$ is the density of the external sources. The terms in (1.28) are sequentially responsible for each one of the processes (1)–(4) which are indicated above, while (1.29) and (1.30) are responsible for processes (5) and (6), respectively; $\Sigma_j(\mathbf{r}, E)$ is the total macroscopic cross section for the interaction of the particles of energy $E$ with the nuclei of the medium near the point $\mathbf{r}$; $\Sigma_j = n(\mathbf{r})\sigma_j(\mathbf{r}, E)/\rho(\mathbf{r})$, where $\rho(\mathbf{r})$ and $n(\mathbf{r})$ are the distribution density and the target density at the point $\mathbf{r}$; $\sigma_j(\mathbf{r}, E)$ is the microscopic cross section; $\Sigma_{jD}(\mathbf{r}, E)$ is the macroscopic cross section for the decay of unstable particles; $\Sigma_{jD}^{-1} = \lambda_{jD} = c\tau_j(p_j/m_j)\rho(\mathbf{r})$ is the range of the particle before its decay, with a lifetime $\tau_j$, mass $m_j$, and momentum $p_j$; $\beta_j(\mathbf{r}, E) = -\frac{1}{\rho(\mathbf{r})} \times \frac{dE_j}{dx}(\mathbf{r}, E)$ is the stopping power of the substance for charged particles at the point $j$; and $\Sigma^s_{ij}(\mathbf{r}, E' \to E, \boldsymbol{\Omega}' - \boldsymbol{\Omega})$ is the macroscopic differential cross section of the inclusive reaction $i + A \to j + X$:

$$\Sigma^s_{ij}(\mathbf{r}, E' \to E, \mu_s) = \frac{n(\mathbf{r})}{\rho(\mathbf{r})} p_j \, f_{ij}(\mathbf{r}, \mathbf{p}_j, S), \tag{1.31}$$

where $f_{ij} = Ed^3\sigma/d^3p$ is the invariant single-particle differential cross section (1.13) (we make use of the invariance of the phase volume, $d^3p/E = pdEd\boldsymbol{\Omega}$). The assumptions used to derive the system of equations (1.26) allow us to restrict the analysis to the azimuthal symmetry of cross section (1.31), where the *scattering indicatrix*

$$\mathscr{P}_{ij} = \Sigma^s_{ij}/\Sigma_i \tag{1.32}$$

depends exclusively on the scalar product

$$\mu_s = \boldsymbol{\Omega}'\boldsymbol{\Omega} = \cos\theta_s,$$

where $\theta_s$ is the scattering angle in the laboratory frame.

In the summation in (1.29) the index $i$ runs through the same values as $j$.

In writing the *nonsteady-state transport equations* we must add on the left side of (1.26) the term

$$\frac{1}{v}\frac{\partial}{\partial t}\Phi_j(\mathbf{r}, E, \mathbf{\Omega}, t),$$

where $v$ is the velocity of the particle of species $j$.

The general boundary value problem (1.26), (1.27) cannot be solved simultaneously for all particle species of the electromagnetic cascade over a broad energy range in the case of arbitrary geometry and systems of arbitrary dimensions.

Methods for the solution of the problem for some special cases are described in Refs. 10–14, for example. If only one type of particles is considered, system (1.26), (1.27) reduces to a single equation. For low energies, well-developed methods are available for the solution of such problems for photons,[10] neutrons,[11,14] and electrons.[13]

The kinetic equations can be simplified appreciably in several cases. In one-dimensional problems, for example, the operator $\mathbf{\Omega}\nabla$ reduces to the operators

$$\mathbf{\Omega}\nabla = \begin{cases} \mu \dfrac{\partial}{\partial x}, & \text{(plane geometry)} \\ \mu \dfrac{\partial}{\partial r} + \dfrac{1-\mu^2}{r}\dfrac{\partial}{\partial \mu}, & \text{(spherical geometry)} \end{cases} \tag{1.33}$$

where $\mu = \cos\theta$.

In azimuthally symmetric problems, the dimensionality of the phase space decreases here by a factor of 2 $(\mathbf{r}, E, \mathbf{\Omega}) \rightarrow (x, E, \mu)$.

If all cross sections are independent of the energy in the range studied in the problem, we can obtain one-velocity equations by integrating (1.26) over the energy. Here we have $(\mathbf{r}, E, \mathbf{\Omega}) \rightarrow (\mathbf{r}, \mathbf{\Omega})$. This also includes the *multigroup approach*, where the cross sections are averaged and the equations are integrated within the limits of the groups into which the energy interval under study is partitioned.

In homogeneous media the spatial dependences vanish in all cross sections and in the stopping power $\beta$ of the substance.

In problems involving a strong scattering anisotropy, it is sometimes possible to use a *simple approximation* in which the scattering indicatrix (1.32) is written in the form

$$\left. \begin{aligned} \mathscr{P}(E', E, \mu_s) &= K(E', E)\,\delta(1-\mu_s)/2\pi; \\ K(E', E) &= \int_{\Delta\Omega} \mathscr{P}(E', E, \mu_s)\, d\Omega, \end{aligned} \right\} \tag{1.34}$$

where $\Delta\mathbf{\Omega}$ is the cone around the direction of motion of the primary particle, in which the momenta of all secondary particles are situated. In the one-dimensional case, the dimensionality of the phase space in such problems can be reduced to two: $(\mathbf{r}, E, \mathbf{\Omega}) \rightarrow (x, E)$. In this particular approximation we obtain the largest number of solutions of the electromagnetic-cascade equations, without the use of the Monte Carlo method.[12]

*The small-angle approximation* is frequently used at high energies. In this approximation the expansion of the distribution functions in the angular variable $(1-\mu \ll 1)$ is restricted to the first-order small terms.

The *spherical-harmonic* method is used extensively at low energies. In this method the differential flux density of the particles and the scattering indicatrix are represented as a Legendre polynomial series

$$\left.\begin{aligned} \Phi(\mathbf{r}, E, \boldsymbol{\Omega}) &= \frac{1}{4\pi}\sum_{l=0}^{\infty}(2l+1)\Phi_l(\mathbf{r}, E)\,P_l(\mu); \\ \mathscr{P}(E', E, \mu_s) &= \frac{1}{4\pi}\sum_{l=0}^{\infty}(2l+1)\,\mathscr{P}_l(E', E)P_l(\mu_s). \end{aligned}\right\} \tag{1.35}$$

In the case of weak anisotropy of the angular distribution we can restrict the analysis to the first two terms in expansions (1.35), which corresponds to the *diffusion approximation*.

The Monte Carlo method is the most general-purpose method among the present methods (see Sec. 1.5 and Chaps. 4, 6, and 7).

The reviews of these methods and of other methods which are used in the transport theory of low-energy radiation ($E < 15$ MeV) are found in Refs. 10–14. The present-day methods which are used in the study of electron–photon showers are described in Chap. 4, those which are used in the cascade calculations are described in Chaps. 5 and 6, and those which are used in the muon transport calculations are described in Chap. 7.

Despite the development and proven success of powerful universal computer programs for radiation transport calculations, we have seen a rapid development of approximate analytic and semianalytic methods which is still continuing. The reason for this trend was clearly explained by Fano, noting that within a reasonably broad range of variation of the phase-space coordinates, none of the approximate methods is considered satisfactory, but each one of them can be used to solve particular transport problems. Analysis of these approximate methods can be used to effectively formulate from them a systematic theory.

The use of physical peculiarities of the problem under consideration is the most radical way of simplifying the procedure by which the solution of the transport problem can be found. This is one of the reasons that considerable attention in this book is given to the physics of particle interactions. Most of the approximate methods described above are based on this approach. In Chaps. 4–7 the reader will find many cases in which the results of Chaps. 2 and 3 are used.

Finally, some of the frequently used *functionals* of the differential flux density of the radiation (the index $j$ is omitted) are the differential flux density of the particle energy

$$I(\mathbf{r}, E, \boldsymbol{\Omega}) = E\Phi(\mathbf{r}, E, \boldsymbol{\Omega}),$$

the spatial-angular flux density of the particles

$$\Phi(r, \boldsymbol{\Omega}) = \int_0^{\infty} \Phi(\mathbf{r}, E, \boldsymbol{\Omega})\, dE, \tag{1.36}$$

the spatial-energy flux density of the particles

$$\Phi(\mathbf{r}, E) = \int_{4\pi} \Phi(\mathbf{r}, E, \boldsymbol{\Omega})\, d\boldsymbol{\Omega}, \tag{1.37}$$

and the spatial distribution of the flux density of the particles

$$\Phi(\mathbf{r}) = \int_0^\infty \Phi(\mathbf{r}, E)\, dE = \int_{4\pi} \Phi(\mathbf{r}, \boldsymbol{\Omega})\, d\boldsymbol{\Omega}. \tag{1.38}$$

Integrating the functional $I(\mathbf{r}, E, \boldsymbol{\Omega})$ in a manner similar to that in (1.36)–(1.38), we find the corresponding definitions for the energy flux density.

Sometimes the *current characteristics* of the radiation field are used. These characteristics are the differential current density of the particles

$$\mathbf{J}_\Phi(\mathbf{r}, E, \boldsymbol{\Omega}) = \boldsymbol{\Omega}\Phi(\mathbf{r}, E, \boldsymbol{\Omega})$$

and the differential current density of the particle energy

$$\mathbf{J}_I(\mathbf{r}, E, \boldsymbol{\Omega}) = \boldsymbol{\Omega} I(\mathbf{r}, E, \boldsymbol{\Omega}).$$

The functionals of the current characteristics are written in a manner similar to Eqs. (1.36)–(1.38).

In application to the specific problems, the differential flux density $\Phi(\mathbf{r}, E, \boldsymbol{\Omega}, t)$ can be interpreted as a path traversed (per unit time) by the particles belonging to a unit volume of the phase space. In particular, this interpretation implies the definition of the *differential collision* (absorption, scattering) *density*

$$F(\mathbf{r}, E, \boldsymbol{\Omega}) = \Sigma(\mathbf{r}, E)\, \Phi(\mathbf{r}, E, \boldsymbol{\Omega}). \tag{1.39}$$

If the differential flux density is known, the reading (response) of any additive detector* can be represented as

$$N = \int\int\int D(\mathbf{r}, E, \boldsymbol{\Omega})\, \Phi(\mathbf{r}, E, \boldsymbol{\Omega})\, d\mathbf{r}\, dE\, d\boldsymbol{\Omega}, \tag{1.40}$$

where $D(\mathbf{r}, E, \boldsymbol{\Omega})$ is the *sensitivity function of the detector*, which is the average contribution to the detector readings from a unit path length of the particle with the coordinates $(\mathbf{r}, E, \boldsymbol{\Omega})$ in the detector volume.

## 5. The Monte Carlo method in radiation-transport problems

The increasing requirements imposed on the mathematical procedures for the solution of problems involving the transport of radiation through matter and the rapid development of computer technology have made the Monte Carlo method the principal, if not the only, method in many fields of application. In its simplest and at the same time most dependable and common modification—direct mathematical modeling—this method involves numerical

*The term "detector" is meant here in the broadest sense (Refs. 13 and 15).

simulation of the interaction and propagation of particles in matter. The use of various modifications of the Monte Carlo method, the so-called dispersion-reduction methods, makes it possible to greatly simplify the solution of this problem in certain cases. Here a very high accuracy can be reached in a limited volume of the phase space. Because the problems of the transport theory are linear, the results of numerical simulation can be used for the real ensembles of particles.

The general theory of the Monte Carlo method and the application of this method to radiation transport problems are dealt with in Refs. 14 and 16–19, for example. The basic concepts of the method, principles used to construct the trajectories, and estimates of the functionals used in Chaps. 4, 6, and 7 to describe the algorithms for simulation of the cascades are discussed below. We assume that the reader is familiar with the principles of the probability theory and that he has read the preceding sections of this chapter.

The random discrete quantity $\xi$ is defined on the discrete set $x_1, x_2,..., x_n$. The probability that $\xi$ will have the value $x_i$ is $P(\xi = x_i) = p_i$, where all $p_i > 0$, and $\sum_{i=1}^{n} p_i = 1$.

The expression $\mathbf{M}\xi = \sum_{i=1}^{n} x_i\, p_i$ is called the *expectation value* (the mean value) of the random quantity $\xi$.

The quantity characterizing the scatter in the values of $\xi$ around the mean value is called *dispersion*, which is defined as

$$D\xi = \mathbf{M}[(\xi - \mathbf{M}\xi)^2] = \mathbf{M}(\xi^2) - (\mathbf{M}\xi)^2. \tag{1.41}$$

The random continuous quantity $\xi$ is determined in the interval $(a,b)$ by the function $p(x)$, which is called the *probability density*. The probability that $\xi$ will be in the interval $(a,x)$ is given by

$$P(a < \xi < x) = \int_a^x p(x')\, dx', \tag{1.42}$$

where $p(x) > 0$ and $\int_a^b p(x)\, dx = 1$.

The expectation value of the random continuous quantity is $\mathbf{M}\xi = \int_a^b xp(x)\times dx$.

It is easy to show[16] that the relation $\mathbf{M}f(\xi) = \int_a^b f(x)\, p(x)\, dx$ holds for the random continuous function $f(x)$.

For a generator of random numbers $\gamma$, which are uniformly distributed in the interval (0,1), the use of the equation

$$\gamma = P(x) = \int_a^x p(x')\, dx' \tag{1.43}$$

allows us to randomly choose the values

$$x = P^{-1}(\gamma). \tag{1.44}$$

Such a selection method is called the *inverse-function method* (or the inversion method). This method is used when the integral in (1.43) can be expressed in terms of the elementary functions. Otherwise, the *Neumann method* (or the rejection method) can be used. This method can be summarized as follows.

In the interval $(a, b)$ we shall redefine the density function of the random quantity $\xi$ as follows:

$$p^*(x) = p(x)/Max[p(x)]. \tag{1.45}$$

We choose two random numbers, $\gamma_1$ and $\gamma_2$, and calculate the quantity $x' = a + \gamma_1(b - a)$. If $\gamma_2 < p^*(x')$, we have $\xi = x'$. If, on the other hand, this condition does not hold, we can discard the $(\gamma_1, \gamma_2)$ pair, choose a new pair of random numbers, and then repeat the procedure.

Direct simulation of the physical picture of the transmission of particles through matter is the most common way in which the Monte Carlo method is used in the radiation transport problems. The range of a particle, $R$, before its interaction can easily be found by solving kinetic equation (1.26), whose right side is zero, $\hat{L}\Phi = 0$. Analyzing this equation in a coordinate system associated with the particle, we find its solution for the neutral particles, $\Phi = \Phi_0 \exp(-\Sigma R)$. The corresponding probability density is $p(r) = \exp(-r)$, where $r = \Sigma R$; $\Sigma$ is the macroscopic cross section for the interaction with the nuclei of the atoms of the medium. Solving Eq. (1.43), we find the algorithm for simulating the range $R$ of neutral particles before their interaction: $r = -\ln(1-\gamma)$ or $R = -\Sigma^{-1}\ln(1-\gamma)$. The appropriate algorithms for charged particles are considered below.

The radius vector of the new interaction point can be determined from the previous coordinates $\mathbf{r}_0$ as follows:

$$\mathbf{r} = \mathbf{r}_0 + R\boldsymbol{\Omega}, \tag{1.46}$$

where $\boldsymbol{\Omega}$ is a unit vector in the direction of motion of the particle.

At this point simulation of the type of interaction and of the interaction itself is carried out by a particular method. Also simulated are the number and type of the newly produced particles, the energy of these particles and their emission angles, and the characteristics of the residual nucleus. The produced particles are transported to the new interaction points, with allowance for the particular features of the system and the possible quasicontinuous effect of the electromagnetic processes. Virtually any functional of the random quantities $\xi$ can be found directly during the simulation. The simulation ends when all particles are absorbed or emitted from the system. The simulation is then repeated $N$ times until the required statistical accuracy of the functionals $\Phi$ is reached.

According to the central limiting theorem of the probability theory,[16] for large values of $N$ the distribution of the sum $\sum_{n=1}^{N} \xi_n$ is approximately normal. The following relation is therefore used to estimate the functional $\Phi$:

$$P\left\{\left|\frac{1}{N}\sum_{n=1}^{N} \xi_n - \Phi\right| < \delta\right\} \approx 0.997. \tag{1.47}$$

This relation shows that the error of the estimate is no greater than $\delta = 3\sqrt{D\xi/N}$ with a probability $P \simeq 0.997$.

The principal *methods of estimating* the differential flux density of particles and the derivatives of the functionals for the simulation of the radiation transport are

(1) Estimate based on the collisions (absorption, scattering)—computation of the number of collisions of the particles of species $j$ in the phase space $V_{klm} = (\Delta\mathbf{r}_k, \Delta E_l, \Delta\boldsymbol{\Omega}_m)$, on the basis of Eq. (1.39):

$$\Phi^{(j)}_{klm} = \frac{1}{NV_{klm}} \sum_{n=1}^{N} \sum_{i} \frac{w^{(j)}_i}{\Sigma^{(j)}_{kl}}, \tag{1.48}$$

where the outer summation is over the number of simulations $N$ and the inner summation in a given simulation is over the number of events of the type under consideration (absorption, scattering, etc.) with the corresponding macroscopic cross section $\Sigma^{(j)}_{kl}$ for particles of species $j$; $w^{(j)}_i$ is the statistical weight of the particles (see the discussion below). In the case of direct simulation, all $w^{(j)}_i = 1$. The nonvanishing term in (1.48) is found only if the phase coordinates of the particles in a given event belong to the phase volume $V_{klm}$.

(2) Estimate from the range—computation of the sum of the lengths $L^{(j)}_i$ of the paths traversed by the particles of species $j$, which belong to the volume of the phase space $V_{klm}$:

$$\Phi^{(j)}_{klm} = \frac{1}{NV_{klm}} \sum_{n=1}^{N} \sum_{i} w^{(j)}_i L^{(j)}_i. \tag{1.49}$$

(3) Expectation-value method—the use of the sum of the expectation values in the estimates presented above. In estimate (1.49), for example, instead of the sum of the lengths of the paths traced out by the particles belonging to $V_{klm}$, we can compute the sum of their expectation values

$$\sim L^{(j)}_i = \frac{1}{\Sigma^{(j)}_{tkl}} [1 - \exp(-\Sigma^{(j)}_{tkl} \Delta R_h)], \tag{1.50}$$

where $\Sigma^{(j)}_{tkl}$ is the total macroscopic cross section for the interaction of particles $j$ of energy $E \in \Delta E_l$ with the nuclei of the atoms of the cell $\Delta\mathbf{r}_k$, and $\Delta R_k$ is the distance from the point of entry of the particle into the cell $\Delta\mathbf{r}_k$ or from the point of interaction of this particle inside the cell $\Delta\mathbf{r}_k$ as it moves toward its boundary. Expression (1.50) is used in the estimates under the assumption that the cross section $\Sigma^{(j)}_{tkl}$ is constant over the entire volume $V_{klm}$. Otherwise, this expression becomes more complicated in the case of charged particles (see, e.g., Sec. 6.2).

(4) Local estimate of the flux—the use of the local detectors to sum the probability density for the arrival from all interaction points of particles which are not scattered:

$$p = r^{-2} \exp(-\Sigma \Delta R), \tag{1.51}$$

where $r$ is the distance from the interaction point to the local detector, $\Delta R$ is a segment of this ray in the substance, and $\Sigma$ is the total macroscopic cross section for the interaction of particles with the nuclei of the atoms of the medium. Expression (1.51) changes again for charged particles which lose their energy in the electromagnetic processes as they pass through matter (see Sec. 6.2).

The quantity $(tD\xi)^{-1}$, where $t$ and $D\xi$ are respectively the counting time and the variance of the estimate per simulation, is called the *effectiveness of the estimate*. An electromagnetic cascade can be calculated as follows. The first and second estimates are effective in the central region of the electromagnetic cascade. Estimate of the range is more suitable for optically thin zones ($d \lesssim \lambda_{in}$), while estimate of the collisions is more suitable for optically thick zones ($d > \lambda_{in}$, where $d$ is the characteristic dimension of the zone). The third estimate is considerably more effective at long distances from the central region of the electromagnetic cascade. The local estimate can be used for small detectors situated outside the system under consideration.

Other methods of reducing dispersion can be used in addition to estimates (3) and (4): essential sampling, method of adjoint walks, superposition method, etc. These methods are described in detail in Refs. 14 and 16–19. We shall consider here only the essential sampling method which is used extensively in the simulation of electromagnetic cascades.

Consider a situation in which it is required to determine the functional $\Phi = \int K(x)dx$, where $K(x)$ is the scattering indicatrix, and $x = (E, \mathbf{\Omega})$. Let us multiply and divide the integrand by the probability density function $p(x)$ with the properties of (1.42):

$$\Phi = \int W(x)\,p(x)\,dx, \tag{1.52}$$

where the quantity $W(x) = K(x)/p(x)$ is called the *statistical weight*.

The integral in (1.52) is found by the Monte Carlo method from the equation $\Phi_N = \dfrac{1}{N}\sum_{k=1}^{N} W(x_k)$, where the points $x_k$ are determined by the method of inverse functions from Eqs. (1.43) and (1.44).

The function $p(x)$ of the form

$$p(x) = K(x)/\Phi \tag{1.53}$$

assures a zero dispersion. Sampling from the function $p(x)$ which approximates (1.53) is the best strategy that can be used in a calculation based on the Monte Carlo method.

The actual algorithms for the realization of random processes by the Monte Carlo method are given in Chaps. 4, 6, and 7.

# Chapter 2 Electromagnetic interaction of particles with matter. Decay of unstable particles

## 1. Bremsstrahlung

The transmission of particles which do not participate directly in the strong interactions (photons, electrons, muons) through matter is determined by their electromagnetic interactions. In this chapter we consider the principal electromagnetic processes which have a strong effect on the development of an electromagnetic cascade (EC) and its muon component. Simple expressions for the cross sections of these processes, which are useful in practical applications, are presented below. Simplification is achieved by replacing exact but cumbersome expressions by approximate equations which are simple in form (in general, accurate to within 1–2%). This remark pertains mainly to the interactions at low energies (on the order of several megaelectronvolts) and to processes involving muons.

The decays of unstable particles are considered here in order to point out the principal sources of the electromagnetic ($\pi^0 \rightarrow 2\gamma$) component and the muon $\pi^{\pm}(K^{\pm}) \rightarrow \mu^{\pm}\nu_{\mu}(\mathrm{j}\nu_{\mu})$ component of the cascade.

Bremsstrahlung is a process in which a charged particle emits a photon in the field of a nucleus and atomic electrons. This process accounts for the main energy loss of relativistic electrons and, at superhigh energies, the main energy loss of muons. Bremsstrahlung, along with the production of electron–positron pairs, is the principal process which determines the development of electromagnetic showers.

### Electron bremsstrahlung

The differential cross section of electron bremsstrahlung in the Coulomb field of a nucleus is described by the equation[20]

$$\sigma_{re}(E,\omega)\, d\omega = 4\alpha Z^2 \ln(183\, Z^{-1/3})\, r_e^2 (d\omega/\omega) F_e(E,u), \qquad (2.1)$$

where $\alpha = e^2/\hbar c$ is the fine-structure constant, $r_e$ is the classical electron radius, $Z$ is the atomic number, and $u = \omega/E$ is the ratio of the emitted photon

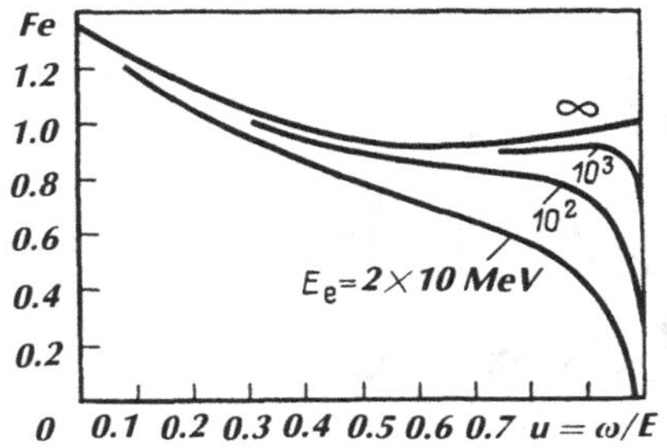


**Figure 2.1. Plots of the function $F_e(E,u)$ at various primary-electron energies $E = E_e$ [the curves were calculated for lead (Ref. 20)].**

energy to the primary electron energy in the laboratory frame. The function $F_e$ (Fig. 2.1) can be written in the form[21]

$$\begin{aligned} F_e(E,u) = {} & [1 - f_c(Z)/\ln(183\, Z^{-1/3})]\{[1 + (1-u)^2] \\ & \times [f_1(\xi)4 - (1/3)\ln Z - f_c(Z)]/ \\ & [f_1(0)/4 - (1/3)\ln Z - f_c(Z)] - (2/3)(1-u) \\ & \times [f_2(\xi)/4 - (1/3)\ln Z - f_c(Z)]/[f_1(0)/4 - (1/3)\ln Z - f_c(Z)]\}, \end{aligned} \tag{2.2}$$

where $f_c(Z) = (\alpha Z)^2 \sum_{k=1}^{\infty} 1/k(k^2 + (\alpha Z)^2)$ is the so-called *Coulomb correction*,[22] which for $\alpha Z \lesssim \frac{2}{3}$ can be approximately described within four significant digits by

$$\begin{aligned} f_c(Z) \simeq {} & (\alpha Z)^2[1/(1 + (\alpha Z)^2) + 0.20206 - 0.0369(\alpha Z)^2 \\ & + 0.0083(\alpha Z)^4 - 0.002(\alpha Z)^6]. \end{aligned}$$

The screening constant $\xi$ is proportional to the ratio of the effective size of the atom to the maximum value of the impact parameter $R_{\max}$. In the Thomas–Fermi model, the effective radius of the atom is given by the expression $a = a_0 Z^{1/3}$, where $a_0 = \hbar^2/m_e e^2$ is the Bohr radius of a hydrogen atom. In this model the screening constant $\xi$ can be calculated from the equation

$$\xi \simeq (136 m_e c^2/E)\, Z^{-1/3} u/(1-u). \tag{2.3}$$

The functions $f_1$ and $f_2$ can be found within an error no greater than 1–2% (with $\xi \leqslant 1$) from the equations[21]

$$f_1(\xi) = 20.867 - 3.242\xi + 0.625\xi^2;$$

$$f_2(\xi) = 20.209 - 1.930\xi - 0.086\xi^2.$$

For $\xi > 1$ we have $f_1(\xi) = f_2(\xi) = 21.12 - 4.184 \ln(\xi + 0.952)$.

If the screening is complete, the screening constant is $\xi \simeq 0$ and $f_1(0) = f_2(0) + \frac{2}{3} = 4 \ln 183$.

The function $F_e$ [see Eq. (2.2)] implies that it depends only slightly on the atomic number $Z$.

In addition to the bremsstrahlung in the nuclear field, which leaves the atom unexcited, there can also be a bremsstrahlung induced by atomic electrons, which causes the atom to undergo a transition to one of its excited states. The cross section for bremsstrahlung of a fast electron induced by a free electron is the same as the cross section for bremsstrahlung of an electron induced by a nucleus with $Z = 1$ in the absence of screening.[23,24] The reason for this correspondence is that the principal contribution to the bremsstrahlung cross section comes from the region of small momentum: $|\mathbf{q}| \sim q_{\min} = m_e^2 c^3 u / 2E(1-u)$. If $|\mathbf{q}| \ll m_0 c$, the electron recoil (and especially nuclear recoil) can be ignored. The condition $|\mathbf{q}| \ll m_e c$ holds if $m_e c^2/E \ll 1$, i.e., in the case of relativistic electrons. The differential cross section for bremsstrahlung of a relativistic electron in the field of atomic electrons (with allowance for their coupling in the atom) is given by the equation[25]

$$\sigma_{ree}(E,\omega)\, d\omega = 4\alpha Z r_e^2 F_{ee}(u)\, d\omega/\omega, \tag{2.4}$$

where

$$F_{ee}(u) = [1 + (1-u)^2]\ln(1444 Z^{-2/3}) - (2/3)(1-u)\ln(1212 Z^{-2/3}).$$

A comparison of expressions (2.1), (2.2), and (2.4) shows that the total cross section for bremsstrahlung of an electron induced by an atomic nucleus or by atomic electrons can be written

$$\sigma_{re}(E,\omega)\, d\omega = 4\alpha Z (Z + \eta_e) \ln(183 Z^{-1/3})\, r_e^2 (d\omega/\omega) F_e(E,u),$$

where $\eta_e = \ln(1444\, Z^{-2/3})/\ln(183 Z^{-1/3})$. Within an error no greater than $|1-\eta_e|/Z \lesssim 0.4/Z$, this formula can be approximated by a simpler expression

$$\sigma_{re}(E,\omega)\, d\omega = 4\alpha Z(Z+1)\ln(183 Z^{-1/3}) r_e^2 (d\omega/\omega)\, F_e(E,u). \tag{2.5}$$

The macroscopic bremsstrahlung cross section (measured in $\text{cm}^{-1}$) for an element is

$$\Sigma_{re}(E,\omega) = (N_A/A)\,\rho\sigma_{re}(E,\omega). \tag{2.6}$$

The radiation length is given by the expression

$$t_r^{-1} = 4\alpha r_e^2 Z(Z+1)\ln(183 Z^{-1/3})\,\rho N_A/A. \tag{2.7}$$

Here, as in Eq. (2.6), $N_A$ is Avogadro's number, $A$ is the atomic mass number of the substance, and $\rho$ is the density of the substance. The macroscopic bremsstrahlung cross section (expressed in units of radiation length) is given by the expression

$$\Sigma_{re}(E,\omega) d\omega = (d\omega/\omega) F_e(E,u). \tag{2.8}$$

Since the function $F_e(E,u)$ is nearly independent of $Z$, the macroscopic bremsstrahlung cross section (expressed in units of radiation length) is essentially the same for various media.

For a homogeneous compound the macroscopic bremsstrahlung cross section is given by

$$\Sigma(E,\omega) = N_A \rho \sum_i w_i \sigma_{re}^i \Big/ \sum_i w_i A_i,$$

where the sum is taken over all elements that comprise the medium. In a compound the radiation length is given by

$$t_r^{-1} = 4\alpha r_e^2 N_A \rho \Big[ \sum_i Z_i (Z_i + 1) \ln(183 Z_i^{-1/3}) w_i \Big] \Big/ \sum_i w_i A_i .$$

Here $w_i$ are the concentrations of the elements that constitute the substance.

In the ultrarelativistic case, the angular distribution of bremsstrahlung is determined by the equation[24]

$$d\sigma_{re} = \frac{\delta d\delta}{(1+\delta^2)^2} \frac{d\omega}{\omega} \Phi(E,\omega,\delta),$$

where $\delta = E\theta / m_e c^2$; $\theta$ is the angle of emission of a bremsstrahlung photon in the laboratory frame. The function $\Phi(E,\omega,\delta)$ depends on $\delta$ only slightly. It can be seen from the expression above that the maximum bremsstrahlung intensity occurs near the angle $\theta = \theta_{\max} = m_e c^2 / E$. Note that the angle $\theta_{\max}$ is independent of the energy of the emitted photon.

It follows from expression (2.1) for the energy spectrum of bremsstrahlung that in the limit $\omega \to 0$ its intensity tends toward infinity and the total bremsstrahlung cross section diverges as $\ln \omega_{\min}^{-1}$. This so-called *infrared catastrophe* is a consequence of the inapplicability of the perturbation theory [in terms of which Eq. (2.1) was derived] to the processes involving long-wave photons. The divergence can be eliminated by introducing radiative corrections of this sort.[26] The radiative corrections must be taken into account in order to describe bremsstrahlung of a single atom. When bremsstrahlung occurs in a condensed medium, the radiation spectrum is modified at low frequencies due to effects which are discussed below.

First, radiation causes the medium to be polarized. This polarization can be taken into account by introducing the factor[21]

$$F_p = [1 + n_e r_e \lambda_e^2 E^2 / \pi \omega^2]^{-1}, \tag{2.9}$$

where $n_e$ is the electron density in the medium, and $\lambda_e$ is the Compton electron wavelength. For $\omega \gg \omega_c = E(n_e r_e \lambda_e^2 / \pi)^{1/2}$ the factor $F_p \simeq 1$, and for $\omega \ll \omega_c$ this factor can be determined from the equation $F_p \simeq \omega^2 / \omega_c^2$. In this case the bremsstrahlung energy spectrum has the form $\sigma_{re} d\omega \sim \omega d\omega$, the infrared catastrophe vanishes, and the total cross section is finite.

Secondly, the Landau–Pomeranchuk effect is seen at very high energy of the primary electron (higher than $10^{13}$ eV) and also during emission of very soft photons.[27] Under these conditions the longitudinal momentum transfer to the nucleus is very small ($q_\parallel \simeq q_{\min}$). Because of the uncertainty principle, the interaction region is extended considerably in the longitudinal direction. As a result, the electrons (both primary and secondary) which participate in the process interact with a group of atoms strung out in the direction of motion of the primary electron, which causes them to undergo multiple Coulomb scattering by these atoms. This scattering disrupts the bremsstrahlung coherence, and hence reduces the radiation cross section.

**Table 2.1. Values of the functions $\varphi(s)$ and $h(s)$.**

| | | | | | | | |
|---|---|---|---|---|---|---|---|
| $s$ | 0.0 | 0.05 | 0.1 | 0.2 | 0.3 | 0.4 | 0.5 |
| $\varphi(s)$ | 0.000 | 0.258 | 0.446 | 0.686 | 0.805 | 0.880 | 0.931 |
| $h(s)$ | 0.000 | 0.094 | 0.206 | 0.475 | 0.695 | 0.800 | 0.875 |
| $s$ | 0.6 | 0.7 | 0.8 | 0.9 | 1.0 | 1.5 | 2.0 |
| $\varphi(s)$ | 0.954 | 0.965 | 0.975 | 0.985 | 0.990 | 0.998 | 0.999 |
| $h(s)$ | 0.917 | 0.945 | 0.963 | 0.975 | 0.985 | 0.994 | 0.998 |

The bremsstrahlung cross section, with allowance for the Landau–Pomeranchuk effect and the polarization of the medium, is[21,27]

$$\sigma_{re}d\omega = 4\alpha r_c^2 Z(Z+1)\frac{d\omega}{\omega}\ln(183Z^{-1/3})[1-f_c(Z)/\ln(183Z^{-1/3})]b(s)$$
$$\times\{[1+(1-u)^2][2\varphi(s)+h(s)]/3-2uh(s)/3\}[1+w_0^2/u^2]^{-1}, \tag{2.10}$$

where

$$\varphi(s) = 12s^2\int_0^\infty \coth\left(\frac{t}{2}\right)\exp(-st)\sin(st)dt - 6\pi s^2;$$
$$h(s) = 24s^2\left[\frac{\pi}{2} - \int_0^\infty \exp(-st)\frac{\sin(st)}{\sin(t/2)}dt\right];$$
$$b(s) = \begin{cases}1, & s\geqslant 1;\\ 1+\ln(s)/\ln(s_0), & 1>s\geqslant s_0;\\ 2, & s_0>s;\end{cases}$$
$$s = 1.37\times10^3\left[\frac{ut_r m_e c^2}{(1-u)Eb(s)}\right]^{1/2};\; s_0=(Z^{1/3}/183)^2;\; w_0^2 = n_e r_e \lambda_e^2/\pi. \tag{2.11}$$

The values of the functions $\varphi(s)$ and $h(s)$ are given in Table 2.1.

At moderately high energies of the primary electron a semiempirical correction must be introduced into the bremsstrahlung cross section[21,28]: $\widetilde{K}(Z) = [1+(\alpha Z)^{1/2}3m_e c^2/E]$. At energy $E\lesssim 50$ MeV the cross section in Eq. (2.5) should be multiplied by the factor $K(Z)$ and $f_c(Z)$ should be set equal to zero. These steps would assure a reasonable calculation error (within several percent).

**Bremsstrahlung of particles heavier than the electron**

Since the bremsstrahlung cross section is inversely proportional to the square of the mass of the emitting particle ($\sigma_{re}\sim r_e^2\sim 1/m_e^2$), for heavy particles this process becomes important only at a very high energy. For definiteness, the corresponding expressions will therefore be given for a muon—the particle which follows the electron on the mass scale. The equation for muon bremsstrahlung in the Coulomb field of a nucleus is derived from the equation for the emission of an electron by substituting the muon mass, $m_\mu$, for the electron mass. Furthermore, the nuclear radius should be used as the minimum value of the impact parameter, since the Compton muon wavelength, $\lambda\!\!\!^{-}_\mu = \hbar/m_\mu c$, is smaller than the nuclear radius. Taking these points into account, we can write the cross section for muon bremsstrahlung in the Coulomb field of a nucleus as[1]

$$\sigma_{r\mu}(E,\omega)\,d\omega = 4\alpha Z^2 \left(\frac{m_e}{m_\mu}\right)^2 r_e^2 \frac{d\omega}{\omega} F_\mu(E,u),$$

where

$$F_\mu(E,u) = \left[1 + (1-u)^2 - \frac{2}{3}(1-u)\right]$$
$$\times \ln\left[\frac{(3/2)183\,(m_\mu/m_e)\,Z^{-2/3}}{(1/2)183\sqrt{e}\,(m_\mu^2c^2/Em_e)\,uZ^{-1/3}/(1-u)+1}\right],$$

$u = \omega/E$, and $\omega$ and $E$ are the photon energy and the primary muon energy, respectively.

For muon bremsstrahlung the screening factor is

$$\xi = 100\left(\frac{m_\mu}{m_e}\right)\frac{m_\mu c^2}{E}\frac{u}{1-u}Z^{-1/3}.$$

In the case of total screening ($\xi \simeq 0$) we have

$$F_\mu(E,u) \simeq \left[1 + (1-u)^2 - \frac{2}{3}(1-u)\right]\ln\left(\frac{3}{2}183\frac{m_\mu}{m_e}Z^{-2/3}\right).$$

The bremsstrahlung of a muon upon its collision with atomic electrons should be considered separately. At a high energy $E \gg m_\mu^2c^2/m_e$, when the momentum transfer to the electron is $|\mathbf{q}| = m_\mu^2c^3u/2E(1-u) \ll m_e c$, and the electron recoil can be ignored, the total differential cross section for bremsstrahlung of a muon induced by a nucleus or by atomic electrons is given by[1]

$$\sigma_\mu(E,\omega)d\omega = 4\alpha\left(\frac{m_e}{m_\mu}\right)^2 r_e^2 Z(Z+\eta_\mu)\frac{d\omega}{\omega}F_\mu(E,u).$$

In the absence of screening, the parameter $\eta_\mu$ is determined from the equation

$$\eta_\mu = \ln\frac{2E(1-u)}{\sqrt{e}m_\mu c^2u} \quad \ln\left[\frac{(3/2)183\,(m_\mu/m_e)Z^{-2/3}}{\dfrac{183\sqrt{e}}{2}\dfrac{m_\mu^2c^2u}{m_eE(1-u)}Z^{-1/3}+1}\right].$$

In the case of total screening we have

$$\eta_\mu = \ln\left(1444\frac{m_\mu}{m_e}Z^{-2/3}\right) \quad \ln\left(\frac{3}{2}183\frac{m_\mu}{m_e}Z^{-2/3}\right).$$

The parameter $\eta_\mu$ in this case can be replaced by 1 within 1.5%. The differential cross section of the process can then be written

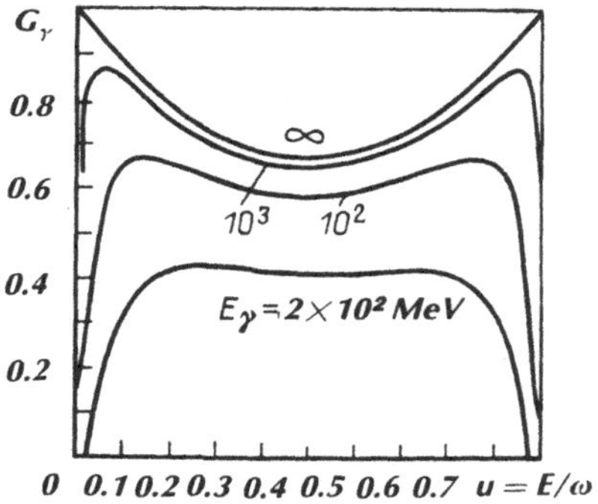


**Figure 2.2. Plots of the function $G_\gamma(\omega,u)$ at various primary-photon energies $\omega = E_\gamma$ [the curves were calculated for lead (Ref. 20)].**

$$\sigma_{r\mu}(E,\omega)d\omega = 4\alpha\left(\frac{m_e}{m_\mu}\right)^2 r_e^2 Z(Z+1)\frac{d\omega}{\omega}F_\mu(E,u).$$

## 2. Electron-positron pair production

**The production of electron–positron pairs by a photon**

The photon production of an electron–positron pair in a Coulomb nuclear field is the reverse of the bremsstrahlung. The differential cross sections of these processes are related by[20] $\sigma_{p\gamma}(\omega,E) = \sigma_{re}(E,\omega)E^2/\omega^2$. By analogy with Eq. (2.1), we can write[20]

$$\sigma_{p\gamma}(\omega,E)\,dE = 4\alpha Z^2 r_e^2 \ln(183Z^{-2/3})\frac{dE}{\omega}G_\gamma(\omega,u), \tag{2.12}$$

where $\omega$ is the energy of the primary photon, $E$ is the energy of one of the electrons of the pair, and $u = E/\omega$.

The screening factor is $\xi = (136m_ec^2/\omega)\times Z^{-1/3}/u(1-u)$, and the function (Fig. 2.2)

$$\begin{aligned}G_\gamma(\omega,u) = {} & [1 - f_c(Z)/\ln(183\,Z^{-1/3})]\\ & \times\{[u^2+(1-u)^2](f_1(\xi)/4 - (1/3)\ln Z - f_c(Z))/(\ln(183\,Z^{-1/3})\\ & - f_c(Z)) + (2/3)u(1-u)(f_2(\xi)/4 - (1/3)\ln Z\\ & - f_c(Z))/(\ln(183\,Z^{-1/3}) - f_c(Z))\}.\end{aligned}$$

Here the functions $f_1(\xi)$, $f_2(\xi)$, and $f_c(Z)$ are the same as those used for electron bremsstrahlung, and the function $G_\gamma(\omega,u)$ is nearly independent of the nuclear charge.

The total cross section for the production of electron–positron pairs by a photon in the field of a nucleus and atomic electrons can be accurately described by

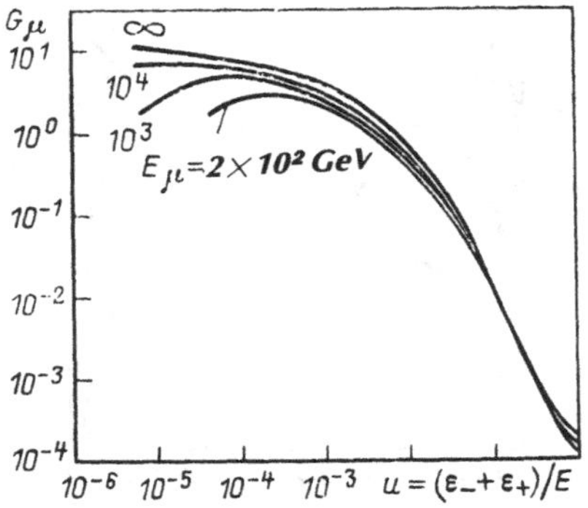


**Figure 2.3. Plots of the function $G_\mu(E,u)$ at various primary-muon energies $E=E_\mu$ [the curves were calculated for lead (Ref. 1)].**

$$\sigma_{p\gamma}(\omega,E)\,dE = 4\alpha Z\,(Z+1)\ln(183\,Z^{-1/3})r_e^2\,(dE/\omega)\,G_\gamma(\omega,u). \tag{2.13}$$

The macroscopic cross section for the production of electron–positron pairs in units of radiation length is given by the expression

$$\Sigma_{p\gamma}(\omega,E)\,dE = (dE/\omega)G_\gamma(\omega,u). \tag{2.14}$$

Since the function $G_\gamma(\omega,u)$ is virtually independent of $Z$, the quantity (2.14) is nearly the same for various media.

In the ultrarelativistic case, the electron and positron are emitted principally in the forward direction: their effective angle of emission is $\theta\sim m_ec^2/\omega$. The Landau–Pomeranchuk effect must be taken into account at superhigh energies and in the case of electron bremsstrahlung. The differential cross section for the production of electron–positron pairs by a photon can then be written as[21,27]

$$\begin{aligned}\sigma_{p\gamma}(\omega,E)\,dE = {}& 4\alpha Z\,(Z+1)\ln(183\,Z^{-1/3})[1-f_c(Z)/\ln(183\,Z^{-1/3})]\\ &\times b(s)\{[u^2+(1-u)^2](1/3)[2\varphi(s)+h(s)]\\ &+(2/3)u(1-u)\times h(s)\}.\end{aligned}$$

Here $s=1.37\times10^3\,[t_rm_ec^2/\omega u(1-u)b(s)]^{1/2}$, and the functions $b(s)$, $\varphi(s)$, and $h(s)$ are the same as those in Eq. (2.10).

At low energies the equation for the cross section of the production of electron–positron pairs must include a semiempirical correction. In Ref. 21, for example, this correction corresponds to the following addition to the total cross section for pair production: $A(Z)\ln(\omega')(\omega'=\omega/m_ec^2)$. The factor $A(Z)$, which depends on the substance, can be determined on the basis of the data of Ref. 29 from the total cross sections for the interaction of photons with the substance at low energies.

### Direct production of electron–positron pairs by a heavy particle

For definiteness, we shall again consider a muon. The differential cross section for the production of an electron–positron pair by a muon is[1]

$$\sigma_{p\mu}(E,u)du = \frac{16}{\pi} Z^2\alpha^2 r_e^2 \frac{du}{u}\left[G_{1\mu}(E,u) + \frac{m_e^2}{m_\mu^2} G_{2\mu}(E,u)\right], \tag{2.15}$$

where $u = \varepsilon/E$, $\varepsilon$ is the total energy of the pair, $E$ is the energy of the primary muon,

$$G_{1\mu}(E,u) = \int_{(x_+)_{\min}}^{1/2} R(E,u,x_+)dx_+;$$

$$R(E,u,x_+) = \int_{t_{\min}}^{t_{\max}} dt/t^2\Big\{[t(1-u+u^2/2) - m_\mu^2c^4u^2]$$
$$\times\left[\frac{1}{\gamma}\left(\frac{1}{2} - x_+x_-\right) + \frac{x_+x_-}{6\gamma^2}(2 - t/m_e^2c^4)\right]$$
$$+ \frac{tx_+^2x_-^2}{3\gamma^2 m_e^2c^4}[t(3(1-u)+u^2/2) - m_\mu^2c^4u^2]\Big\};$$

$$G_{2\mu}(E,u) = \frac{1}{12}\left[\left(\frac{4}{3} - \frac{4}{3}u + u^2\right)\left(\ln\frac{m_\mu^2u^2}{m_e^2(1-u)} - \frac{5}{3}\right) + 1 - \frac{1}{3}(1+u)^2\right]$$
$$\times\ln\left(\frac{183Z^{-1/3}m_\mu/m_e}{1 + \dfrac{183\sqrt{e}}{2}\dfrac{m_\mu^2c^2uZ^{-1/3}}{Em_e(1-u)}}\right);$$

$x_\pm = \varepsilon_\pm/\varepsilon, \gamma = 1 + (t/m_e^2c^4|)x_+x_-; \varepsilon_\pm$ is the positron (electron) energy; and $t$ is the square of the 4-momentum transfer to the nucleus.

The function $G_\mu(E,u) = G_{1\mu}(E,u) + (m_e^2/m_\mu^2)G_{2\mu}(E,u)$ (Fig. 2.3) depends only slightly on the nuclear charge.

The direct production of electron–positron pairs by a muon in an electromagnetic field of the atomic electrons is taken into account through substitution of the factor $Z(Z+1)$ for $Z^2$ in Eq. (2.15). Such a substitution should be made at high energies ($E \gg m_\mu^2c^2/m_e$), when the electron recoil can be ignored (at high muon energies, when the direct pair production is important, this condition always holds).

The equation which was proposed in Ref. 30 and which approximates the results of numerical calculations based on Eq. (2.15) can be used in actual calculations:

$$\sigma_{p\mu}(E,u) = B(Z)F(E)C(u),$$

where

$$B(Z) = 6.3(m_e/m_\mu)\alpha^2r_e^2Z(Z+1.3)[\ln(189\,Z^{-1/3}) + 0.6051];$$
$$F(E) = 1 - \exp(-d^2/40); \quad d = \ln(E/m_\mu c^2); \; C(u)$$
$$= a(1+a)/u(u+a)^2; \quad a = 0.0071.$$

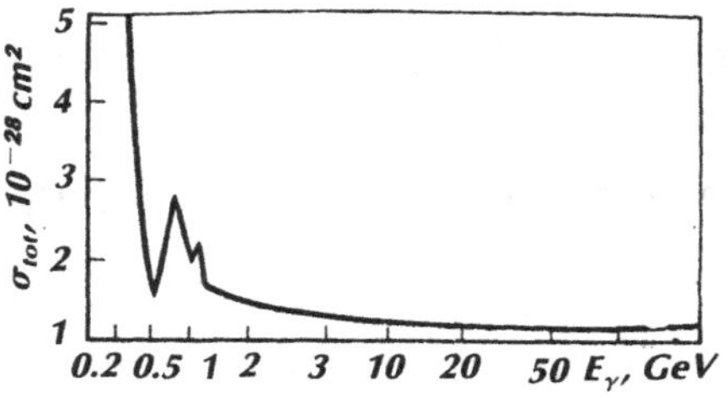


**Figure 2.4. Cross section for the interaction of photons with nucleons measured experimentally versus the photon energy (Ref. 31).**

# 3. Inelastic interaction of muons with atomic nuclei

In the loss of energy of a muon in a medium, an important role is played by the interaction of the muon with atomic nuclei at high energies, when the multiple production of secondary hadrons is possible from the standpoint of the energy. The production of hadrons as a result of interaction of a muon with the nucleus can be viewed as the photoproduction of hadrons as a result of interaction with the nucleus of a high-energy virtual photon which is emitted by the muon. Here we are considering the region of small 4-momentum transfer to the nucleons of the nucleus, where the properties of the virtual photon are similar to those of the real photon.

Since the photoproduction cross section has been determined (in experiments with photon beams; Fig. 2.4) over a broad energy range, and since the energy spectrum of virtual photons can be calculated exactly, the cross section for interaction of a muon with a nucleus can be linked with the cross section for the $\gamma A$ interaction. The differential cross section of this process can be written as[1]

$$\frac{d^2\sigma_{\mu n}}{dq^2 dE_h} = \frac{\alpha}{8\pi^2}\frac{1}{p^2 q^4}\left\{L_\perp\left[(E^2+(E-E_h)^2)q^2 - 2m_\mu^2 c^4 E_h^2 - \frac{1}{2}q^4\right]\right.$$
$$\left. + L_{\parallel}(2m_\mu^2 c^4 c^4 - q^2)\, q^2\right\},$$

where $E$ and $p$ are the energy and momentum of the primary muon, $E_h$ is the energy of the secondary hadrons, $q^2$ is the square of the 4-momentum of the virtual photon, and $L_\perp$ and $L_\parallel$ are functions of $q^2$ and $k$, the photon energy. These quantities are related to the cross section for the multiple production of hadrons by a real photon by $\sigma_\gamma(k) = (k/4\pi)(L_\perp - L_\parallel q^2/k^2)|_{q^2\to 0}$. This relation implies that the second term, which is proportional to the function $L_\parallel$ and which describes the contribution of the longitudinal component of the electric field, does not contribute to the photoproduction of hadrons (pions) as $q^2 \to 0$. Assuming that the interaction of a virtual photon is similar to the interaction of a real photon, we can consider in the expression for the differential cross section only the term proportional to $L_\perp$, which is the transverse field component.

The $q^2$ dependence of the quantity $L_1 k/4\pi\sigma_\gamma(k)$ is usually parametrized in the form[1]: $[\Lambda^2/(q^2+\Lambda^2)]^2$ ($\Lambda^2 = 0.365$ GeV$^2$). Assuming that $\sigma_\gamma$ = (const within good accuracy) (see Fig. 2.4), after integrating the differential cross section over $q^2$ in the range from $q^2_{\min} = m_\mu^2 c^4 E^2/(E-E_h)$ to $q^2_{\max} = 2ME_h$ ($M$ is the nucleon mass), we find[1]

$$\sigma_{\mu n}(E,u)du = \frac{\alpha}{2\pi}\sigma_\gamma \frac{du}{u} F_{\mu n}(E,u),$$

where

$$F_{\mu n}(E,u) = [2-2u+(1+4/a)u^2]\ln\frac{u^2+a(1-u)}{u(u+b)}$$
$$-[2-2u+(1+2/a)u^2+\Lambda^2/2E^2]$$
$$\times\left[\frac{a(1-u)}{u^2+a(1-u)}-\frac{b}{u+b}\right]-2[1-(1+b/a)u];$$
$$u = E_h/E;\quad a = \Lambda^2/m_\mu^2 c^4;\quad b = \Lambda^2/2ME.$$

These expressions disregard the contribution from the region of large values of $q^2$ ($q^2 \gtrsim 1$ GeV$^2$), which is called the deep inelastic region. This region is now under active study.[30] However, since its contribution to the total energy loss of a muon is small, it is not considered here in the first approximation.

## 4. Interaction of electrons and photons with the atomic electrons

Let us consider the interaction of photons, electrons, and positrons with the atomic electrons. This interaction, which is important at low energies, determines the development of the low-energy component of the electron–photon shower.

### Compton scattering of a photon

The differential cross section for Compton scattering of a photon by a free electron is determined by the equation[20]

$$\sigma_c(\omega_0,\omega)d\omega = \pi r_e^2(m_e c^2 d\omega/\omega_0)G_c(\omega_0,u), \qquad (2.16)$$

where $\omega_0$ and $\omega$ are the energies of the primary and secondary photons, respectively, and $u = \omega/\omega_0$. The function $G_c(\omega_0,u)$ has the form

$$G_c(\omega_0,u) = \frac{1}{u}\left(1+u^2-2\frac{1+k_0}{k_0^2}+\frac{1+2k_0}{k_0^2}u+\frac{1}{k_0^2 u}\right).$$

Here $k_0 = \omega_0/m_e c^2$ and $1/(1+2k_0) < u < 1$.

The angle of emission of the secondary photon relative to the direction of the momentum of the primary photon is uniquely related to the photon ener-

Table 2.2. The parameter values in Eq. (2.17).

| $\alpha Z$ | $b_0$ | $b_1$ | $b_2$ | $f(\alpha Z)$ | $\alpha Z$ | $b_0$ | $b_1$ | $b_2$ | $f(\alpha Z)$ |
|---|---|---|---|---|---|---|---|---|---|
| 0.00 | 1.008 | 1.926 | 2.107 | — | 0.40 | 0.323 | 1.265 | 0.753 | 1.115 |
| 0.10 | 0.704 | 1.647 | 1.592 | 1.079 | 0.45 | 0.293 | 1.247 | 0.636 | — |
| 0.15 | 0.604 | 1.547 | 1.411 | — | 0.50 | 0.268 | 1.234 | 0.528 | 1.134 |
| 0.20 | 0.522 | 1.460 | 1.258 | 1.092 | 0.55 | 0.248 | 1.234 | 0.407 | — |
| 0.25 | 0.455 | 1.392 | 1.114 | — | 0.60 | 0.232 | 1.243 | 0.278 | 1.162 |
| 0.30 | 0.402 | 1.339 | 0.985 | 1.102 | 0.65 | 0.218 | 1.263 | 0.134 | — |
| 0.35 | 0.358 | 1.297 | 0.866 | — | 0.70 | 0.207 | 1.293 | −0.041 | 1.201 |

gy: $\cos\theta_\gamma = 1 - 1/uk_0 + 1/k_0$. The angle of emission of the electron in this case is $\cos\theta_e = \sqrt{1-u}(1+1/k_0)/\sqrt{(1-u)+2/k_0}$.

If the binding energy of the electron in the atom is ignored, the cross section for the scattering of a photon by atomic electrons can be found from Eq. (2.16) by multiplying it by $Z$.

### Photoelectric effect

The cross section for the photoelectric effect induced by electrons of the $K$, $L$, and higher atomic shells is described by an approximate equation[21]:

$$\sigma_{ph}(\omega) \approx 4\pi r_e^2\,(\alpha Z)\,(4Z/k)\,(b_0 + b_1/k + b_2/k^2)\,f(\alpha Z), \qquad (2.17)$$

where $k = \omega/m_e c^2$, and the parameters are given in Table 2.2.

### The scattering of electrons by atomic electrons

The differential cross section for scattering of an electron by a free electron (the Møller scattering) is given by[20]

$$\sigma_M(E_0,E)\,dE = \frac{2\pi r_e^2}{\beta^2}\, m_e c^2\, \frac{du}{T_0}\left[\frac{1}{u^2} + \frac{1}{(1-u)^2} + \frac{T_0^2}{E_0^2} - \frac{2T_0+1}{E_0^2}\,\frac{m_e c^2}{u(1-u)}\right]. \qquad (2.18)$$

The angle between the incident electron and one of the secondary electrons is uniquely related to their energies: $\cos^2\theta = T(T_0 + 2m_e c^2)/T_0(T + 2m_e c^2)$. Here $u = T/T_0$, where $T$ and $T_0$ are the kinetic energies of the secondary and primary electrons, respectively. The cross section for the Møller scattering in the form given in (2.18) takes into account the identity of the secondary electrons. In particular, this is indicated by the fact that $0 < u < \frac{1}{2}$.

If in the case of scattering of a free electron by an electron bound in the atom the energy transferred to the latter electron is much higher than its binding energy, the atomic electron may be considered a free electron. The cross section for scattering of an electron by atomic electrons in this case can be found from Eq. (2.18) by multiplying it by $Z$.

**The scattering of positrons by atomic electrons**

By analogy with the scattering of electrons when a large amount of energy is transferred to the atomic electron, the differential cross section for the scattering of a positron is determined by the Bhabha formula[20] for free electrons,

$$\sigma_a(E_0,E)\,dE = \frac{2\pi r_e^2}{\beta_0^2} m_e c^2 \frac{du}{T_0}\left\{\frac{1}{u^2} - \beta_0^2\left[(2-y^2)\frac{1}{u} - (3-6y+y^2-2y^3) + (2-10y+16y^2-8y^3)u - (1-6y+12y^2-8y^3)u^2\right]\right\}, \quad (2.19)$$

where $u = T/T_0$ ($T_0$ and $T$ are the kinetic energies of the primary positron and secondary electron, respectively), $y = 1/(T_0/m_e c^2 + 2)$, and $0 < u < 1$.

**The positron and electron annihilation**

The cross section for annihilation of a free positron and free electron with the emission of two photons is given by the equation[20]

$$\sigma_a(E) = \frac{\pi r_e^2}{(\varepsilon+1)}\left[\frac{\varepsilon^2+4\varepsilon+1}{(\varepsilon^2-1)}\ln(\varepsilon+\sqrt{\varepsilon^2-1}) - \frac{\varepsilon+3}{\sqrt{\varepsilon^2-1}}\right], \quad (2.20)$$

where $\varepsilon = E/(m_e c^2)$, and $E$ is the positron energy. At $\varepsilon > 2$ the annihilation cross section can be accurately described by a simple equation[21]: $\sigma_a(E) = 1.6\pi r_e^2 \varepsilon^{-7/9}$.

It can be seen from Eq. (2.20) that two-photon annihilation is important only when the positron energy is very low. A positron can be annihilated by a bound atomic electron with the emission of a single photon, but the probability for such annihilation is no greater than 5% of the two-photon annihilation.

## 5. Energy loss of charged particles due to the electromagnetic interaction

**Energy loss by ionization**

The mean energy loss of a charged particle which is heavier than an electron, with a charge $ze$, due to ionization and excitation of the atoms of the medium, ignoring the *density effect*, is given by the Bethe–Bloch equation[34]

$$\frac{1}{\rho}\frac{dE_i}{dx} = z^2\frac{L}{\beta^2}\left[\ln\left(\frac{2m_e c^2\beta^2 W_{max}}{I^2(1-\beta^2)}\right) - 2\beta^2\right]. \quad (2.21)$$

Here $L = 2\pi r_e^2 m_e c^2 N_A (Z/A) = 0.1535(Z/A)$ MeV cm$^2$/g, $\beta = v/c$, $Z$ and $A$ are the atomic number and atomic mass number of the substance, $\rho$ is its density, $I$ is the average ionization potential of the atom ($I \simeq 20Z$ eV for $Z \sim 1$ and $I \simeq 10Z$ eV for $Z \gtrsim 20$), and $W_{max}$ is the maximum detectable energy or the maximum energy transferred to the $\delta$ electron

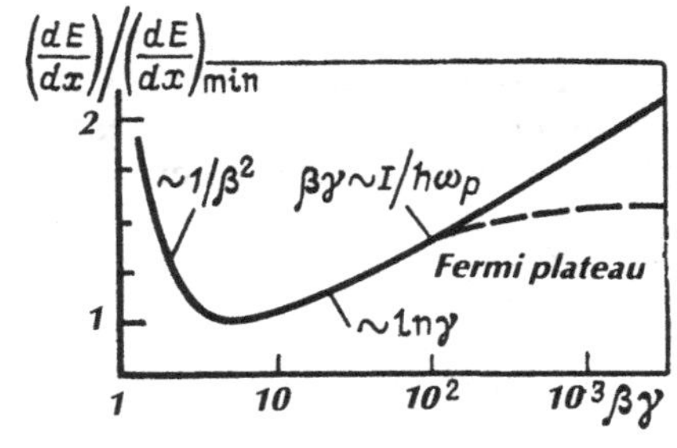


**Figure 2.5. The average energy loss of a charged particle due to ionization versus $\beta\gamma = p/Mc$ ($p$ is the particle momentum).**

$$W_{max} = 2\beta^2\gamma^2 m_e c^2/(1 + 2\gamma m_e/M + m_e^2/M^2), \qquad (2.22)$$

where $\gamma = E/M$, and $E$ and $M$ are the energy and mass of the primary particle.

For a compound the energy loss by ionization can be accurately determined by[32]

$$\left(\frac{1}{\rho}\frac{dE}{dx}\right) = \sum_i \left(\frac{1}{\rho}\frac{dE}{dx}\right)_i w_i,$$

where $w_i$ is the concentration of the elements of the substance.

For small values of $\beta$ (a nonrelativistic particle) the energy loss due to ionization depends strongly on $\beta$ ($\sim 1/\beta^2$), reaches a minimum at $\beta\gamma \simeq 4$, and then increases linearly with $\ln \gamma$ (the so-called logarithmic increase), as shown in Fig. 2.5. Such a behavior of the ionization loss can be explained qualitatively as follows. In the nonrelativistic region, the function $1/\beta^2$ reflects the fact that the lower the particle velocity, the longer the particle remains near the atom, and hence the greater is its probability for interaction. The logarithmic increase of the ionization loss is caused by the relativistic broadening of the transverse electric field of the particle, which gives rise to a logarithmic ($\ln \gamma$) increase in the radius of the region surrounding the particle track, where the atoms of the medium are excited or ionized. On the other hand, the relativistic broadening of the transverse electric field of a charged particle gives rise to a polarization of the medium (the density effect) and to screening of the atoms which are situated a considerable distance from the particle track. As a result, the logarithmic increase slows down and the ionization loss reaches the so-called *Fermi plateau*.[33] The behavior of the energy loss due to ionization near the density effect and the Fermi plateau is illustrated by the dashed curve in Fig. 2.5. The depth to which the transverse electric field penetrates effectively is given by the equation[32] $r = c/\omega_p$, where $\omega_p = [4\pi N_A(Z/A)\rho r_e]^{1/2}$ is the plasma frequency. The logarithmic increase begins to slow down at $\gamma \simeq I/\hbar\omega_p$. The polarization of the medium can be approximately taken into account by substituting the difference $I^2/\gamma^2 - \hbar^2\omega_p^2$ for $I^2(1 - \beta^2)$ in Eq. (2.21).

In the measurement of the energy loss due to ionization, the slowing down of the logarithmic increase and the plateauing of the loss can be attributed simply to the incapability of the detector to detect an energy loss greater than $W_D < W_{max}$ ($W_{max}$ is the maximum kinematically permissible energy of

**Table 2.3. Parameter values in Eqs. (2.23) and (2.24) for the energy loss of charged particles due to ionization (Ref. 35).**

| Substance | $Z$ | $I$, eV | $L$, MeV·cm²/g | $-C$ | $a$ | $b$ | $y_1$ | $y_0$ |
|---|---|---|---|---|---|---|---|---|
| C | 6 | 78.1 | 0.0768 | 3.22 | 0.531 | 2.63 | 2 | −0.05 |
| Al | 13 | 163 | 0.0740 | 4.21 | 0.0906 | 3.51 | 3 | 0.05 |
| Ar* | 18 | 218 | 0.0692 | 12.27 | 0.0255 | 4.36 | 5 | 2.02 |
| Fe | 26 | 337 | 0.0715 | 4.62 | 0.127 | 3.29 | 3 | 0.10 |
| Cu | 29 | 376.7 | 0.0701 | 4.74 | 0.119 | 3.38 | 3 | 0.20 |
| Sn | 50 | 709 | 0.0647 | 6.28 | 0.404 | 2.52 | 3 | 0.20 |
| W | 74 | 991 | 0.0618 | 6.03 | 0.0283 | 3.91 | 4 | 0.30 |
| Pb | 82 | 1180 | 0.0608 | 6.93 | 0.0652 | 3.41 | 4 | 0.40 |
| U | 92 | 1325 | 0.0594 | 6.69 | 0.0652 | 3.37 | 4 | 0.30 |
| Water | | 74.1 | 0.0853 | 3.47 | 0.519 | 2.69 | 2 | 0.23 |
| Scintillator (plastic) | | 63.8 | 0.0826 | 3.15 | 0.429 | 2.85 | 2 | 0.13 |
| Nuclear emulsion | | 373 | 0.0698 | 5.55 | 0.022 | 4.01 | 4 | 0.23 |
| NaI | | 562 | 0.0643 | 6.66 | 0.525 | 2.32 | 3 | 0.08 |

*GaAs under normal conditions.

the $\delta$ electron). This circumstance occurs, for example, when the ionization loss of a particle that passes through a relatively thin layer of gas is measured. The energetic $\delta$ electron in this case may escape from the sensitive volume of the detector, losing in it only a part of its energy. To account for an effect of this sort, $W_{max}$ should be replaced by $W_D$ in Eq. (2.21).

In practical calculations the following equations for the average ionization energy loss can be used[34]:

$$\frac{1}{\rho}\frac{dE_i}{dx}=z^2\frac{L}{\beta^2}(B_M+0.69+2\ \ln\beta\gamma-2\beta^2-\delta) \tag{2.23}$$

for a particle heavier than an electron and

$$\frac{1}{\rho}\frac{dE_i}{dx}=\frac{L}{\beta^2}(B_e+0.43+2\ \ln\beta\gamma-\beta^2-\delta) \tag{2.24}$$

for an electron. Here $B_M=\ln(m_ec^2W_{max}/I^2)$, $B_e=\ln(m_ec^2E/I^2)$, and $\delta$ is the correction for the density effect:

$$\delta=\begin{cases}0, & y<y_0,\\ 2y\ln 10+C+a(y_1-y)^b, & y_0<y\leqslant y_1,\\ 2y\ln 10+C, & y>y_1,\end{cases}$$

where $y=\log\beta\gamma$, and the parameter values are given in Table 2.3.

The energy loss of a charged particle due to ionization is comprised of discrete elementary events. In each event the particle loses a part of its energy with a particular probability. Since the probability that the particle will lose a large part of its energy is low ($\sim 1/\varepsilon^2$, where $\varepsilon$ is the energy lost by the particle), a relatively few events can occur upon the passage of the particle

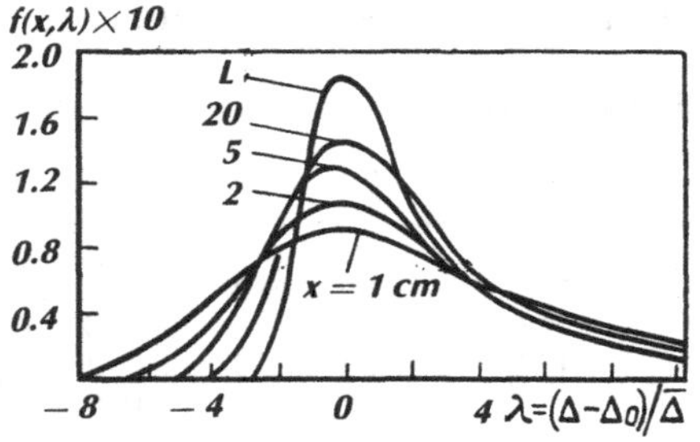


**Figure 2.6. Fluctuation function of the ionization energy loss of charged particles (β = 0.9) in argon layers of various thicknesses *x* under normal conditions (Ref. 31) [*L* is the Landau function (2.25)].**

through a thin layer of the substance. These rare events account for the appreciable fluctuations in the energy loss of a particle relative to its most probable value. The fluctuations are described by the function for the probability density of the ionization loss $f(x,\Delta)$, where $x$ is the path length of a particle in the substance (the layer thickness of the substance), and $\Delta$ is the energy lost by the particle along this path length. The fluctuations of the ionization losses relative to their most probable value, $\Delta_0$, are very high in the case of a thin layer ($x$) of the substance (Fig. 2.6). With increasing thickness of the layer, the fluctuations decrease, approaching a distribution determined by the *Landau function*.

The fluctuations in the energy loss by ionization of a relativistic charged particle upon passage through a thin layer of the substance were initially studied by Landau. The equation for the probability density of the energy loss by ionization in a thin layer of a substance (of thickness $x$) is[36]

$$f(x,\Delta) = (1/\xi)\varphi_L(\lambda), \tag{2.25}$$

where $\varphi_L(\lambda) = \dfrac{1}{2\pi i}\displaystyle\int_{-i\infty+\sigma}^{+i\infty+\sigma} du\, \exp(u \ln u + \lambda u)$, $\varphi_L(\lambda)$ is the universal Landau function (Fig. 2.7), $\xi = Lx/\beta^2$, and $\lambda = (\Delta - \Delta_0)/\xi$ [here $L$ is the same as that in Eq. (2.21)]. The most probable energy loss, $\Delta_0$, is related to the average energy loss in the same layer [$\overline{\Delta} = (dE/dx)x$] by $(\overline{\Delta} - \Delta_0)\beta^2/Lx = 0.37$.

It is evident from Eq. (2.25) that the Landau function for the energy-loss distribution has the property of self-similarity; i.e., the distribution does not depend on the layer thickness of the substance or on the primary particle energy in the range of applicability of the equation. Specifically,

$$I \ll \overline{\Delta} \ll W_{\max} \quad \text{or} \quad I/(dE/dx) \ll x \ll W_{\max}/(dE/dx).$$

Let us consider the behavior of the Landau function for large values of the variable $\lambda$ (i.e., for a large energy loss). We shall rewrite the expression for $\varphi_L(\lambda)$ in the form

$$\varphi_L(\lambda) = (1/\pi)\int_0^{\infty} \exp(-u \ln u - \lambda u)\sin(\pi u)u\,du.$$

**Figure 2.7. A plot of the universal Landau function (Ref. 35).**

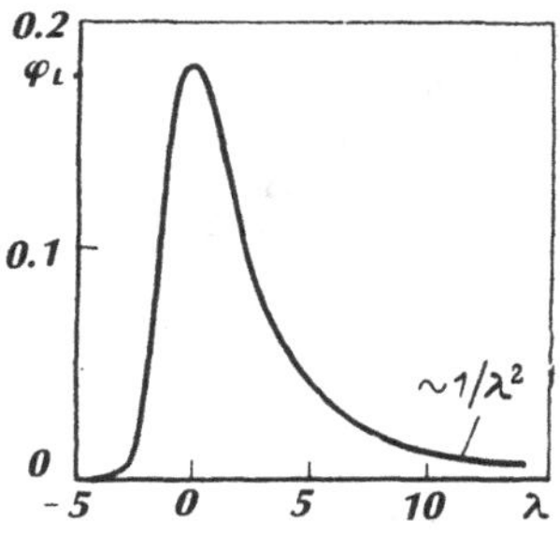


Using this expression, we can easily determine the asymptotic form of the Landau function for $\lambda \gg 1$ ($u_{\text{eff}} \lesssim 1/\lambda$): $\varphi_L(\lambda) \sim 1/\lambda^2$. It follows from this equation that the probability for a large energy loss decreases as $1/\Delta^2$, i.e., in the same way as the probability for the formation of a $\delta$ electron with energy $\varepsilon$ decreases in the interaction of a charged particle with the atomic electrons ($\sim 1/\varepsilon^2$). This conclusion is confirmed by the fact that the fluctuations of a large energy loss are determined by the infrequent (single) collisions, which cause the primary particle to lose an appreciable part of its energy. The peak near the maximum of the ionization-loss distribution (see Fig. 2.7) is caused by a large number of long-range collisions which excite the atoms of the substance.

The equation for the probability density of the ionization loss in a thick layer of the substance ($x$ is the layer thickness) was obtained by Vavilov. This equation is[37]

$$f(x,\Delta) = (1/\xi)\varphi_V(\lambda_V,\varkappa,\beta^2), \tag{2.26}$$

where

$$\varphi_V(\lambda_V,\varkappa,\beta^2) = \frac{1}{\pi}\exp[\varkappa(1+\beta^2 c)] \times \int_0^\infty \exp[\varkappa f_1(y)]\cos(y\lambda_V + \varkappa f_2(y))dy;$$

$$\lambda_V = (\Delta - \bar{\Delta})/W_{\max} - \varkappa(1+\beta^2 - C); \quad C = 0.577216;$$

$$f_1(y) = \beta^2[\ln y + \text{Ci}(y)] - \cos y - y\,\text{Si}(y);$$

$$f_2(y) = y[\ln y + \text{Ci}(y)] + \sin y + \beta^2\,\text{Si}(y);$$

$$\text{Si}(y) = \int_0^y \frac{\sin u}{u}du; \quad \text{Ci}(y) = \int_{-\infty}^y \frac{\cos u}{u}du; \quad \varkappa = \xi/W_{\max}.$$

Here $W_{\max}$ is determined from Eq. (2.22) and the quantity $\xi$ is the same as that in Eq. (2.25).

For small values of the parameter $\varkappa$ ($\varkappa \lesssim 0.01$) the function $\varphi_V(\lambda_V,\varkappa,\beta^2)$ becomes the Landau function $\varphi_L(\lambda)$. At $\varkappa \gg 1$ the distribution (2.26) is approximately equal to the Gaussian distribution.

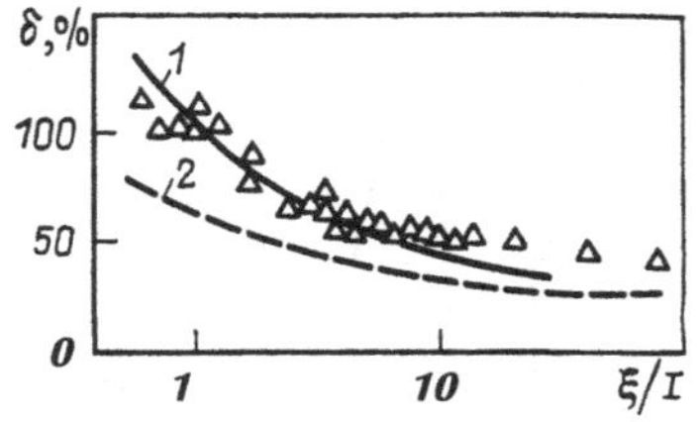


**Figure 2.8. The distribution half-width δ of the fluctuations of the energy loss by ionization of a relativistic charge particle in inert gases vs the ratio ξ/I (Ref. 38). Δ, experiment; 1, calculation based on Eq. (2.27); 2, calculation based on the Landau equation Eq. (2.25).**

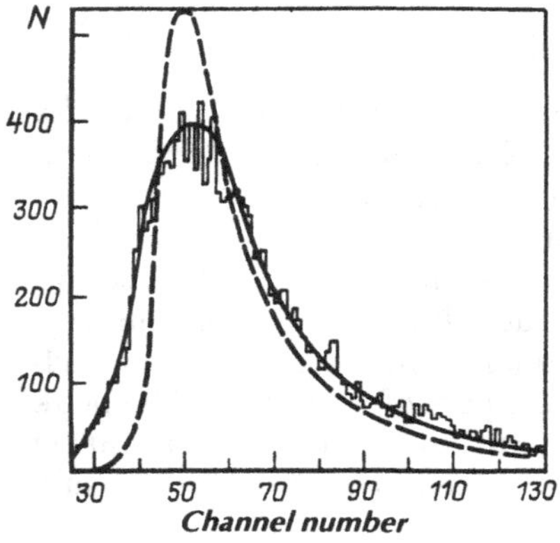


**Figure 2.9. Fluctuation function of the energy loss by ionization of protons with a momentum $p = 2.1$ GeV/c in a gas layer (95% Ar + 5% $CH_4$ under normal conditions) of thickness 1.5 cm (Ref. 38). Histogram, experimental data; —, calculation based on Eq. (2.27); - - -, calculation based on the Landau equation Eq. (2.25).**

The Landau equation cannot be used for the passage of a particle through a very thin layer of the substance (see Fig. 2.6), since a small energy transfer ($\sim I$) contributes substantially to the fluctuations of the ionization loss in the case of a thin layer of the substance. In the Landau approach the corresponding energy loss does not fluctuate and is equal to the average loss. The equation for fluctuations of the energy loss by ionization in very thin layers of a substance was obtained by Ermilova and Chechin[38]

$$f(x,\Delta) = \sum_{i=0}^{\infty} \delta(\Delta - nI)C_n, \tag{2.27}$$

where the coefficient $C_n$ can be calculated by using the recurrence relations

$$C_0 = \exp(-\xi A/I);\ C_1 = C_0\xi(A-1)/I;$$
$$C_{n+1} = (\xi/I)\left[(A-1)C_n + \sum_{m=0}^{n-1} C_m/(n-m)\right];$$
$$A = \ln\left(\frac{2mc^2\beta^2}{I(1-\beta^2)}\right) - \beta^2 + 1 - 0.577 - \delta.$$

The fluctuations of the ionization loss which are calculated from Eq. (2.27) are in good agreement with the experimental data (Figs. 2.8 and 2.9).

**Figure 2.10. Relative energy loss of an electron due to ionization (1) and due to radiation (2) in lead per unit radiation length (Ref. 31).**

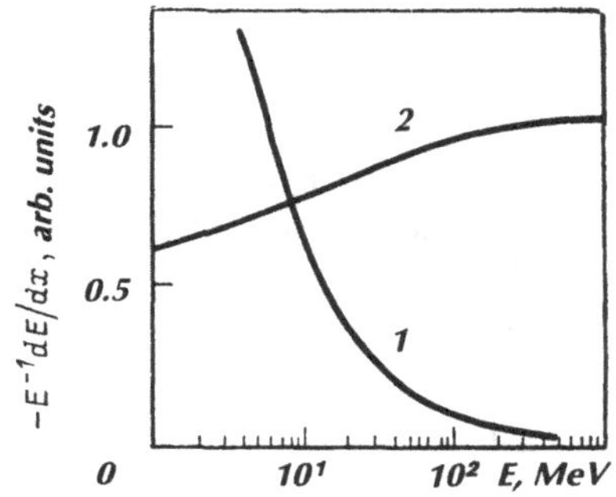


## Energy loss of an electron due to radiation

The average energy loss of an electron due to bremsstrahlung, per unit path length in a substance is given by the equation

$$\left(\frac{dE}{dx}\right)_r = \int_0^{E - m_e c^2} \frac{N_A \rho}{A} \sigma_{re}(E,\omega)\omega \, d\omega, \tag{2.28}$$

where $N_A$ is Avogadro's number, $\rho$ is the density of the substance, and $A$ is the atomic mass number. In the case of total screening ($\xi \simeq 0$) the integral in (2.28) can easily be evaluated:

$$(dE/dx)_r = 4ar_e^2 Z (Z+1)(N_A \rho / A) E \,[\ln(183\, Z^{-1/3}) + 1/18].$$

Or, introducing the radiation length (2.7), we obtain

$$\frac{t_r}{E}\frac{dE_r}{dx} = 1 + b, \quad b^{-1} = 18 \ln(183\, Z^{-1/3}). \tag{2.29}$$

Equation (2.29) clarifies *the physical meaning of radiation length*: Radiation length is the range of an electron in the material, at which the electron energy decreases by approximately a factor of *e* as a result of bremsstrahlung.

The loss of energy due to radiation (Fig. 2.10) surpasses the loss of energy due to ionization at a certain critical energy $\varepsilon_c$ given by

$$\left.\frac{dE}{dx}(\varepsilon_c)\right|_r = \left.\frac{dE}{dx}(\varepsilon_c)\right|_i .$$

## The loss of energy by a heavy particle

The total mean energy loss of a particle is given by the equation

$$\frac{dE_{\text{tot}}}{dx} = K(E) + [\,b_r(E) + b_p(E) + b_n(E)\,]E, \tag{2.30}$$

where $K$, $b_r$, $b_p$, and $b_n$ are functions which are nearly independent of the particle energy $E$. The term $K(E)$ describes the energy loss due to ionization [Eq. (2.23)]; $b_r$, $b_p$, and $b_n$ are the specific energy loss due to radiation, direct pair production, and the loss due to the multiple production of hadrons:

$$b_{r,p,n} = \left(\frac{1}{E}\frac{dE}{dx}\right)_{r,p,n} = \frac{1}{E}\frac{N_A}{A}\rho\int_0^E \omega\sigma_{r,p,n}(E,\omega)d\omega,$$

where $\sigma_{r,p,n}$ are the differential cross sections of the corresponding processes, and $\omega$ is the energy of the photon, the electron–positron pair, or the hadronic jet. The quantities $b_r$, $b_p$, and $b_n$ ($cm^2/g$) can be described* in an approximate way by[1,3]

$$b_r = \begin{cases} L_r(Z(Z+1)/A)\ln\chi\left[1+\dfrac{6\varepsilon^2\ln\varepsilon-\varepsilon^2-10\varepsilon\ln\varepsilon+6\ln\varepsilon+1}{6(1-\varepsilon)^3\ln\chi}\right], & \varepsilon\neq 1; \\ L_r(Z(Z+1)/A)\,[\ln\chi-17/18], \quad \varepsilon=1; & \end{cases}$$

$$b_p = L_p(Z(Z+1)/A)\left(0.97\ln\frac{183\,Z^{-1/3}}{183\,Z^{-1/3}M^2c^2/2m_eE+1}+2.15\right);$$

$$b_n = \frac{2\alpha}{\pi}\frac{N_A}{A}\sigma_0[\ln(E/Mc^2)-0.5],$$

where

$$L_r = \alpha N_A(2r_e m_e/M)^2,$$

$L_r = 3.27\times10^{-8}$ for muons,

$$L_p = (16/\pi)(m_e/M)\alpha^2 r_e^2 N_A,$$

$L_p = 6.01\times10^{-8}$ for muons,

$$\varepsilon = E\bigg/\left(\frac{183\sqrt{e}}{2}\frac{M^2c^2}{m_e}Z^{-1/3}\right),$$

$$\chi = \tfrac{3}{2}\,183\,(M/m_e)\,Z^{-2/3},$$

$\sigma_0 = 1.8\times10^{-29}A^{0.9}$, $M$ is the particle mass, and $\alpha = \frac{1}{137}$. If the particle is not a muon, we have $b_n = 0$.

The energy loss of heavy particles versus their energy is shown in Fig. 2.11. The value of (2.30) increases markedly with increasing atomic number of the material and with decreasing particle mass. In each case the energy losses due to the bremsstrahlung and direct pair production dominate at $E \gtrsim 1000$ GeV. At particle energies no greater than several hundred GeV, when the energy loss due to the ionization is still the principal energy loss, the fluctuation of the energy loss is described well by the Landau or Vavilov function (depending on the layer thickness of the material). At a higher energy, the energy loss of the heavy particle and its fluctuation are determined by the bremsstrahlung, the direct pair production, and the production of secondary

*The energy loss due to the pair production for particles heavier than the muon can be described more correctly according to A. I. Nikishov and N. V. Pichkurov, "Production of an $e^-e^+$ pair in the collision of a fast nucleus with a heavy atom," Sov. J. Nucl. Phys. **35**, 964 (1982).

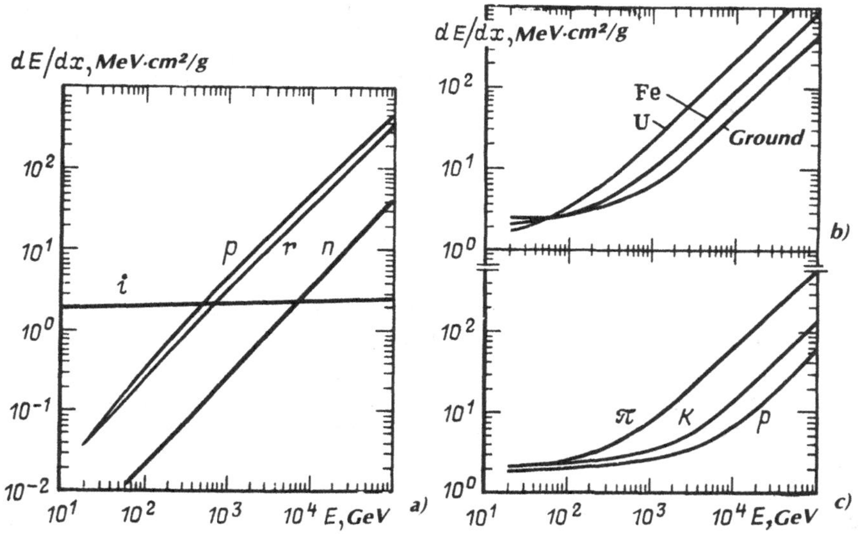


**Figure 2.11. Average specific energy loss of particles due to the electromagnetic radiation versus the energy of these particles. a, The energy loss of a muon in iron during the ionization (*i*), the pair production (*p*), the bremsstrahlung (*r*), and the inelastic nuclear interaction (*n*); b, the total energy loss of a muon in various materials; c, the energy loss of hadrons in iron.**

hadrons in muon–nucleus interactions. A simple analytical expression was obtained for the probability density of the energy loss of a high-energy muon (see Sec. 7.2). The distribution of the energy loss of a muon in a particular energy range can be calculated relatively simply by the Monte Carlo method. This topic is discussed in greater detail in Chap. 7, which deals with the transmission of muons through matter.

## 6. Multiple Coulomb scattering

### Scattering of charged particles in a thin layer of the material

A charged particle, as it passes through a finite layer of the material, interacts through the electromagnetic forces with the electrons and nuclei of the atoms. As a result of this interaction, the particle experiences multiple scattering, which causes its direction of motion at the exit from the layer to change from the initial direction. The point at which the particle emerges from the material shifts with respect to the point at which it would emerge if it had maintained its initial direction of motion until it crossed the surface of the layer at its exit point (Fig. 2.12). In the final state (at the exit from the layer) the particle is characterized by the coordinates $\Delta x$, $\Delta y$, $t$ and by the direction of motion $\sin\theta\cos\varphi$, $\sin\theta\sin\varphi$, $\cos\theta$. Clearly, the path $l$ traversed by the particle in the material in this case is greater than the thickness $t$ of the

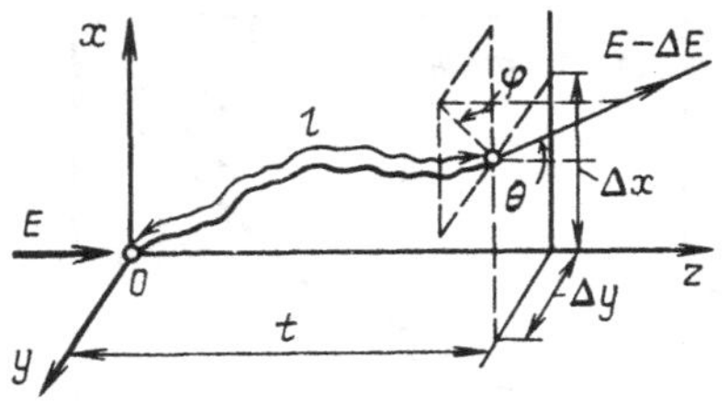


**Figure 2.12. Schematic representation of multiple Coulomb scattering of a charged particle which initially moves along the direction perpendicular to the layer of the material of thickness *t*.**

layer. Upon the traversal of the particle through the material, its energy decreases through various energy-loss mechanisms (the effect of the energy loss by ionization on the Coulomb scattering is considered below). Since the nature of scattering of the particle depends on the energy of the particle, the scattering will be different at different points along its path in the layer of the material.

We shall first ignore the transverse displacement of the particle, the change in energy, and the differences between $l$ and $t$; i.e., we assume that the layer of the material is thin. In the simplest form, the problem of multiple Coulomb scattering of a fast particle as it passes through a layer of material was solved by Moliere.[39] A simple derivation of Moliere's equation was given by Bethe.[40] We consider this problem on the basis of these two studies and also on the basis of Scott's[41] and Ford and Nelson's[42] studies.

The distribution of charged particles in the polar angle after the passage through a layer $t$ of the material is given by the function[40]

$$f(\theta)\theta d\theta = v\,dv[\,f^{(0)}(v) + (1/B)f^{(1)}(v) + (1/B^2)\,f^2(v) + \dots], \tag{2.31}$$

where

$$\left.\begin{aligned} f^{(0)}(v) &= 2\exp(-v^2);\\ f^{(n)}(v) &= \frac{1}{n!}\int_0^\infty u\,du\,J_0(vu)\exp\left(-\frac{1}{4}u^2\right)\left[\frac{1}{4}u^2\ln\left(\frac{u^2}{4}\right)\right]^n. \end{aligned}\right\} \tag{2.32}$$

The variables $\theta$ and $v$ in Eq. (2.31) are related by $\theta = \chi_c\sqrt{B}v$. Here $\chi_c$ and $B$ are the parameters which depend on the material, the particle energy, and the layer thickness, and functions (2.32) are shown in Fig. 2.13. For large values of $v$ the functions $f^n(v)$ $(n\geqslant 1)$ behave as $v^{-2n-2}$.

The parameter $B$ is determined from the equation

$$B - \ln B = b, \tag{2.33}$$

where

$$\exp(b) = \frac{6702.33\rho t}{\beta^2}\,\frac{Z_S}{M}\exp\left(\frac{Z_E - Z_X}{Z_S}\right).$$

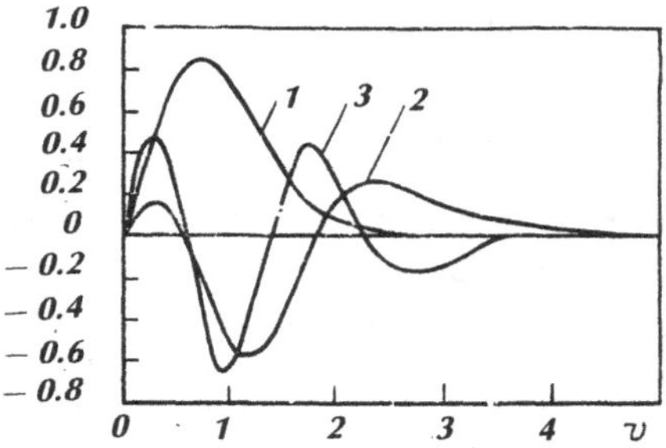


**Figure 2.13. Plots of the following functions: 1, $\nu f^{(0)}(\nu)$; 2, $\nu f^{(1)}(\nu)$; and 3 $\nu f^{(2)}(\nu)$.**

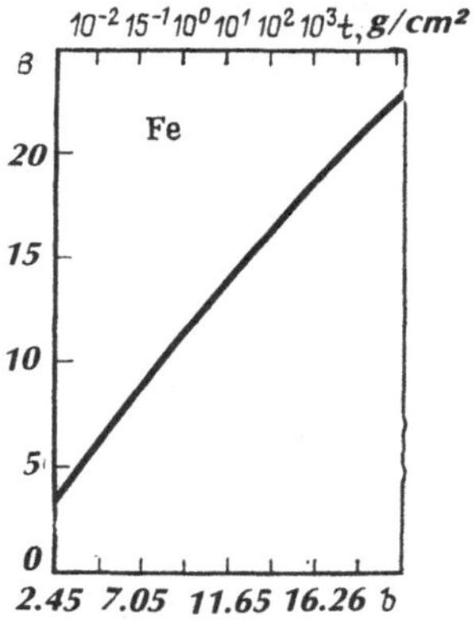


**Figure 2.14. The parameter *B* vs *b* [see Eq. (2.33)]. The upper scale shows the thickness of the layer of iron.**

Here $M = \sum_{i=1}^{N} w_i A_i$ is the molecular mass of the compound, $Z_S = \sum_{i=1}^{N} w_i Z_i (Z_i + 1)$, $Z_E = \sum_{i=1}^{N} w_i Z_i (Z_i + 1) \ln Z_i^{-2/3}$, $Z_X = \sum_{i=1}^{N} w_i Z_i (Z_i + 1) \times \ln(1 + 3.34(\alpha Z_i)^2)$, $\rho$ is the density of the material (g/cm$^3$), $t$ is the thickness of the layer of the material (cm), and $w_i$ are the relative concentrations of the constituent elements of the material.

The quantity exp($b$) may be regarded as the average number of scattering events in the material. In a very thin layer, where there are only a few scattering events, Eq. (2.31) is inapplicable. The condition for the applicability of Moliere's equation is generally assumed to be the inequality: exp($b$) $\gtrsim 20$, from which it follows that $b \gtrsim 3$ or $B \gtrsim 4.5$. These inequalities impose a lower limit on the thickness of the layer of material, allowing Eq. (2.31) to be used. The quantity $B$ versus the layer thickness of the material is shown in Fig. 2.14.

The quantity $\chi_c$ is calculated from the equation

$$\chi_c^2 = 4\pi r_e^2 (\rho N_A / M) Z_S \, t m_e^2 c^4 / E^2 \beta^4. \qquad (2.34)$$

Here, as in the previous case, $\beta = v/c$, $v$ is the particle velocity, and $c$ is the velocity of light. The physical meaning of $\chi_c$ is that the probability for a single scattering at an angle greater than $\chi_c$ is equal to unity. If the thickness of the layer is measured in radiation lengths, then $\tilde{\chi}_c = \chi_c\sqrt{t}$ and $\exp(\tilde{b}) = t_r \exp(b)$.

Equation (2.31) can be used for moderately large scattering angles. The corresponding condition imposes an upper limit on the layer thickness of the material, for which the indicated equation is still valid: $\chi_c^2 B \lesssim 1$. It follows from relations (2.31)–(2.33) that the distribution in the given angle $v$ has an important feature: It does not depend on the primary energy of the incident particle, but is rather determined by the nuclear charge of the atoms of the material, the atomic mass, and the layer thickness $[B = B(Z, A, t)]$. If the scattering occurs in a relatively thick layer of the material and $B$ is large, we can restrict expansion (2.31) to the first term $f^0(v)$. The distribution in $v$ in this case is no longer dependent on the layer thickness, on $Z$, or on $A$.

As was already noted, the particle loses its energy as a result of passing through the material. Since the parameter $B$ is essentially independent of the particle energy, the effect of the energy loss of the particle on its scattering can be taken into account by substituting in Eq. (2.34) the integral $\int_0^t dt'/E^2(t')\beta^4(t')$ for $t/E^2\beta^4$. For a relativistic particle we have $\beta \simeq 1$. In the continuous-energy-loss approximation we shall then have $E(t') = E - \varepsilon t$ ($\varepsilon$ is the average ionization loss of a particle per unit path length in the material). The integral can easily be evaluated and the expression for $\chi_c$ becomes

$$\chi_c^2 = 4\pi r_e^2(\rho N_A/M)\, Z_S(m_e^2c^4/E^2)\, t/(1 - \varepsilon t/E),$$

which is equivalent to extending the particle path in the material by a factor of $(1 - \varepsilon t/E)^{-1}$ (the energy loss in a thin layer of the material is small in comparison with the initial particle energy $\varepsilon t \ll E$).

Apart from the fictitious lengthening of the particle path in the material due to the energy loss, a multiple change in the direction of motion of the particle causes its actual path $l$ in the material to exceed the thickness of the layer $t$ (see Fig. 2.12). At a small scattering angle the average particle path is related to the layer thickness by $l = (E_s/2E\beta^2)^2\,(l^2/t_r) + t$ or $(l - t)/l = (E_s/2E\beta^2)\,^2(l/t_r)$, where $E_s = m_e c^2\sqrt{4\pi/\alpha} = 21.2$ MeV. For high-energy particles and moderately thick layers of the material the difference between $l$ and $t$ can thus be ignored within error determined by the equation given above.

## Scattering of charged particles in a thick layer of the material

If the scattering occurs in a thick layer of the material and the numerical value of $B$ is large, the dominant term in expansion (2.31) is the quantity $f^0(v) = 2\exp(-v^2)$. As a result of multiple scattering in a thick layer of the material, the particle is displaced appreciably in the transverse direction $(\Delta x, \Delta y)$. Therefore, in addition to the distribution in the scattering angle, it is necessary to consider the spatial distribution of particles in the transverse direction, which is strongly correlated with the scattering angle. Since the scattering in two mutually perpendicular planes, $xz$ and $yz$ (see Fig. 2.12), occurs independently because of the homogeneity of the medium, the distribution in the projection scattering angle $\varphi_x$ and the displacement $\Delta x$ are equal to the distribution in $\varphi_y$ and $\Delta y$, respectively. We shall consider below the distribution in the multiple scattering angle and in the displacement in a particular plane, in the $xz$ plane, for example, and we shall denote $\varphi_x$ in terms of $\varphi$ and $\Delta x$ in terms of $x$.

The distribution of particles in the scattering angle $\varphi$ and in the transverse displacement $x$ at the exit from the layer of the substance of thickness $t$ is given by[43]

$$F(\varphi,x) = \frac{1}{4\pi\sqrt{D}} \exp\{-(\varphi^2 A_2 - 2x\varphi A_1 + x^2 A_0)/4D\}, \tag{2.35}$$

where

$$\left.\begin{aligned} &D = A_0 A_2 - A_1^2; \quad A_0(t) = \int_0^t dz/W^2(z); \\ &A_1(t) = \int_0^t dz(t-z)/W^2(z); \\ &A_2(t) = \int_0^t dz(t-z)^2/W^2(z); \end{aligned}\right\} \tag{2.36}$$

$W(z) = 2E(z)\beta^2(z)/E_s$; $E(z)$ is the particle energy (MeV) of the particle at a distance $z$ from the beginning of the flat layer of the material; $\beta(z) = v(z)/c \simeq 1$ for relativistic particles; and $E_s = 21.2$ MeV; the distances $z$ and $t$ are measured in units of radiation length $t_r$.

In the continuous-loss approximation, the particle energy at the point with the longitudinal coordinate $z$ is $E(z) = E - \varepsilon z$. Integrals (2.36) can easily be evaluated in this case. For particles of reasonably high energy and for moderately thick layer of the material ($\varepsilon t/E \ll 1$) we can assume that $E(z) \simeq \text{const} \simeq E$. Using the notation $\sigma_\varphi = \sqrt{2A_0}$, $\sigma_x = \sqrt{2A_2}$, and $\rho = A_1/\sqrt{A_0 A_2}$, we can write expression (2.35) in the form

$$F(\varphi,x) = \frac{1}{2\pi\sigma_x\sigma_\varphi\sqrt{1-\rho^2}} \exp\left\{-\frac{1}{2(1-\rho^2)}\left(\frac{\varphi^2}{\sigma_\varphi^2} - \frac{2\rho\varphi_x}{\sigma_x\sigma_\varphi} + \frac{x^2}{\sigma_x^2}\right)\right\}, \tag{2.37}$$

where $\sigma_\varphi = E_s\sqrt{t/t_r}/E\sqrt{2}$, $\sigma_x = E_s\sqrt{t^3/3t_r}/E\sqrt{2}$, and $\rho = \sqrt{3}/2$ if $E(z) = \text{const} = E$. The energy loss can be approximately taken into account by substituting $\sqrt{EE_t}$ for $E$ in the last equations, where $E_t$ is the energy of the particle at the exit from the layer ($E_t = E - \varepsilon t$).

As can be seen from expression (2.37), transverse displacement of the particle at the exit from the layer of the material is correlated with the multiple scattering angle $\varphi$; here the correlation coefficient is $\rho = \sqrt{3/2}$.

Integrating expression (2.35) over the transverse displacement $x$, we find the distribution in the projection scattering angle $\varphi$ in one of the two mutually perpendicular planes

$$F_\varphi(\varphi)=\frac{1}{\sqrt{2\pi\langle\varphi^2\rangle}}\exp\left\{-\frac{1}{2}\frac{\varphi^2}{\langle\varphi^2\rangle}\right\}$$

with a mean-square scattering angle $\sqrt{\langle\varphi^2\rangle}=\sigma_\varphi=E_s\sqrt{t/t_r}/E\sqrt{2}$. The distribution in the displacement $x$ can be obtained by integrating expression (2.35) over the projection scattering angle $\varphi$:

$$F_x(x)=\frac{1}{\sqrt{2\pi\langle x^2\rangle}}\exp\left\{-\frac{1}{2}\frac{x^2}{\langle x^2\rangle}\right\},$$

where $\sqrt{\langle x^2\rangle}=\sqrt{\langle\varphi^2\rangle}t/\sqrt{3}$.

Since the scattering in two mutually perpendicular planes occurs independently, the combined distribution in the projection scattering angles $\varphi_x$ and $\varphi_y$ can be written as

$$F(\varphi_x,\varphi_y)=F(\varphi_x)F(\varphi_y)=\frac{1}{2\pi\langle\varphi^2\rangle}\exp\left\{-\frac{1}{2}\left(\frac{\varphi_x^2}{\langle\varphi^2\rangle}+\frac{\varphi_y^2}{\langle\varphi^2\rangle}\right)\right\}. \quad (2.38)$$

The polar scattering angle $\theta$ is related to the projection angles $\varphi_x$ and $\varphi_y$ by (in the small-angle approximation of $\theta$, $\varphi_x$ and $\varphi_y \ll 1$) $\varphi_x=\theta\cos\psi$ and $\varphi_y=\theta\sin\psi$, where $\psi$ is the azimuthal angle reckoned from the $x$ axis. Making the replacement of the variables $(\varphi_x,\varphi_y)\to(\theta,\psi)$ in expression (2.38) and integrating over the azimuthal angle $\varphi$, we find the distribution in the polar scattering angle

$$F_\theta(\theta)=\frac{2\theta}{\langle\theta^2\rangle}\exp\{-\theta^2/\langle\theta^2\rangle\},$$

where the mean-square scattering angle is

$$\sqrt{\langle\theta^2\rangle}=\sqrt{2\langle\varphi^2\rangle}=E_s\sqrt{t/t_r}/E. \quad (2.39)$$

## 7. Decay of unstable particles

### Unstable particles and their role in the development of the internuclear cascade.

Most of the massive elementary particles (except the electron and proton) are unstable, i.e., they are capable of spontaneously decaying into several secondary particles. According to the energy-momentum conservation laws, an unstable particle with a mass $m_0$ may decay to any number of particles with masses $m_1,\dots,m_n$ if the following condition is satisfied: $m_0>m_1+m_2+\cdots+m_n$. The possible sets of secondary particles are, however, limited by the laws of conservation of the angular momentum, the isotopic spin, and various discrete quantum numbers, for example, by the laws of conservation of the electric charge, the lepton number, the baryon number, etc.[44]

Let us consider the principal relations which describe the decay of unstable particles and the role of these particles in the development of the internuclear cascade. The probability that a free unstable particle would move a

**Table 2.4. Lifetime and decay modes of muons, pions, and kaons.**

| Particle | Mass, MeV | Lifetime, s | Decay mode | |
|---|---|---|---|---|
| $\mu^+$ | 105.65946 | $2.197120\times10^{-6}$ | $e^+\nu_e\tilde{\nu}_\mu$ | 98.6% |
| $\mu^-$ | | | $e^-\tilde{\nu}_e\nu_\mu$ | 98.6% |
| $\pi^0$ | 134.9626 | $0.828\times10^{-16}$ | $\gamma\gamma$ | 98.79% |
| | | | $\gamma e^-e^+$ | 1.21% |
| $\pi^+$ | 139.5669 | $2.6030\times10^{-8}$ | $\mu^+\nu_\mu$ | 100% |
| $\pi^-$ | | | $\mu^-\tilde{\nu}_\mu$ | 100% |
| $K_L^0$ | 497.67 | $5.183\times10^{-8}$ | $\pi^\pm e^\mp\tilde{\nu}_e(\nu_e)$ | 38.5% |
| | | | $\pi^\pm\mu^\mp\tilde{\nu}_\mu(\nu_\mu)$ | 27.0% |
| $K_S^0$ | | $0.8923\times10^{-10}$ | $\pi^+\pi^-$ | 68.61% |
| | | | $\pi^0\pi^0$ | 31.39% |
| $K^+$ | 493.669 | $1.2371\times10^{-8}$ | $\mu^+\nu_\mu$ | 63.50% |
| | | | $e^+\nu_e\pi^0$ | 4.82% |
| $K^-$ | | | $\mu-\tilde{\nu}_\mu$ | 63.50% |
| | | | $e^-\tilde{\nu}_e\pi^0$ | 4.82% |

distance $x$ from the site of its production and would decay over an interval from $x$ to $x+dx$ is given by the equation $dW(x)=(1/L)\exp(-x/L)dx$, which is a simple corollary of the exponential (with respect to time) law of any radioactive decay. It is useful to use here the following notation: the mean free path of an unstable particle before its decay is $L=c\tau\gamma\beta=c\tau p/mc$; $\gamma=E/mc^2$ is the Lorentz factor of the disintegrating particle; $\beta=p/E$; $E$, $p$, and $m$ are the energy, momentum, and mass of the disintegrating particle, respectively; and $\tau$ is the lifetime of an unstable particle in its own rest frame. The coefficient $\gamma$ takes into account the relativistic slowing down of time in the laboratory frame in comparison with the reference frame in which the disintegrating particle is at rest, $c\beta=v$ is the velocity of the unstable particle in the laboratory frame, and $c$ is the velocity of light which is assumed equal to unity, unless stated otherwise. For brevity, the quantity $L$ is frequently called the decay length.

Table 2.4 gives the principal characteristics of the unstable particles which participate in the development of the nuclear-electromagnetic cascade.

The role of unstable particles in the development of the internuclear cascade in condensed matter and in the formation of the various components of this cascade is determined not only by the types of interaction (strong, electromagnetic, weak) in which these particles participate but also by their lifetimes. Table 2.5 gives the characteristic decay lengths of some unstable particles. The range of a hadron before it undergoes a strong inelastic interaction $\lambda_{in}$ is approximately 100–200 gf/cm$^2$. For materials with different densities the variation range of $\lambda_{in}$ is from 100 cm (for water) to 10 cm (for uranium

**Table 2.5. Decay length of some unstable particles at the momentum $p = 1$ GeV/$c$.**

| Particle | $\mu^{\pm}$ | $\pi^0$ | $\pi^{\pm}$ | $K_L^0$ | $K_S^0$ | $K^{\pm}$ |
|---|---|---|---|---|---|---|
| $L$, m | $6.3\times10^3$ | $1.9\times10^{-7}$ | 55 | 31 | $5.4\times10^{-2}$ | 7.5 |

and tungsten). Let us consider the role of unstable particles in the development of high-energy nuclear-electromagnetic cascades.

### The $\pi^0$ meson

The decay of a $\pi^0$ meson, which occurs as a result of electromagnetic interaction, is characterized by a very short lifetime ($\tau_{\pi^0} \simeq 10^{-16}$ s)—8–10 orders of magnitude shorter than the lifetime of $\pi^{\pm}$, $K^{\pm}$, and $\mu$ mesons which decay due to the weak interaction (see Table 2.4). Accordingly, the decay length of a $\pi^0$ meson is also very short ($L \sim 10^{-4}$ cm at an energy of several GeV). Since $L/\lambda_{in} \ll 1$, the $\pi^0$ meson does not have time to interact with the material in which the cascade is developed. It can be assumed, therefore, that it decays at almost the point where it was produced. Although the $\pi^0$ meson does not participate in the development of the nuclear-active component of the cascade, it is nevertheless the principal source of the electromagnetic component, since the two photons produced as a result of its decay initiate the electron-photon showers.

### Charged pions and kaons

The decay lengths of charged $\pi$ and $K$ mesons are much greater than their mean free paths before the inelastic nuclear interaction ($L_{\pi^{\pm},K^{\pm}}/\lambda_{in} \gg 1$; (see Table 2.5). Because of this difference, many of the charged pions and kaons actively participate in the production and development of the hadronic component of the cascade. A few of the charged $\pi$ and $K$ mesons nevertheless have time to decay before experiencing a strong interaction. The decay $\pi^{\pm}(K^{\pm}) \rightarrow \mu\nu_{\mu}$ is the main source of the muonic component of the cascade and also of the muonic neutrinos. The decay of charged kaons along the channel $K \rightarrow \pi e \nu_e$, called the $K_{e3}$ decay, is one of the principal sources of electronic neutrinos (antineutrinos).

### Neutral kaons

Since the decay length of a $K_s^0$ meson is comparable to its path length before the nuclear interaction, a large fraction (50% at $p_{K_s^0} \simeq 4$ GeV/$c$) of the $K_s^0$ mesons has time to decay before the interaction with the nuclei of the material. In 70% of the cases the $K_s^0$ meson decays into two charged pions, which combine with the nuclear-active component of the cascade, and in 30% of the cases this meson decays into two neutral pions, which are the sources of the electromagnetic component of the cascade. The role of the $K_s^0 \to \pi^0\pi^0$ decay as a source of electron–photon showers is, however, insignificant in comparison with the decay of $\pi^0$ mesons which are produced directly in the hadron–nucleus interactions. The decay length of the $K_L^0$ meson is much greater than its path length before the strong inelastic interaction. As a result, the $K_L^0$ meson participates in the development of the hadronic component of the cascade. Only a small part of $K_L^0$ mesons decays before experiencing a strong interaction. Noteworthy among the $K_L^0$-meson decays are those along the channel $K_L^0 \to \pi e \nu_e$, which are the source of electronic neutrinos. The importance of $K_{\mu 3}$ decays ($K_L^0 \to \pi\mu\nu_\mu$) as the sources of muons and muonic neutrinos is slight in comparison with the decay $\pi^\pm(K^\pm) \to \mu\nu_\mu$.

**Muons**

The lifetime of a muon is so large in comparison with charged pions and kaons ($\tau_\mu/\tau_\pi \simeq 10^2$) that the probability for its in-flight decay in condensed matter is very small. Since the muon does not participate in strong interactions, it slows down principally due to the ionization loss in the material and decays at the end of its range, having come to a virtual stop. Forming the most deeply penetrating component, the muons thus play a dominant role in the cascade at large depths in the material. The small fraction of the in-flight decaying muons ($\mu \to e\nu_e\nu_\mu$) is of interest primarily because it is a source of relatively high-energy electronic neutrinos.

**Neutrino-producing decays**

The neutrino-producing pion, kaon, and muon decays are of particular interest for the following reasons. First, these decays are the principal sources of muonic and electronic neutrinos which are used to produce neutrino beams in the present-day proton accelerators. Secondly, the charged pion and kaon decays with the production of neutrinos are one of the sources of the energy loss in the region in which an internuclear cascade is developed. Having a very small cross section for interaction with the material [$\sigma_{\nu N} \simeq 0.7 \times 10^{-38} E_\nu$ (cm$^2$), where $E_\nu$ is the neutrino energy (in GeV)], the neutrino escapes virtually unimpeded from the effective region of development of the nuclear-electromagnetic cascade.

**Decay kinematics of the principal components of the cascade. Two-body decays**

In the rest frame of a particle which decays into two secondary particles ($0 \to 1 + 2$) the laws of conservation of the momentum energy have a very simple form[45]

$$E_1^* + E_2^* = m_0, \quad \mathbf{p}_1^* + \mathbf{p}_2^* = 0. \tag{2.40}$$

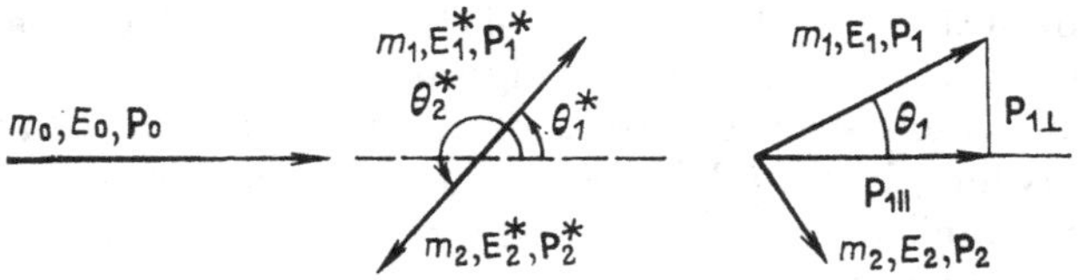


**Figure 2.15. Two-body decay scheme [see Eq. (2.41)].**

Here the subscript 0 pertains to the decaying particle with mass $m_0$ and the subscripts 1 and 2 refer to the decay products of this particle, with masses $m_1$ and $m_2$. It follows from Eq. (2.40) that in the rest frame of the decaying particle the momentum vectors of the decay products are the same in magnitude but opposite in direction. It also follows from Eq. (2.40) that the secondary-particle energies are

$$E_1^* = (m_0^2 + m_1^2 - m_2^2)/2m_0; \quad E_2^* = (m_0^2 + m_2^2 - m_1^2)/2m_0.$$

The absolute values of the momenta of these particles can be calculated from the equation

$$|\mathbf{p}_1^*| = |\mathbf{p}_2^*| = \sqrt{E_1^{*2} - m_1^2} = \sqrt{[m_0^2 - (m_1 + m_2)^2][m_0^2 - (m_1 - m_2)^2]}/(2m_0).$$

The energy and the momentum-vector components in the laboratory frame are found from the Lorentz transformation equations[45]

$$\left.\begin{aligned} E_i &= \gamma_0(E_i^* + \beta_0 p_0^* \cos\theta_i^*); \\ p_{i\parallel} &= p_i \cos\theta_i = \gamma_0(\beta_0 E_i^* + p_i^* \cos\theta_i^*); \\ p_{i\perp} &= p_i \sin\theta_i = p_i^* \sin\theta_i^*. \end{aligned}\right\} \tag{2.41}$$

Here $E_i$ is the energy of the secondary particle $i$ ($i = 1,2$), $p_{i\parallel}$ and $p_{i\perp}$ are the longitudinal and transverse components of the momentum of the particle $i$ relative to the momentum direction of the decaying particle, $\theta_i$ is the angle of emission of the particle $i$ relative to the momentum of the decaying particle, $\gamma_0 = E_0/m_0$, $\beta_0 = p_0/E_0$, $E_0$ and $p_0$ are the energy and momentum of the decaying particle, and $\theta_i^*$ is the angle of emission of the decay product $i$ in the rest frame of the decaying particle relative to the direction along which the Lorentz transformation occurs (Fig. 2.15).

The angle of emission $\theta_i$ is uniquely related to the angle of emission $\theta_i^*$ (or, equivalently, to the energy of the particle $i$) in the rest frame by

$$\tan\theta_i = p_i^* \sin\theta_i^* / \gamma_0(\beta_0 E_i^* + p_i^* \cos\theta_i^*).$$

The relationship between the angle of emission $\theta_i$ and the energy (momentum) of the particle $i$ in the laboratory frame is described by the equation

$$\cos\theta_i = (E_0 E_i - m_0 E_i^*)/p_0 p_i. \tag{2.42}$$

The back coupling, i.e., the coupling between the energy and the angle of emission of the particle $i$ is given by[45]

$$E_i = (m_0 E_0 E_i^* \pm p_0 \cos\theta_i R_i)/(E_0^2 - p_0^2 \cos\theta_i), \tag{2.43}$$

where

$$R_i = +\sqrt{R_i^2} \text{ and } R_i^2 = m_0^2 p_i^{*2} - m_i^2 p_0^2 \sin^2\theta_i.$$

For $\beta_i^* > \beta_0$ the solution with a plus sign in the numerator of Eq. (2.43) has a physical meaning. The energy in this case is uniquely related to the angle of emission. If $\beta_i^* < \beta_0$, the solutions with both a plus and a minus sign are possible. Two different energies of the particle in this case correspond to one angle of emission; i.e., the energy is nonuniquely related to the angle of emission. The condition $\beta_i^* < \beta_0$ leads to the appearance, as will be shown below, of a limiting (maximum possible) angle of emission for the particle $i$.

The condition $R_i^2 \geqslant 0$ must be satisfied for Eq. (2.43) to have a meaning. It follows, therefore, that the angles of emission must satisfy the inequality $m_0 p_i^* - m_i p_0 \sin\theta_i \geqslant 0$. If the mass $m_i = 0$, this inequality will always be satisfied, which means that the particle $i$ can have any angle of emission (from 0 to $\pi$). It can also have any angle of emission when $m_0 p_i^*/m_i p_0 = \beta_i^* \gamma_i^*/\beta_0\gamma_0 \geqslant 1$. Otherwise, a restriction is imposed on the angle of emission: $\theta_i \leqslant \theta_i^{\max} = \arcsin(\beta_i^*\gamma_i^*/\beta_0\gamma_0)$. In the decay $\pi(K) \to \mu\nu_\mu$, for example, the neutrino may be emitted at any angle (from 0 to $\pi$) relative to the momentum of the $\pi(K)$ meson in the laboratory frame, since the neutrino mass is zero. The angle of emission of a muon in the same reference frame is limited by

$$\sin\theta_\mu \leqslant \sin\theta_\mu^{\max} = (1/2)[(m_{\pi(K)}/m_\mu)^2 - 1]\, m_\mu/p_{\pi(K)},$$

if $p_{\pi(K)} \geqslant m_{\pi(K)} p_\mu^*/m_\mu$. In the decay of the $\pi$ and $K$ mesons with identical momenta ($p_{\pi(K)} \gtrsim 1.2$ GeV/$c$) along the channel $\pi(K) \to \mu\nu_\mu$, the maximum angle of emission of the muon produced as a result of the decay of a $K$ meson is larger than the maximum angle of emission of the muon from the decay of a $\pi^\pm$ meson. For high-energy pions and kaons with momenta $p_{\pi(K)}/m_{\pi(K)} \geqslant p_\mu^*/m_\mu$ we find

$$\theta_{\mu(K)}^{\max}/\theta_{\mu(\pi)}^{\max} \simeq (m_K^2 - m_\mu^2)/(m_\pi^2 - m_\mu^2) \sim 20.$$

The divergence angle $\psi$ of the decay products in the laboratory frame is[45]

$$\cos\psi = [E_1E_2 - (1/2)(m_0^2 - m_1^2 - m_2^2)]/p_1p_2.$$

For the decay $\pi^0 \to 2\gamma$ the angle of divergence between the $\gamma$ rays is $\cos\psi = 1 - m_\pi^2/2E_1E_2$. The minimum angle of divergence (the minimum possible angle is $\psi_{\min} \leqslant \psi \leqslant \pi$) occurs when the photons in the rest frame of a $\pi^0$ meson escape along the direction perpendicular to the motion of the $\pi^0$ mesons in the laboratory frame. The minimum divergence angle for photons is $\sin(\psi_{\min}/2) = m_{\pi^0}/E_{\pi^0}$. The maximum divergence angle, $\psi_{\max} = \pi$, occurs when the photons in the rest frame of a $\pi^0$ meson move strictly in the direction of motion of the $\pi^0$ meson or opposite to it in the laboratory frame.

Let us determine the permissible range of variation of the energy of the decay products in the laboratory frame. It follows from Eqs. (2.41) that the secondary particle, which is emitted in the rest frame of the primary particle in the forward direction ($\theta_i^* = 0$), has a maximum energy,

$$E_i^{\max} = \gamma_0(E_i^* + \beta_0 p_i^*). \tag{2.44}$$

When the secondary particle moves in the rest frame of the primary particle in the backward direction ($\theta_i^* = \pi$), the secondary particle has a minimum energy in the laboratory frame,

$$E_i^{\min} = \gamma_0(E_i^* - \beta_0 p_i^*). \tag{2.45}$$

Let us now determine how the products of a two-body decay of unstable particles are distributed in terms of energy, the angle of emission, and the angle of divergence. If the decaying particle is not polarized or if its spin is zero, then there is no preferred direction in its rest frame.* As a result, the decay products of this particle are distributed isotropically:

$$\left.\begin{aligned} &dW(\theta_i^*,\varphi_i^*) = (1/4\pi)\, d\cos\theta_i^*\, d\varphi_i^*;\\ &-1 \leqslant \cos\theta_i^* \leqslant 1, 0 \leqslant \varphi_i^* \leqslant 2\pi;\\ &dW_{z^*}(z^*) = (1/2)\, dz^*, z^* = \cos\theta^*. \end{aligned}\right\} \tag{2.46}$$

It is immediately evident from the isotropic nature of the angular distribution that the energy distribution of the decay products in the laboratory frame is uniform within the permissible kinematic limits [see Eq. (2.41)]:

$$\left.\begin{aligned} dW_E(E_i) = dE_i/(E_i^{\max} - E_i^{\min}) = dE_i/(2\gamma_0\beta_0 p_i^*);\\ E_i^{\min} \leqslant E_i \leqslant E_i^{\max}. \end{aligned}\right\} \tag{2.47}$$

The minimum and maximum energies are determined from Eqs. (2.44) and (2.45).

The general procedure for obtaining $dW_y(y)$ from any variable is as follows. A relation $x = x(y)$ is established between the quantity $y$, the distribution along which must be found, and the quantity $x$, the distribution along which, $dW_x(x)$, is known. A simple replacement of variables then makes it possible to obtain $dW_y(y)$ from $dW_x(x)$:

$$dW_y(y) = \frac{dW_x(x)}{dx}\frac{dx(y)}{dy}\, dy. \tag{2.48}$$

An energy distribution, for example, can be found from distribution (2.46) in $\cos\theta^*$ by making use of the relation between $\cos\theta^*$ and the energy of the decay product in the laboratory system, which is given by Eq. (2.41). Applying transformation (2.48) to (2.46), we obtain $dW_E(E_i) = dE_i/2\gamma_0\beta_0 P_i^*$, in complete agreement with (2.47).

To determine the distribution in the angle of emission of the decay product in the laboratory system, we shall make us of the relation between the angle of emission $\theta$ and the energy $E_i$ of the particle, which is given by Eq. (2.42). Differentiating expression (2.42) with respect to $z_i = \cos\theta_i$ with allowance for the relation $E_i^2 = p_i^2 + m_i^2$, we obtain

*It is also assumed that since the polarization of secondary particles is not measured, there is also no preferred direction in space in the final state.

$$dE_i(z_i)/dz_i = p_0 p_i^2/|E_0 p_i - p_0 E_i z_i|.$$

If $\beta_0 \leqslant \beta_i^*$, Eq. (2.43) determines the single-valued relation between the energy and the angle of emission of one of the decay products, and the $z_i$ function becomes

$$dW_z(z_i) = \frac{dW_E(E_i)}{dE_i}\frac{dE_i(z_i)}{dz_i} = \frac{dz_i}{2\gamma_0\beta_0 p_i^*}\frac{p_0 p_i^2}{|E_0 p_i - p_0 E_i z_i|}.$$

If $\beta_0 > \beta_i^*$, each angle of emission corresponds to two values of the energy function and two values of the density function. The combined function can be expressed in the form[45]

$$dW_z(z_i) = \frac{dz_i}{\gamma_0\beta_0 p_i^*}\frac{p_0}{R_i}\frac{m_0^2 E_i^{*2} p_0^2 z_i^2 + E_0^2 R_i^2}{(E_0^2 - p_0^2 z_i^2)^2}. \tag{2.49}$$

Here the quantity $R_i^2$ is the same as that in Eq. (2.43).

If $\beta_0 > \beta_i^*$, the particle $i$ has a maximum angle of emission $\theta_i^{\max}$. If the angle of emission $\theta_i$ of the particle tends to the maximum value ($\theta_i \to \theta_i^{\max}$), the density function (2.49) becomes infinite. This means that the most probable angles of emission are those near the maximum angle of emission. At $\theta_i = \theta_i^{\max}$ the square-root singularity of the density function (2.49) is integrable.

The distribution in the angle of divergence $\psi$ for the decay $\pi^0 \to 2\gamma$ is

$$dW_\eta(\eta) = (m_{\pi^0}^2/p_{\pi^0}E_{\pi_0})\left[(1-\eta)^2\sqrt{1-(1-\eta_{\max})/(1-\eta)}\right]^{-1}d\eta.$$

Here $\eta = \cos\psi$, $-1 \leqslant \eta \leqslant \eta_{\max}$, and $\eta_{\max}$ corresponds to the minimum divergence angle of the two gamma rays, $\eta_{\max} = 1 - 2m_{\pi^0}^2/E_{\pi^0}^2$.

The density function in the angle of divergence has a characteristic feature at $\eta = \eta_{\max}$ or at $\psi = \psi_{\min}$): $W_\eta(\eta \to \eta_{\max}) \to \infty$ in a square-root manner. This means that the most probable angles of divergence of photons are those near the minimum angle of divergence.

**Three-body decay**

Let us consider the decay $\mu^- \to e^- \tilde{\nu}_e \nu_\mu$. In the $m_e/E_e \ll 1$ approximation, the energy and angle-of-emission function of an electron in the rest frame of a muon has the form[44]

$$dW_e = (1/2\pi)[(3-2\varepsilon) + \xi\mathbf{n}(1-2\varepsilon)]\varepsilon^2 d\varepsilon\, ds\Omega^*$$

Here $\xi$ is the vector of the (spin-induced) muon polarization: $\xi = \langle\mathbf{s}\rangle/(1/2) = 2\langle\mathbf{s}\rangle$, $\langle\mathbf{s}\rangle$ is the mean value of the muon-spin vector, $\mathbf{n} = \mathbf{p}_e^*/|\mathbf{p}_e^*|$ is the unit vector in the direction of the electron 3-momentum, $\varepsilon = E_e^*/E_e^{\max}$, $E_e^{\max} = m_\mu/2$, and $\cos\theta^* = \xi\mathbf{n} = z^*$. For the decay $\mu^+ \to e^+\nu_e\tilde{\nu}_\mu$ the plus sign in front of $\xi\mathbf{n}$ should be replaced by a minus sign. The muons from the decay $\pi^\pm(K^\pm) \to \mu^\pm + \nu_\mu(\tilde{\nu}_\mu)$ are, as we know[44] (more on this below), completely longitudinally polarized: for $\mu^+$ muons we have $\xi\mathbf{p}_{\mu^+}/|\mathbf{p}_{\mu^+}| = -1$ and for $\mu^-$ muons we have $\xi\mathbf{p}_{\mu^-}/|\mathbf{p}_{\mu^-}| = +1$. In the absence of depolarization effects, the muon polarization remains until the muons decay upon stopping.

The angle-of-emission function of an electron (positron) during the decay of a completely polarized stopped muon is[44]

$$dW_z(z^*) = (1/2)[1 \mp (1/3)z^*]dz^*. \tag{2.50}$$

Here the minus sign corresponds to the electron and the plus sign corresponds to the positron. It is evident from Eq. (2.50) that the muon polarization can be determined experimentally by measuring the asymmetry of the angular distribution of electrons (positrons) which are produced as a result of the decay of the stopped muons. If the longitudinal polarization of the muon is incomplete, the angle-of-emission function of the electron (positron) becomes

$$dW_z(z^*) = (1/2)[1 \mp (1/3)Pz^*]dz^*,$$

where $P$ is the coefficient which characterizes the degree of longitudinal polarization of a muon ($P \leqslant 1$). One of the principal causes of muon depolarization—the so-called kinematic depolarization—will be considered below.

The electron energy distribution can be found by integrating the expression for $dW_e$ over the angles of emission[44]

$$dW_e(\varepsilon) = 2(3 - 2\varepsilon)\,\varepsilon^2 d\varepsilon.$$

We wish to emphasize again that three-body muon decay and $K_{e3}$ decay are the main sources of the electronic neutrinos. They are therefore of interest from the standpoint of the production of electronic neutrino beams in proton accelerators. In the production of muonic neutrino beams the $K_{\mu 3}$ decay plays a minor role in comparison with the two-body decay, $\pi(K) \to \mu\nu_\mu$. The $K_{e3}$ decay determines the admixture of electronic neutrinos in the muonic neutrino beam. In many cases this factor may appreciably affect the interpretation of the results of neutrino experiments carried out on proton accelerators.

The energy and angle-of-emission function of the electronic neutrino, which is produced as a result of the decay of an in-flight muon ($\mu \to e\nu_e\nu_\mu$), in the laboratory frame, is[46]

$$dW = (3/2\pi)\, f_\mu(\varepsilon_\nu, \theta_\|)\, \varepsilon_\nu^2 d\varepsilon_\nu d\Omega_\nu,$$

where

$$f_\mu(\varepsilon_\nu, \theta_\nu) = (\gamma - \sqrt{\gamma^2 - 1}\cos\theta_\nu)[1 - (2/3)\varepsilon_\nu \times (\gamma - \sqrt{\gamma^2 - 1}\cos\theta_\nu)] \text{ and } \varepsilon_\nu = 2E_\nu/m_\mu, \quad \gamma = E_\mu/m_\mu.$$

For a three-body decay of an in-flight $K$ meson—the $K_{e3}$ decay—the energy and angle-of-emission function of neutrinos in the laboratory frame is[46]

$$dW = (4\pi j(a))^{-1} f_K(\varepsilon_\nu, \theta_\nu)\varepsilon_\nu^2 d\varepsilon_\nu d\Omega_\nu,$$

where

$$f_K(\varepsilon_\nu, \theta_\nu) = (\gamma - \sqrt{\gamma^2 - 1}\cos\theta_\nu) \times [1 - \varepsilon_\nu(\gamma - \sqrt{\gamma^2 - 1}\cos\theta_\nu)^2] / [1 - (\varepsilon_\nu/a)(\gamma - \sqrt{\gamma^2 - 1}\cos\theta_\nu)].$$

The quantity $j(a)$, which is determined from the normalization condition, is

$$j(a)=\frac{a^3}{2}\left[2(a-1)^2\ln\left(\frac{a}{a-1}\right)-\frac{1}{6a^2}-\frac{2}{3a}+3-2a\right],$$

where $a=m_K^2/(m_K^2-m_\pi^2)$, $\varepsilon_\nu=2aE_\nu/m_K$, and $\gamma=E_K/m_K$. Here, as in the case of the muon decay, the electron mass is ignored.

The energy-momentum conservation law imposes a constraint on the energy and emission angle of the neutrino in the three-body decay: $\varepsilon_\nu[\gamma-\sqrt{\gamma^2-1}\cos\theta_\nu]\leqslant 1$.

**Kinematic depolarization**

In the rest frame of a decaying particle the longitudinal polarization of a muon is complete in the two-body decay $\pi(K)\to\mu\nu_\mu$. This forced muon polarization is a consequence of the fact that only left-handed neutrinos participate in the weak interactions. The massless-neutrino spin is always directed opposite to its momentum, i.e., the longitudinal polarization of the neutrino is complete; $\xi_\nu\mathbf{n}=-1$, where $\xi_\nu=2\langle\mathbf{s}_\nu\rangle$ is the neutrino polarization vector, and $\mathbf{n}$ is the unit vector along the neutrino-momentum direction (for an antineutrino $\xi_\nu\mathbf{n}=+1$). By virtue of the law of conservation of the angular-momentum projection the muon helicity is $\lambda(\mu^\pm)=2(\mathbf{s}_{\mu^\pm}\mathbf{n}_{\mu^\pm})=\mp 1$ in the rest of the pion (kaon). In other words, the sign and magnitude of the longitudinal polarization (helicity) of the muon are the same as the sign and magnitude of the polarization of the neutrino (antineutrino) which is produced along with the muon.

In the transformation to the laboratory frame, the neutrino helicity is conserved, whereas the muon experiences a kinematic depolarization. The kinematic depolarization mechanism can be explained as follows. The Lorentz transformations, which are used to effect a transition from the rest of the decaying particle to the laboratory frame, do not change the spin direction of the decay products, whereas their momentum may reverse direction. In the rest frame of the decaying pion or kaon $[\pi(K)\to\mu\nu_\mu]$ the muon polarization is always the same [for example, $\lambda(\mu^+)=-1$ for the decay $\pi^+(K^+)\to\mu^+\nu_\mu$], regardless of the direction of emission of the muon with respect to the momentum of the $\pi(K)$ meson. The muons which are emitted in the rest frame of the $\pi(K)$ meson in the backward direction relative to the direction of motion of the primary $\pi(K)$ meson may be emitted in the forward direction in the laboratory frame. The longitudinal polarization of these muons in this case will change sign (for $\beta_{\pi(K)}>\beta_\mu^*$ the muons in the laboratory frame always move in the forward direction, regardless of the direction in which they were emitted in the rest frame of the decaying particle). Since in the laboratory frame the muon beam consists of particles which are emitted both in the forward and backward directions in the rest frame of the $\pi(K)$ meson, the average longitudinal polarization of muons will differ from the polarization of muons which they had in the rest frame of the decaying particle.

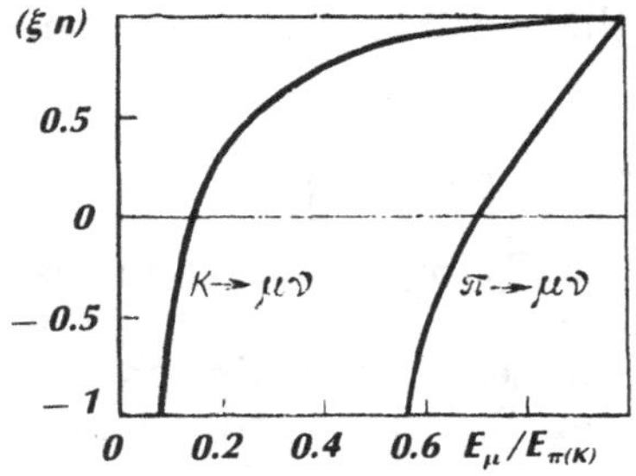


**Figure 2.16. The degree of longitudinal polarization of a muon produced as a result of the decay $\pi(K)\to\mu\nu$ versus $E_\mu/E_{\pi(K)}$.**

The degree of longitudinal polarization of a muon due to the decay $\pi(K)\to\mu\nu_\mu$ in the laboratory frame is given by the equation[47]

$$(\xi_\mu \mathbf{n}_\mu) = (E_\mu E_\mu^*/p_\mu p_\mu^*) - (E_{\pi(K)} m_\mu^2/m_{\pi(K)} p_\mu p_\mu^*)$$

or for $m_\mu/E_\mu \ll 1$ by the equation

$$(\xi_\mu \mathbf{n}_\mu) = \left[\varepsilon - \frac{m_\mu^2(1-\varepsilon)}{m_{\pi(K)}^2}\right] \left[\varepsilon + \frac{m_\mu^2(1-\varepsilon)}{m_{\pi(K)}^2}\right],$$

where

$$\varepsilon = (E_\mu - E_\mu^{\min})/(E_\mu^{\max} - E_\mu^{\min}), 0 \leqslant \varepsilon \leqslant 1.$$

The depolarization of a muon produced in the decay $\pi\to\mu\nu_\mu$ is stronger than the depolarization of a muon produced in the decay $K\to\mu\nu_\mu$ (Fig. 2.16), because the muon in the decay $\pi\to\mu\nu_\mu$ has in the rest frame of the pion a velocity $\beta^*_{\mu(\pi)}$ approximately one-third the velocity it has in the decay of a kaon, $\beta^*_\mu(K)$.

# Chapter 3
# Description of multiple production of hadrons at high energies

## 1. Preliminary remarks

In this chapter we briefly describe the results of an analysis of the interaction of particles with nucleons and nuclei. More detailed information may be found in the monographs of Refs. 7–9 and 48 and in the reviews of Refs. 49–52.

A theoretical description of the transmission of a hadron beam through a condensed medium must necessarily be based on the experimental data on the reactions involving the collision of various types of hadrons with nucleons (a hydrogen target) and with various nuclei. The large variety of reactions involving hadron–nucleon and hadron–nucleus interactions and a large multiplicity of secondary hadrons and their many characteristics preclude, however, the exclusive use of experimental data in the calculations, since at high energies it would hardly be possible to obtain comprehensive information about all reaction channels in the entire allowable range of variation of the kinematic variables and over a broad range of primary collision energies.

Under these conditions we can use the information only on the inclusive spectra of various types of secondary hadrons produced in various hadron–nucleus reactions. This approach allows us to obtain all the information about the inclusive single-particle distributions of a particular type of hadrons which are produced in a medium during the development of an electromagnetic cascade (EC). Simulation of EC on the basis of inclusive spectra of hadron–nucleus collisions greatly simplifies the problem and leads to results which are quite adequate for the planning and interpretation of experiments with thick targets, for the calculation of a radiation shield and channels that contain a large amount of matter, etc.

From a theoretical point of view, such an approach gives an inclusive description of the spectra of hadrons inside condensed matter at a certain depth. The resulting inclusive distributions of various types of hadrons produced as a result of several successive interactions with the nuclei of the medium yield minimum information necessary for the prediction of virtually all average characteristics of EC. It should be remembered, however, that such an approach ignores the correlation between hadrons produced in a single interaction event and consequently sets aside the corresponding class of problems. It can be assumed, nevertheless, that these correlations cancel out to a large extent as a result of multiple interactions and therefore can be ignored below.

It is not realistic from the practical standpoint to expect that comprehensive experimental data on inclusive single-particle hadron spectra in various hadron–nucleus collisions can be obtained in the energy range which is now experimentally accessible or which will be accessible in the near future. It is therefore necessary to use highly incomplete data which must be extrapolated to the region of kinematic variables where no measurements have been carried out. It is also necessary to use data obtained with use of certain nuclear targets and to extend the observed systematic features to all other nuclei, whose masses take on values intermediate between those that were measured.

Such extrapolations are based on model-based representations of the multiple production of hadrons by nucleons and nuclei, which make it possible to generalize the data obtained in special cases to a broad range of values of primary energies, to the kinematic characteristics of secondary particles, to other types of targets, etc. The agreement of the predictions based on a theoretical model with parameters found from experimental data obtained under one set of conditions and those obtained under another set of conditions [at another energy and (or) with another target] justify these extrapolations to the energy region accessible to measurements.

The experimental data and the basic theoretical models which describe the multiple production at the initial energies, $E_0 \gtrsim 10$ GeV, are briefly analyzed below. This analysis focuses principally on the qualitative behavior of the inclusive spectra of secondary hadrons produced by nucleon and nuclear targets and a quantitative description of experimental data in terms of the present-day theoretical models.

## 2. Fundamental experimental data

### Hadron-nucleon interactions

The interaction of elementary particles is characterized by their total cross sections $\sigma_{tot}$ and inelastic interaction cross sections $\sigma_{in}$. For hadron–nucleon interactions in the range of available energies in the laboratory frame of reference ($E = 5$–$2000$ GeV) these cross sections amount to tens of millibarns ($\sim 10^{-26}$ cm$^2$). In this energy range $\sigma_{tot}$ and $\sigma_{in}$ vary only slightly. In the $pp$ interaction $\sigma_{tot}$ initially decreases slowly down to $E \sim 30$ GeV, acquires a broad minimum in the region $E \sim 30$–$70$ GeV, and then increases slowly by approximately 10–12% at $E = 1500$ GeV (Fig. 3.1). For the $\pi^{\pm} p, K^{\pm} p$, and $\tilde{p}p$ interactions, the $\sigma_{tot}$ have been measured to energy $E = 280$ GeV and they exhibit the same behavior at high energies as they do in $pp$ interactions, but their low-energy behavior and the position of the minima differ from the case of $pp$ interaction and from each other (Fig. 3.1).

**Figure 3.1. The energy dependence of the total cross section of hadron–nucleon interactions.**

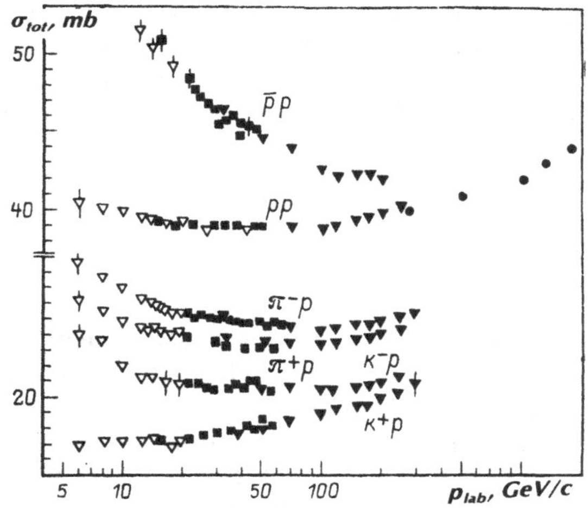


**Figure 3.2. The average multiplicity of charged secondary hadrons in hadron–nucleon collisions vs the energy required to produce new particles, $Q = \sqrt{S} - m_a - m_b$.**

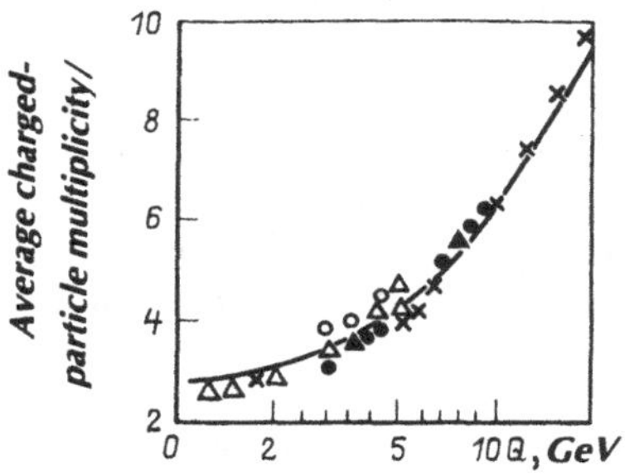


The average *multiplicity of secondary hadrons* produced in the collision of hadrons and other particles with nucleons increases with increasing energy. It should be mentioned here that this increase is much slower than the maximum permissible increase [$\langle N \rangle_{\max} \simeq (\sqrt{S} - m_a - m_b)$; see Sec. 1.3]. The experimentally observed increase in the average multiplicity of charged secondary hadrons as a function of energy (Fig. 3.2), including the data obtained in the cosmic radiation, is described by the equation

$$\langle N_{ch} \rangle = A + B \ln S + C(\ln S)^2 ,$$

where $S$ is the square of the total energy in the c.m. frame (GeV$^2$); $3 < \sqrt{S} < 150$ GeV; and $A$, $B$, and $C$ are the adjustable parameters which differ in different reactions. At $\sqrt{S} > 10$ GeV the equation $\langle N_{ch} \rangle = A + B \ln S$, where $A = -(1\text{–}2)$ and $B = 1.5\text{–}2$ for various reactions, gives a completely satisfactory description. The data obtained at $\sqrt{S} > 10$ GeV are also described well by a power function: $\langle N_{ch} \rangle = \alpha S^{\beta}$, where $\alpha =$ const and $\beta \simeq 0.2$.

Remarkably, the dependence of $\langle N_{ch} \rangle$ on the energy evolution of the reaction, $Q = \sqrt{S} - m_a - m_b$, is nearly universal for a wide variety of interactions (different beams and different targets) (see Fig. 3.2). This result was verified experimentally for the beams $p$, $\pi^{\pm}$, $K^{\pm}$, $\gamma$, $\mu^{+}$, $e^{-}$, $\nu$, and $\tilde{\nu}$ using a proton target and for the $e^{+}e^{-}$ annihilation. For the photon and lepton beams we have $Q = W - m_N$, where $W$ is the effective mass of secondary hadrons.

The majority ($\sim 80\%$) of secondary (newly produced) hadrons are pions. It should be noted that the average multiplicity of kaons, antiprotons, hyperons, and other heavy particles in the energy range which is studied increases more rapidly than the total multiplicity.

**Inclusive longitudinal-momentum distributions (see also Sec. 3 of Chap. 1)**
Most of the secondary hadrons are produced in the region of values of the Feynman variable $|x_F| \simeq 2|p_{\parallel}^*|/\sqrt{S} \lesssim 0.1$. This region of values of $x_F$ is conventionally known as the central region or the pionization region, since the bulk of the secondary particles consists of pions. A relatively few particles are produced in the fragmentation region of the primary hadrons $0.1 \leqslant |x_F| \leqslant (0.8\text{–}0.9)$ and in the diffraction region $0.8\text{–}0.9 \lesssim |x_F| \lesssim 1$. The fastest particles, which are of the same type (with possible exception of the charge) as the incident particle (or the target particle) are produced in this longitudinal-momentum region. These particles are generally produced in the inelastic diffraction processes, in which one or both of the primary hadrons are excited as a result of the interaction and in which a group of particles with a small multiplicity, which is characterized by the same quantum numbers as the primary hadron, with possible exception of the spin and parity, is produced. In particular, such a particle may be simply a hadron of the same type as the primary hadron.

In a very crude approximation, the experimental data on the invariant inclusive cross sections are in agreement with the Feynman-scaling hypothesis, i.e.,

$$E_c \frac{d^3\sigma}{d^3p_c} = f(x_F, p_\perp) = f(y^*, p_\perp), \tag{3.1}$$

and the dependence on the primary energy is not very strong. Specifically, in the central region (with $x_F = 0$) the function $f(x_F, p_\perp, \sqrt{S})$ increases in the $pp$ collisions by a factor of 1.5–2 for charged secondary pions at energies between $\sqrt{S} = 5$ and 23 GeV. In the $pp$ collisions at energies $\sqrt{S} = 23$–63 GeV, this function increases again by 30–40% when $x_F \simeq 0$. In considering the function $\rho(x_F, p_\perp; \sqrt{S}) = f/\sigma_{\text{in}}$, we see that when $x_F \simeq 0$, it increases by only 15–25% at energies $\sqrt{S} = 23$–63 GeV, i.e., almost half as rapidly.

The plot of $f(y^*, p_\perp, \sqrt{S})$ as a function of the rapidity in the central region has the shape of a broad bell (Fig. 3.3) with a maximum near the small values of $|y^*|$. This variable has no plateau in the central region. A rigorous Feynman scaling so far has not been observed. The uncertainties that arise, however, are no greater than 1.5–2, which is reasonable in the case of strong-interaction phenomenology.

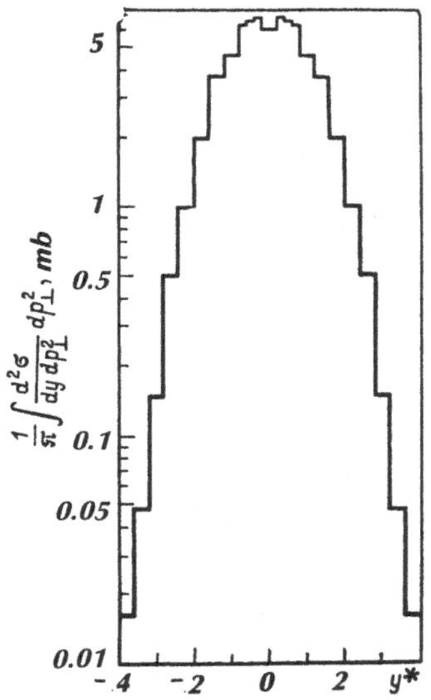


**Figure 3.3. Inclusive rapidity distribution in the c.m. frame of charged secondary pions produced in *pp* collisions at energy $E_0 = 69$ GeV in the laboratory frame.**

In the case of numerical calculations for specific cases, using equations in which the Feynman scaling holds, the slight increase of function (3.1) with the energy in the central region can be taken into account by introducing a correction factor (or factors) into the fitting formulas. As for the fragmentation region, here the function $\rho(x_F, p_\perp; \sqrt{S})$ is virtually independent, beginning with the energy $\sqrt{S} \simeq 8$ GeV, of $\sqrt{S}$ with about a 10% error. This property is known as *early scaling* in the fragmentation region.

In *pp* collisions the inclusive spectra are naturally symmetric with respect to the point $x_F = 0$ ($y^* = 0$) and in the $\pi p^-$, $Kp^-$, and other processes, they are asymmetric.

## Inclusive transverse-momentum secondary-hadron distributions

At $p_\perp \lesssim 1.5$–$2.0$ GeV/$c$ the $p_\perp$ dependence of the secondary-hadron distributions, a nearly universal dependence, is given by

$$f(x_F, p_\perp) \sim \exp(-bm_\perp), \tag{3.2}$$

where $m_\perp = \sqrt{m^2 + p_\perp^2}$, $m$ is the mass of secondary particles, and the parameter $b \simeq 6$ (GeV/$c$)$^{-1}$. A more thorough analysis of the data shows that the parameter $b$ depends on $x_F$ and also has a more complex functional dependence on $p_\perp$. Equation (3.2) can be used only for a ballpark estimate. It is important to note that the average value of $\langle m_\perp \rangle$ is essentially independent of energy for particles of the given type. Figure 3.4 is a plot of $dN/dp_\perp^2$ versus $m_\perp$ for an energy of the CERN *pp* colliding beams ($\sqrt{S} \simeq 60$ GeV) for secondary $\pi$ and $K$ mesons.

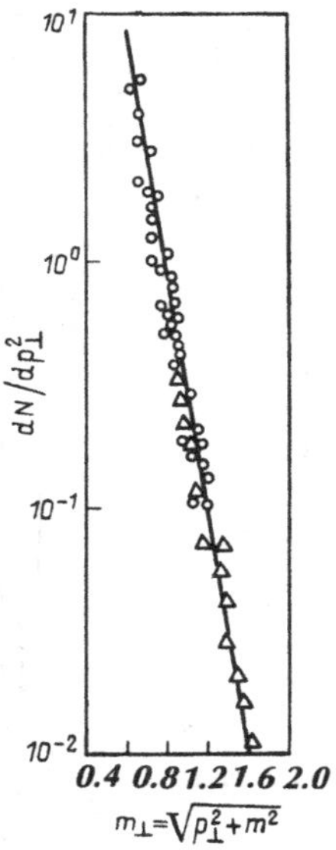


**Figure 3.4. Distribution of secondary hadrons in the transverse mass $m_\perp$. Solid line, data on $\pi^\pm$ mesons; $\Delta$, $\pi$ mesons; O, $K$ mesons.**

At $p_\perp \gtrsim 1.5$–$2$ GeV/$c$ expression (3.2) is replaced by a slower (a power-law) decrease of $f(x_F, p_\perp)$ as a function of $p_\perp$, where the rate of decrease depends on the primary collision energy. In this range of values the following type of fits are used:

$$f(x_F, p_\perp) = \varphi(x_R)\,(p_\perp^2 + m^2)^{-n(S)}, \tag{3.3}$$

where $x_R = 2p^*/\sqrt{S}$, and $m$, $n(S)$ are certain parameters.

## Multiple production hadron–nucleus processes

The collision of particles with nuclei is usually compared with reactions involving an interaction with quasifree nucleons. The experimental data are generally given as a ratio of differential cross sections, of multiplicities, and of other characteristics of the processes involving nuclei and nucleons.

The secondary hadrons are conventionally divided into three groups. This convention arose in the study of the interaction of particles with nuclei by the photographic-emulsion method. The first group includes particles with a velocity $\beta = v/c \geqslant 0.7$. These particles, called *shower particles*, leave light tracks in the photographic emulsion. The *average multiplicity* of these particles is given by $\langle N_s \rangle$. The particles with a velocity $0.3 < \beta < 0.7$ leave gray tracks in the photographic emulsion. These particles are primarily protons which were knocked out of the nucleus as a result of an intranuclear cascade. The average multiplicity of these protons is given by $\langle N_g \rangle$. The particles with velocities $\beta \leqslant 0.3$ leave black tracks in the photographic emulsion. These particles are primarily the nuclear decay products and the so-called *evaporative*

**Table 3.1. Average multiplicity of secondary particles in the collision of protons with the nuclear-emulsion nucleus, *Em*.**

| $E_0$, GeV | $\langle N_s\rangle$ | $\langle N_g\rangle$ | $\langle N_b\rangle$ | $\langle N_h\rangle$* | $E_0$, GeV | $\langle N_s\rangle$ | $\langle N_g\rangle$ | $\langle N_b\rangle$ | $\langle N_h\rangle$* |
|---|---|---|---|---|---|---|---|---|---|
| 6.2 | 3.2 | 3.58 | 5.68 | 9.26 | 3500 | 22.0 | 2.06 | — | 6.80 |
| 17.2 | 5.4 | 2.10 | 5.69 | 7.79 | 200 | 12.7 | — | — | 7.68 |
| 22.5 | 6.5 | 3.38 | 5.22 | 8.60 | 200 | 13.0 | — | — | 7.30 |

*$\langle N_h\rangle = \langle N_g\rangle + \langle N_b\rangle$.

protons. The average multiplicity of black tracks is given by $\langle N_b\rangle$ and the average multiplicity of gray and black tracks is given by $\langle N_h\rangle = \langle N_g\rangle + \langle N_b\rangle$. These particles are the nuclear fragmentation products, while the shower particles are principally the fragmentation products of the incident particle and the secondary hadrons from the central rapidity region. The experimental data on the average multiplicities are given in Table 3.1.

Measurements involving the use of nuclear targets were carried out in the laboratory frame and all data are presented for this frame of reference.

The inclusive secondary-hadron spectra are studied as functions of the variable $p_\perp$ and of one of the longitudinal variables (see Sec. 1.3):

$$y = (1/2)\ln[(E+p_\parallel)/(E-p_\parallel)]; \quad \eta = -\ln\tan(\theta/2); \quad x = p_\parallel/p_0 .$$

The *pseudorapidity* variable $\eta$ is similar to the rapidity variable $y$ for particles with longitudinal momenta $p_\parallel \gg m_\perp$. The use of the variable $\eta$ stems from the fact that momenta of secondary particles are not measured in many experiments involving the use of the nuclear targets (in these experiments only the angles of emission with respect to the momentum direction of the primary-beam particles are measured in the laboratory frame).

The experimental data for the multiple processes induced by nuclei are presented in the form $R = \langle N_s\rangle_A/\langle N_s\rangle_N$; $R_z = (dN_s/dZ)_A/(dN_s/dZ)_N$, etc., where the subscripts $A$ and $N$ refer to the nuclear target and the nucleon target ($A$ is the number of nucleons in the nucleus), and $Z = y$, $\eta$, or $x$.

A useful measure of the target thickness is the quantity

$$\bar{\nu} = A\sigma^N_{\rm abs}/\sigma^A_{\rm abs} , \qquad (3.4)$$

where $\sigma^{N,A}_{\rm abs}$ is the cross section for the absorption of beam particles in the interaction with the nucleon or nucleus.* The parameter $\nu$ characterizes the average thickness of the nucleus in units of the mean free path of a hadron in the nucleus before the absorption $\lambda_{\rm abs} = \langle(\sigma^N_{\rm abs}\rho_A)^{-1}\rangle$, where $\rho_A$ is the nucleon density in the nucleus. The mean value of $\langle\rho_A^{-1}\rangle$ is contained in this expression.

*The absorption cross section is taken to mean the cross section of all inelastic processes, with the exception of the cross section for the excitation of a target which does not produce any new particles.

In a model for sequential collisions of an incident particle with the nucleons of the nucleus, the quantity $\bar{\nu}$ is interpreted as the average number of inelastic intranuclear collisions during which the primary particle remains the leading particle (particle with the highest energy).

The two- and three-body correlation functions and other characteristics of the process are also measured in experimental studies of the multiple production by nuclei. These are important data for the selection of the specific target, but after the target is chosen, only the inclusive single-particle spectrum of secondary hadrons produced by the nuclei must be known. The fine details of the process gleaned from the experimental data will therefore be disregarded here.

*The cross sections for the absorption* of charged hadrons are measured by knocking them out of the beam. The measured values of the cross sections $\sigma^A_{\rm abs}$ are described reasonably well by the optical-model equation[8,51]

$$\sigma^A_{\rm abs} = 2\pi \int_0^\infty dbb\{1 - \exp[-\nu(b)]\}, \tag{3.5}$$

where $b$ is the impact parameter, $\nu(b) = \tilde{\sigma}^N_{\rm abs} T(b)$, $\tilde{\sigma}^N_{\rm abs}$ is the cross section for the absorption of the incident particle by the nucleon of the nucleus,* and $T(b)$ is the optical thickness of the nucleus,

$$T(b) = \int_{-\infty}^{\infty} dz\rho_A(z,b); \quad \int \rho_A(z,b)dV = A. \tag{3.6}$$

The integration in Eq. (3.6) is carried out over the nuclear volume. If the expectation value over the nuclear volume $\langle\nu(b)\rangle \gg 1$, then we would have $\sigma^A_{\rm abs} \approx \pi R^2$, where $R$ is the mean-square radius of the nucleus: $R \simeq r_0 A^{1/3}$, where $r_0 = 1.2$ fm. If, on the other hand, $\langle\nu(b)\rangle \ll 1$, then $\sigma^A_{\rm abs} \simeq A\sigma^N_{\rm abs}$.

Experimental data imply that for the hadron–nucleus interactions the parameter $\alpha$ in the expression $\sigma^A_{\rm abs} = \sigma_0 A^\alpha$ is approximately equal to $\frac{2}{3}$. The parameter $\alpha \simeq 0.69$ for protons, and it is 0.75 for $\pi^\pm$ mesons and 0.74–0.77 for $K^+$ mesons. In the case of incident hadrons the dependence of $\sigma^A_{\rm abs}(E)$ on the energy $E$ in the laboratory frame is much weaker than $\sigma^N_{\rm abs}(E)$. The cross sections $\sigma^A_{\rm abs}$ can be assumed essentially constant to energies $E = 200$–$300$ GeV and evidently at higher energies as well (Table 3.2). For photons $\langle\nu(b)\rangle \ll 1$, but the experimental value of the parameter $\alpha(\gamma A) \simeq 0.9$, which is evidence for the existence of the hadronic component in the photon wave function. For virtual photons (the electroproduction of hadrons by nuclei) we have $\alpha(\gamma^* A) \simeq 1$ in the deep inelastic region, where the effective mass of secondary hadrons, $W \gtrsim 2$ GeV, is the square of the 4-momentum transfer to the hadrons, $Q^2 \gtrsim 1$–$2$ GeV$^2$.

*It should be noted that $\tilde{\sigma}^N_{\rm abs}$ differs from the cross section for the absorption of hadrons by a free nucleon (Ref. 48). The use of $\sigma^N_{\rm abs}$ for free nucleons accounts for an up to 20% error in (3.5).

Table 3.2. Cross section for the absorption of protons and $\pi^-$ mesons by various nuclei, $\sigma^A_{abs}$, mb.

| $E_0$, GeV | Be | C | Al | Fe | Cu | Pb | U |
|---|---|---|---|---|---|---|---|
| | | | Protons | | | | |
| 10 | 208 | 256 | 456 | 753 | 822 | 1804 | 1974 |
| 20 | 205 | 252 | 450 | 746 | 814 | 1793 | 1963 |
| 50 | 204 | 251 | 448 | 743 | 812 | 1790 | 1960 |
| 100 | 205 | 252 | 450 | 746 | 814 | 1793 | 1963 |
| 200 | 207 | 255 | 454 | 750 | 816 | 1800 | 1970 |
| 500 | 211 | 260 | 460 | 759 | 828 | 1813 | 1984 |
| 1 000 | 215 | 264 | 466 | 766 | 836 | 1824 | 1996 |
| 2 000 | 219 | 268 | 472 | 774 | 844 | 1835 | 2008 |
| 5 000 | 224 | 274 | 481 | 785 | 855 | 1851 | 2025 |
| 10 000 | 228 | 279 | 487 | 793 | 864 | 1863 | 2037 |
| 20 000 | 232 | 283 | 493 | 801 | 872 | 1875 | 2049 |
| | | | $\pi^-$ mesons | | | | |
| 10 | 164 | 204 | 380 | 651 | 715 | 1646 | 1812 |
| 20 | 157 | 196 | 367 | 633 | 696 | 1619 | 1782 |
| 50 | 152 | 190 | 358 | 621 | 683 | 1597 | 1761 |
| 100 | 151 | 189 | 357 | 619 | 681 | 1584 | 1757 |
| 200 | 153 | 191 | 360 | 623 | 685 | 1601 | 1764 |
| 500 | 158 | 197 | 369 | 636 | 699 | 1623 | 1787 |
| 1 000 | 163 | 204 | 380 | 651 | 715 | 1646 | 1812 |
| 2 000 | 171 | 213 | 393 | 669 | 734 | 1675 | 1841 |
| 5 000 | 183 | 226 | 412 | 695 | 762 | 1716 | 1884 |
| 10 000 | 192 | 237 | 428 | 717 | 784 | 1749 | 1919 |
| 20 000 | 202 | 249 | 445 | 739 | 808 | 1783 | 1953 |

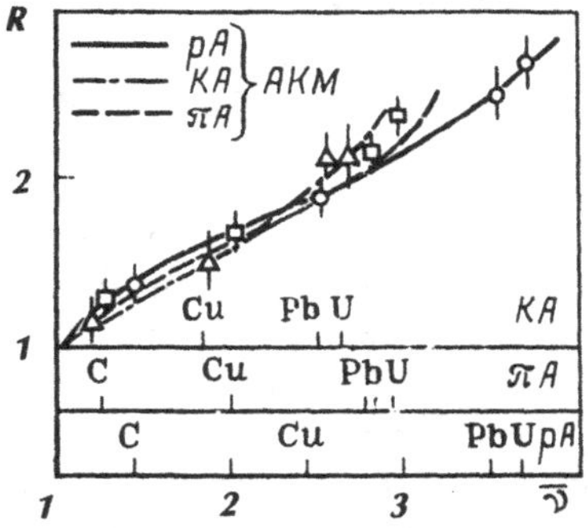


**Figure 3.5.** ***R*** **vs the parameter $\bar{\nu}$ in *KA* (△), $\pi A$, (□), and *pA*(○) interactions at primary-beam energy of 100 GeV.**

**Average multiplicity of hadrons produced by nuclei**

At incident hadron energies $E > 50$ GeV in the laboratory frame, the observed ratio of the average multiplicity of secondary shower hadrons in hadron-nucleus collisions to the quantity $\langle N_s \rangle_N$ is described by an approximate equation[51] $R = a + b\bar{\nu}$, where $a \simeq 0.4$ and $b \simeq 0.6$–$0.7$ (Fig. 3.5).

It should be emphasized that the values of $R$ are, according to this equation, much smaller than the values obtained from the cascade model for a

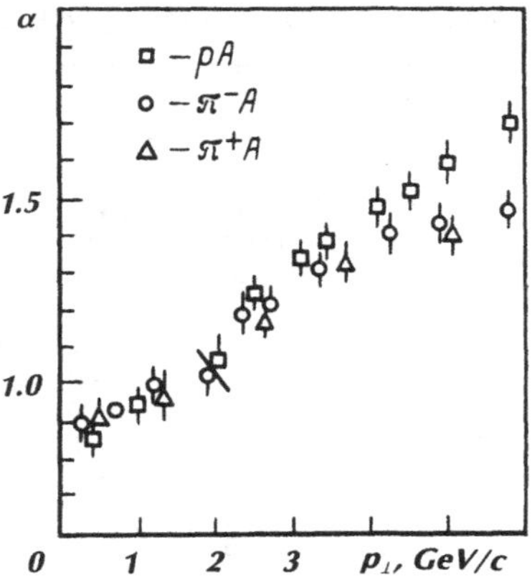


**Figure 3.6.** The exponent $\alpha(p_\perp)$ in a function of the type $E\,D^3\sigma/d^3p \sim A^{\alpha(p_\perp)}$ for the inclusive cross section for the production of hadrons and jets with large $p_\perp$ by nuclei.

hadron–nucleus collision.[48] This is a crucial point in choosing a model for the description of a hadron–nucleus collision.

For virtual primary photons and neutrinos $\bar{\nu} = 1$, although $R(\gamma^* A) > 1$ and $R(\nu A) > 1$, which is evidence that there is a hadron component in the wave function of the primary beam particles.

**Inclusive spectra of secondary hadrons in hadron–nucleus interactions**

Experimental data show that the spectra of secondary hadrons in hadron–nucleus interactions exhibit the following systematic features[51]:

(1) In the region of primary-beam fragmentation ($\eta \gtrsim \eta_{\max} - 2$) the ratio $R_\eta < 1$ and decreases with increasing parameter $\bar{\nu}$.

(2) In the region of nuclear fragmentation ($\eta \leqslant 1$–$2$) the ratio $R_\eta$ is described by an approximate phenomenological equation $R_\eta = 1 + S(\eta) \times (\bar{\nu} - 1)$, where $S(\eta) \simeq 2.5$–$3$. In this case $R_\eta$ does not depend on the primary energy, beginning with $E \gtrsim 30$–$50$ GeV.

(3) In the region which lies between cases (1) and (2), the ratio $R_\eta$ begins to plateau at $E \gtrsim 100$ GeV.

(4) The data on the electro- and neutrino production of hadrons by nuclei show that $R_\eta \approx 1$ for large values of $\eta$ (larger than 2), and $R_\eta > 1$ for $\eta \simeq 0$.

(5) The experimental data on nuclear fragments show that the multiplicities $\langle N_g \rangle$ and $\langle N_b \rangle$, isotopic yield, etc., are virtually independent of $E$ at incident-hadron energies $E \gtrsim 10$–$20$ GeV. These results are evidence that at high energies the average number of intranuclear interactions is no longer dependent on $E$. This phenomenon is called the *limiting nuclear fragmentation regime.*[51]

(6) In the production of hadrons with large $p_\perp$ ($\gtrsim 2$ GeV/$c$) the cross section depends nontrivially on the number of nucleons in the nucleus (Fig. 3.6).

(7) In the case of shower particles, the spectra for transverse momenta behave identically to those in the hadron–nucleon collision.

In concluding the short review of the experimental data on the multiple production resulting from the collisions with nucleon and nuclear targets we note that these data allow us to reduce the number of theoretical models that can be used in the calculation of the passage of hadrons through condensed matter.

## 3. Theoretical models for multiple-production processes

### Hadron–nucleon collisions

We restrict the discussion here to a short review of the two qualitatively and quantitatively most developed theoretical models of multiple-production processes: the statistical hydrodynamic model and the multiperipheral Regge model. (The basic types of models of multiple-production processes are discussed in Refs. 7, 8, and 48–51. The quark-parton model conventionally falls in the same category as the multiperipheral models and will be discussed below.) These models are based to some degree on opposite points of view concerning the multiple hadron production. The statistical hydrodynamic models are based on the concept of collective hadron interaction which is described by statistical methods and hydrodynamic equations of motion of a relativistic hadronic fluid. The multiperipheral Regge models are based on the field theory picture of a quasi-independent production of hadrons from multiple centers which produce either a single hadron (a hadronic resonance) or a group of hadrons (a cluster), which appears in the initial stage of the process as a single production and which subsequently decays in accordance with certain laws (including the statistical-hydrodynamic laws). We discuss below several simple examples of these models which agree reasonably well with the data on the more important characteristics of the multiple-production processes, which permit the experimental systematic features to be represented in simple analytic form and which contain a reasonable number of adjustable parameters that are nearly independent of the primary energy and other factors. This slightly simplified approach will make it possible to later introduce relatively simple, universal semiphenomenological equations which can describe secondary-hadron spectra over a broad range of kinematic characteristics of primary and secondary particles.

### The statistical hydrodynamic model

Let us consider the most popular version of this model which was initially developed by Landau.[53] This model is based on the following qualitative assumptions concerning the multiple-production mechanism[7,8]:

(1) The first stage of the process involves the production of a single, highly excited hadron system (a fluid), which quickly reaches a state of thermodynamic equilibrium as a result of a strong interaction between its elements.

(2) In the second stage, the hadron system is assumed to experience an isoentropy expansion in accordance with the thermodynamic and relativistic hydrodynamics laws.

(3) In the final stage of the process, when the temperature of the elements of the system falls to a level on the order of the pion mass, we see the production of real hadrons, and the elements of the system decay to secondary particles.

The volume element of the hadronic fluid in the proper frame of reference is assumed to decay in accordance with the statistical laws; i.e., the momentum distribution of secondary hadrons in this reference frame is given by the Fermi or Bose standard statistical distribution for hadrons with half-integer and integer spins, respectively.

$$d^3N/d^3p = (2\pi)^{-3} gV[\exp(E/T) \pm 1]^{-1},$$

where $g$ is the number of spin states, charge states, and other states of a particle of a given type, $V$ is the finite volume of the system at the end of the hydrodynamic expansion stage, $E = \sqrt{p^2 + m^2}$ is the energy ($p$ is the momentum in the indicated reference frame, and $m$ is the particle mass), and $T$ is the temperature of an element of the hadronic fluid at the time of the decay to the real hadrons.

The equation of state for a hadronic fluid $P = c_0^2 \varepsilon$, where $P$ is the pressure, $\varepsilon$ is the energy density, and $c_0$ is the velocity of sound in the hadronic fluid, plays an important role in predicting various characteristics of secondary particles in the statistical hydrodynamic model. When the state of the system is similar in its properties to the ultrarelativistic gas, we have $c_0^2 = \frac{1}{3}$. Other values of $c_0^2$, which allow a better agreement to be obtained between the statistical hydrodynamic model and the experiment, are also discussed in the literature. To illustrate the basic predictions of the model, we consider the value $c_0^2 = \frac{1}{3}$, which is frequently used to compare the model with experiment. In the statistical hydrodynamic model the number of secondary particles is not specified (the chemical potential is $\mu = 0$) and is determined from the condition for thermodynamic equilibrium (the thermodynamic potential vanishes).

The statistical hydrodynamic model further assumes that (a) at the time immediately preceding the onset of hydrodynamic dispersal (expansion) of the system, the hadronic fluid is at rest as a whole (the velocity of the fluid elements is $v = 0$ at the time $t = 0$). (In the case of nucleon-nucleon collision, this condition is satisfied in the c.m. frame. In the case of a collision of hadrons with different masses, the analysis is carried out in an equal-velocity frame); (b) at the time of collision, the nucleon is a thin disk of thickness $z \sim (1/m_\pi) m_N/E_0^*$ ($m_\pi$ and $m_N$ are the pion and nucleon masses, and $E_0^*$ is the energy of the primary nucleon in the c.m. frame) and radius $\sim m_\pi^{-1}$. In the statistical hydrodynamic model, the initial volume of the system is generally a free parameter, whose value is chosen from physically obvious considerations or by comparing the predictions of the model with the experiment.

The following are the principal physical predictions of the Landau statistical hydrodynamic model[7,8]:

(1) The multiplicity of secondary hadrons increases with the collision energy according to a power law $\langle N\rangle \sim S^{1/4} \sim E^{1/4}$, where $E$ is the energy of the incident hadron in the laboratory frame.

(2) The rapidity spectrum of secondary hadrons (primarily pions) is determined completely by the hydrodynamic-expansion laws and has an approximately Gaussian shape

$$\frac{1}{N}\frac{dN}{dy^*} \approx (2\pi B)^{-1/2} \exp[-(y^*)^2/2B] , \tag{3.7}$$

where $B \simeq 0.6\ln(E/m_N) + 1.6$ for the $NN$ collision. The width of the rapidity distribution increases slowly with increasing primary energy in accordance with $\Delta y \sim (\ln E/m_N)^{1/2}$. The statistical hydrodynamic model actually describes only the central part of the rapidity spectrum. In this region the rapidity spectrum increases slowly as a function of energy, i.e., there is no scaling.

(3) The transverse-momentum secondary-hadron distribution, which is determined in the model by nearly thermal motion of the system's elements, is described approximately by the equation

$$dN/dp_\perp \sim p_\perp \exp(-m_\perp/T) . \tag{3.8}$$

The hydrodynamic expansion accounts for the slight increase of the average transverse momentum with increasing primary energy. This momentum increase can be approximated by the equation $\langle p_\perp \rangle \sim E^\beta$, where the value of $\beta$ is estimated to be $\beta = \frac{1}{14}-\frac{1}{12}$ (based on several estimates).

Predictions (1)–(3) are in reasonably good qualitative agreement with the experimental data. The distributions of the type (3.7) and (3.8) are frequently used to approximate experimental data. The predicted $p_\perp$ distribution [Eq. (3.8)] in the region $p_\perp \lesssim 1.0$–$1.5$ GeV/$c$ is, in general, virtually the only clear theoretical prediction of the $p_\perp$ distribution of secondary hadrons. Furthermore, the parameter value $T^{-1} \approx m_\pi^{-1} \approx 7$ GeV$^{-1}$ is approximately equal to the experimentally measured value $T^{-1} \approx 6$–$7$ GeV$^{-1}$ for hadrons of various types.

The statistical hydrodynamic models are poorly adapted for use in the region of fragmentation (of leading particles), although the experimental data can be described well when different versions of this model are used to accomplish this goal. The statistical hydrodynamic model is sometimes used to calculate the production of hadrons with large transverse momenta, when agreement with the experiment is achieved by fitting the model parameters. The use of the statistical hydrodynamic model in the kinematic regions indicated above is not yet a universal or standard procedure. We shall therefore discuss here the models which effectively describe the qualitative behavior of the hadron spectra in a reasonably broad range of kinematic variables.

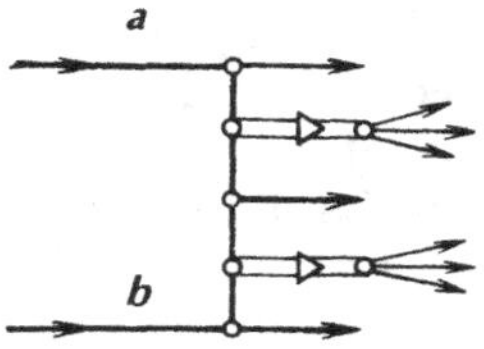


**Figure 3.7. Multiperipheral diagram of multiple production.**

### The multiperipheral Regge model[7,8]

The multiperipheral approach to the multiple production of hadrons is based on the concept of the formation of many quasiindependent centers for the emission of secondary hadrons (one or several) as a result of collision of high-energy hadrons. This concept is based on the assumption that the effective momentum transfer between the secondary-hadron-formation centers is small. This assumption is basically in fair agreement with the experimental data on the multiple-production processes.

Theoretical analysis of hadron-hadron interactions in terms of the multiperipheral model is based on the use of the multiperipheral-diagram technique (Fig. 3.7). The lines labeled *a* and *b* refer to the primary hadrons, the inner lines between the emission centers correspond to the virtual states which transfer the energy and momentum from the primary to the secondary particles, and the rays or bundle of rays which are emerging from the emission centers represent the secondary hadrons.

The hypothesis on the dominant role of such diagrams in the description of multiple production of hadrons with small $p_\perp$ (less than 1.0–2.0 GeV/$c$) proved to be very useful. Several versions of the multiperipheral model have been developed. These models were used to explain physical phenomena such as the power-law (Regge) behavior of a hadron–hadron interaction at energies of the elastic scattering amplitude and quasi-two-body processes, the logarithmic increase of the secondary-hadron multiplicity, scaling and its violation at relatively low energy in the fragmentation region, a smooth (plateau-like) rapidity variation of the spectrum in the central region, etc.[7,8,50]

The complex-momentum method (Reggistics) was developed in parallel with the multiperipheral method.[54] This method is based on the idea that the partial hadron scattering amplitudes behave analytically as functions of the angular momentum which are extended analytically to the complex-momentum plane. At high energies the behavior of the total amplitude of the process in the scattering channel was found to be determined by the rightmost partial-amplitude poles in the annihilation reaction channel, which are situated in the upper half-plane of the complex momenta.[54]

In addition to the poles, the more complex singularities in the complex-momentum plane also play an important role in the behavior of the scattering amplitude which is asymptotic with respect to energy. These singularities are the various type of branch points which appear as a result of pole-amplitude iterations. The amplitude poles in the complex-momentum plane are called *Regge poles*. These poles account for the power-law behavior of the

**Figure 3.8. (a) Reggeon diagram of elastic scattering and (b) ladder diagram.**

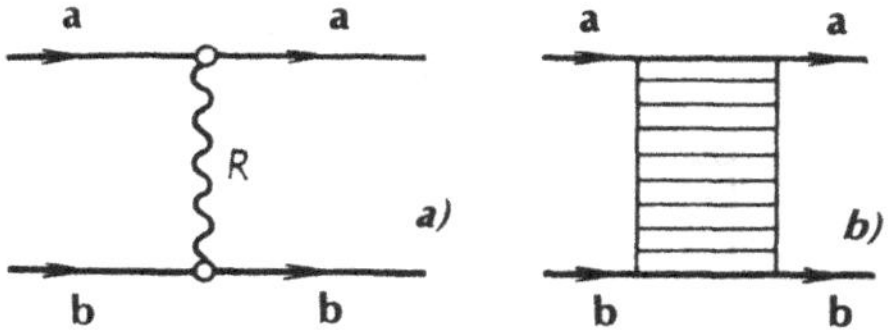


amplitude with increasing energy. A special diagram technique, which unifies the computational procedure in terms of the complex-momentum method, has been developed. The amplitude for the scattering of two hadrons, $a + b \rightarrow a + b$, is described by the diagram in Fig. 3.8a, where the wavy line $R$ represents the Regge pole (in the annihilation channel the reaction is of the form $a + \bar{a} \rightarrow b + \bar{b}$). This virtual state, which the interacting hadrons exchange, is called a *Reggeon*. As in the case of an ordinary virtual particle, the emission and absorption of a Reggeon are described by a propagation function or a *Reggeon propagator*, which is proportional to the power function of the energy: $D_R(S,t) \propto (S/S_0)^{\alpha(t)}$, where $S$ is the square of the collision energy in the c.m. frame; $S_0 = \text{const}$; $t$ is the square of the 4-momentum transfer between the interacting hadrons; and $\alpha(t)$ is the $t$-dependent position of the Regge pole in the complex-momentum plane. At $t < 0$ the function $\alpha(t)$ is a real function of $t$ (in the scattering channel). The function $\alpha(t)$ is called a *Reggeon trajectory*.

As a virtual entity a Reggeon has quantum members: parity ($P$), isospin ($T$), $G$-parity, etc. At arbitrary values of $t$ a Reggeon does not have a specific physical spin value. At $t = m^2$ ($m$ is the mass of a physical particle, whose quantum numbers are the same as those of the Reggeon), however, one can postulate a relation $\text{Re}\alpha(m^2) = J$, where $J$ is the spin of the physical particle. The physical particle is said to be situated on the Regge trajectory, which is called the $\rho$ trajectory, $\pi$ trajectory, $N$ trajectory, etc., depending on the particular particle.

On the basis of the data on the mass spectrum, hadron spins, and the behavior of $\alpha(t)$ at $t < 0$ the Regge trajectories were found to be linear:

$$\text{Re}\alpha\,(t) = J + \alpha'(0)\,(t - m^2)\,,$$

where $\alpha'$ is the trajectory slope. Another distinctive feature of Reggeons is their special quantum number, the signature $\sigma$: $\sigma = (-1)^J$ for bosons and $\sigma = (-1)^{J \pm 1/2}$ for fermions. Trajectories with different signatures but the same other quantum numbers are different.

Aside from the trajectories on which the known physical particles are situated, the vacuum-trajectory hypothesis plays a key role in the theory. This hypothesis, which corresponds to a Reggeon with the vacuum quantum numbers, is characterized by the value $\alpha(0) = 1$ (or a value approximately equal to unity). The parameter $\alpha(0)$ is called an *intercept*. A vacuum Reggeon is called a *pomeron* (in honor of I. Ya. Pomeranchuk, who proved the theorem on the asymptotic equality of the cross sections for the interaction of particles and antiparticles with the same target). The intercepts of the other Reggeons are $\alpha(0) \lesssim 0.5$.

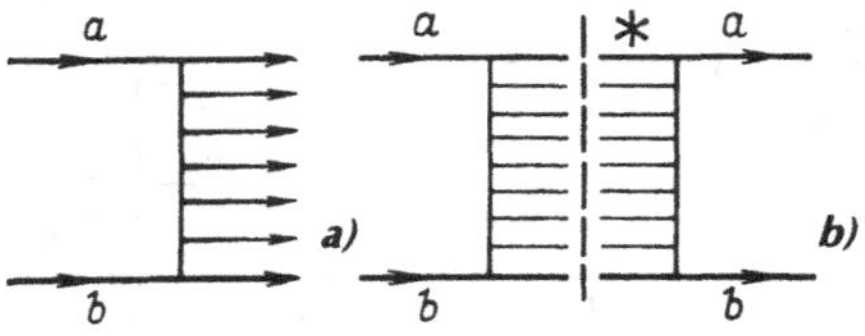


**Figure 3.9. (a) Multiperipheral "comb" and (b) the result of its squaring (the asterisk means complex conjugation).**

The amplitude for the elastic scattering $T(S,t)$ through zero angle ($t = 0$) is related to the total interaction cross section by the optical theorem $\operatorname{Im}T(S,0) = S\sigma_{\text{tot}}(S)$. Since the contribution from a given Reggeon to the imaginary part of the amplitude depends on the energy in accordance with $\operatorname{Im}T_R(S,0) \propto D_R(S,0) \propto S^{\alpha(0)}$, the corresponding contribution to the cross section is $\sigma_{\text{tot}}^{R}(S) \sim S^{\alpha(0)-1}$. In the limit $S \to \infty$, the pomeron plays the dominant role. We therefore have $\sigma_{\text{tot}}(S) \approx \sigma_{\text{tot}}^{P}(S) \simeq \text{const}$. This behavior of $\sigma_{\text{tot}}(S)$ can be described by taking into account the corrections to the mechanism for the pomeron exchange due to the exchange of other Reggeons, due to multiple Reggeon exchange (in particular, multiple pomeron exchange), and due to the interaction between Reggeons (pomerons). The multiple Reggeon exchange and the interaction between Reggeons correspond to singularities of the branch-point type in the complex-momentum plane.

A comparison of the predictions of the complex-momentum method and predictions of the multiperipheral models shows that they are closely related. The reason for this close relationship is that the total contribution of *multiperipheral ridges* (diagram 3.9a), with different number of hadron-production centers, to the total cross section $\sigma_{\text{tot}}(S)$ also leads to a power function $\sigma_{\text{tot}}(S) \propto S^{\alpha(0)-1}$, where $\alpha(0)$ is determined by the interaction vertex at the hadron-emission center. On the one hand, the contribution from a single ridge to $\sigma_{\text{tot}}$ is a kind of a cut *ladder-type diagram* (Figs. 3.8b and 3.9b), which corresponds to the multiplication of the height of the ridge by a complex-conjugate quantity.

On the other hand, the imaginary part of the uncut ladder diagram is in agreement, to within a numerical factor, with the result of calculation of the contribution of a given ridge to $\sigma_{\text{tot}}$. It can thus be concluded that ladder diagrams behave in a Regge-like (power-law) manner. An exchange of ladders with a different number of steps is equivalent to an exchange of Reggeons. Conversely, a calculation of the imaginary part of the diagram with an exchange of Reggeons is equivalent to taking the sum of the ridge-type diagrams which are squared in the modulus.[7,8]

Because of this relationship between the Regge model (the complex-momentum method) and the multiperipheral model, the properties of these models can be effectively used to make theoretical predictions regarding the inclusive spectra of secondary hadrons in the multiple-production processes.[7,8,50]

Analysis of the properties of the multiperipheral ridge showed that this process is characterized by the following properties of secondary-particle distribution[7,8]:

(1) In the central rapidity region the rapidity spectrum does not depend on the primary energy at high energies [$\ln(E/m_N) \geqslant 4$–5] and has the shape of a plateau: $dN/dy \simeq \text{const}$.

(2) In the fragmentation regions the rapidity spectra have the scaling property: $Ed^3\sigma/d^3p = F(y_a - y, p_\perp)$; $Ed^3\sigma/d^3p = F(y - y_b, p_\perp)$. These equations are for the fragmentation regions of particles $a$ and $b$, respectively.

(3) The average transverse momentum of the inclusive hadron does not depend on the energy, $\langle p_\perp \rangle \simeq \text{const}$. This conclusion is a corollary of the hypothesis that the momentum transfer between the emission centers is effectively small (energy independent).

(4) In view of properties (1) and (2), the average particle multiplicity at the ridge increases logarithmically with the energy: $\langle N \rangle = a + b \ln(E/m)$.

(5) The emission of hadrons from various centers is essentially an independent process. If each center emits only a single hadron, then the multiplicity distribution of the secondary events is a Poisson distribution: $p(N) = \sigma_N / \sigma_{\text{in}} = \exp(-N/\langle N \rangle)\langle N \rangle^N / N!$, where $\sigma_N$ is the cross section for the production of $N$ hadrons (the topological cross section). This is a valid cross section for both the total number of particles and only the charged particles.

The properties (1) and (2) are consistent with the Regge model in view of the above-noted qualitative equivalence of each approach. The Regge model makes it possible to predict more accurately the longitudinal-momentum spectra of the fastest secondary particles and it also suggests that the components which are responsible for the disruption of the scaling behavior of the spectra at moderately high energies are energy-dependent components. A modified version of the Regge model with a pomeron crossing, $\alpha_p(0) = 1 + \Delta$ ($\Delta = 0.06$–$0.15$) can be used to explain the increase of $\sigma_{\text{tot}}(S)$ and $\sigma_{\text{in}}(S)$ as functions of energy and to explain the rapidity spectra in the central region.[8,50]

### The central rapidity region

Let us consider the diagram in Fig. 3.10a and isolate the inclusive particle $c$. At the multiperipheral ridge the longitudinal momenta of the c.m. particles, on the average, decrease systematically from $p^*_{1\parallel} \approx p^*_a$ to $p^*_{2\parallel} \approx -p^*_a = p^*_{b\parallel}$. We assume that the primary energy and the effective masses of the particles with $p^*_\parallel > p^*_{c\parallel}$ and $p^*_\parallel < p^*_{c\parallel}$ are adequately large (the condition for the applicability of the Regge model in the estimates below). It is easy to show that the square of the effective mass of the particles with $p^*_\parallel > p^*_{c\parallel}$ is $S_{ac} \approx |(p_a - p_c)^2| \approx m_a m_{c\perp} \exp(y_a - y_c)$. It is assumed here that $y_a \gg |y_c|$, since the particle $c$ is taken from the central rapidity region which is developed sufficiently only when the condition indicated above is satisfied.

The square of the effective mass of particles with $p^*_\parallel < p^*_{c\parallel}$ is $S_{bc} \approx |(p_b - p_c)^2| \approx m_b m_{c\perp} \exp(y_c - y_b)$. Calculation of the invariant inclusive differential cross section (squaring of the diagram modulus in Fig. 3.10a) reduces to the calculation of the diagram in Fig. 3.10b or 3.10c, where the

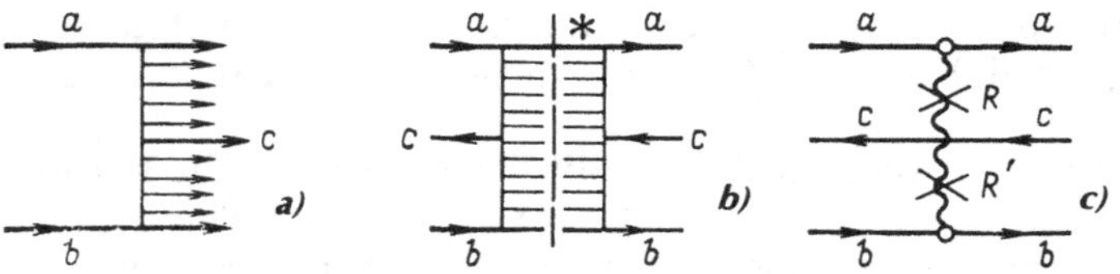


**Figure 3.10. A "comb"-type diagram with (a) a clearly identifiable inclusive $c$ particle and (b) and (c) the result of its squaring.**

crossed-out wavy Reggeon lines replace the cut ladders. This procedure is based on the above-mentioned equivalence of the ladder and Reggeon exchange. A crossing out of the Reggeon means that in the calculations of the diagram in Fig. 3.10 only the imaginary part of the Reggeon propagator must be taken into account at $t=0$: $\text{Im}D_R(S',0)\sim(S')^{\alpha_R(0)}$, where $S'$ is the square of the effective mass of the particles—of the ladder steps that constitute the Reggeon. The vertices $aaR$ and $bbR$ are constants which can be determined independently from the data on the behavior of the total and differential cross sections for elastic scattering of hadrons.[7,8,55] The vertex $RR'cc$ depends exclusively on $m_{c\perp}$ by virtue of the relativistic invariance and is independent of the primary energy (central-region scaling).

The invariant differential cross section can be calculated from the equation[7,8] $E_c\,d^3\sigma/d^3p_c\sim S^{-1}\Delta T$, where $\Delta T$ is the amplitude corresponding to the diagram in Fig. 3.10c with cut Reggeons (ladders). Since this diagram contains two Reggeon propagators $R$ and $R'$, we have $\Delta T\sim(S_{ac})^{\alpha_R(0)-1}(S_{bc})^{\alpha_{R'}(0)-1}$. It is easy to show that $S_{ac}S_{bc}\approx m_a m_b m_{c\perp}^2\exp(y_a-y_b)\approx m_\perp^2 S$. We then have

$$E_c\frac{d^3\sigma}{d^3p_c}=\sum_{R,R'}\varphi^{c}_{RR'}(m_{c\perp})\left(\frac{S_{ac}}{S_0}\right)^{\alpha_R(0)-1}\left(\frac{S_{bc}}{S_0}\right)^{\alpha_{R'}(0)-1},$$

where the summation is over all kinds of exchangeable Reggeons with compatible quantum numbers. The function $\varphi^{c}_{RR'}(m_{c\perp})$ is a collection of all the $m_{c\perp}$-dependent factors which have not been predicted theoretically. Taking into account only the principal contributions, we find, as $S\to\infty\,[\alpha_p(0)=1;\ \alpha_R(0)\approx0.5]$

$$E_c\frac{d^3\sigma}{d^3p_c}=\varphi^{c}_{PP}(m_{c\perp})+\varphi^{ac}_{PR}(m_{c\perp})\exp\left(-\frac{y_a-y_c}{2}\right)$$
$$+\varphi^{bc}_{RP}(m_{c\perp})\exp\left(-\frac{y_c-y_b}{2}\right).$$

If $a=b$, then

$$E_c\frac{d^3\sigma}{d^3p_c}=\varphi^{c}_{PP}(m_{c\perp})+2\varphi^{ac}_{RP}\cosh\frac{y_c}{2}\Big/\left(\frac{S}{m_a^2}\right)^{0.25}.$$

**Figure 3.11.** (a) Ladder diagram and (b) Reggeon diagram for the production of an inclusive $c$ particle in the region of fragmentation of hadron $a$.

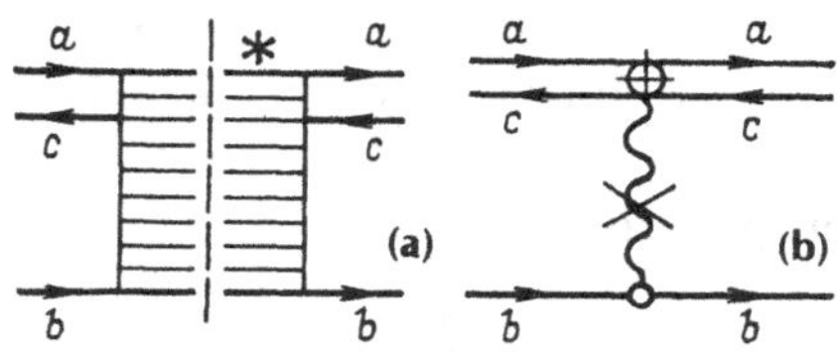


It follows from this expression that the pomeron-correction terms die out rather slowly with the increase in energy.

The experimentally observed increase in the hadron yield with the energy in the control region can be attributed to different factors (see Refs. 8, 50, and 53). It may occur because of the dying out of the negative-sign corrections to the purely pomeron contribution [when $\alpha_P(0) = 1$] or because of the increase in the resonance yield whose decay increases the yield of the observable stable hadrons (with respect to the strong interactions) in the central region, or because $\alpha_P(0)$ deviates from unity [$\alpha_P(0) = 1 + \Delta, \Delta > 0$]. In the last case, a power-law increase in the secondary-particle yield $E_c d^3\sigma/d^3p_c \sim S^\Delta \varphi(m_\perp)$ and in the average multiplicity $\langle N \rangle \sim S^\Delta \ln S/(a + b \ln S + c \ln^2 S)$ is expected. One mechanism cannot be given a preference over another because of the absence of convincing unambiguous arguments in the investigated energy region in favor of a particular mechanism.[50,52]

As $S \to \infty$ the corrections due to the simultaneous emission of several ridges in the central rapidity region cancel out because of the AGK rules,[56] along with the contributions from the elastic rescattering and diffractive production of hadronic showers with a small mass ($S \lesssim 5$–$6$ GeV$^2$). At these energies, however, the cancellation is far from being complete: Because the energy conservation law is in effect, the primary energy is distributed equally, on the average, among the several emitted ridges. At final energies, the average number of ridges is limited. At a fixed energy the particles, whose rapidities become increasingly more limited, are emitted with increasing number of ridges. This effect can also simulate an increase in the plateau height with increasing energy. The broadening of the multiplicity distribution in comparison with the Poisson distribution is attributed to the emission of several ridges.[8,57]

## Fragmentation regions

For brevity, let us consider the region of fragmentation of an incident particle $a$. The fragmentation of a target hadron $b$ can be considered in a similar way. In the fragmentation region the difference in the rapidities of the incident particle and the inclusive particle $c$ is negligible ($y_a - y_c \sim 1$). The effective mass $S_{ac}$, therefore, is also negligible, so that the upper cut part of the ladder in the diagram in Fig. 3.11a cannot be reduced to the cut Reggeon. The mass $S_{bc}$ is assumed, as before, to be large ($S_{bc} \approx S$). As a result, the inclusive differential cross section is described by the amplitude $\Delta T$ which can be calculated from the diagram in Fig. 3.11b. The upper vertex ($ac\,Rac$) in this diagram is an unknown function $\Phi_R^{ac}(y_a - y_c, m_{c\perp})$, while the imaginary part of the Reggeon propagator $R$ is $\mathrm{Im}D_R\ (S_{bc},0) \sim (S_{bc})^{\alpha_R(0)} \sim S^{\alpha_R(0)}$. In this case we have

$$E_c \frac{d^3\sigma}{d^3p_c} = \sum_R \Phi_R^{ac}(y_a - y_c, m_{c\perp}) g_R^b \left(\frac{S}{S_0}\right)^{\alpha_R(0)-1}.$$

Retaining, as $S \to \infty$, the pomeron contribution with $\alpha_P(0) = 1$ and the poles with $\alpha_R(0) \approx 0.5$, we find

$$E_c \frac{d^3\sigma}{d^3p_c} = \Phi_P^{ac}(y_a - y_c, m_{c\perp}) g_P^b + \Phi_R^{ac}(y_a - y_c, m_{c\perp}) g_R^b \left(\frac{S}{S_0}\right)^{-1/2}.$$

The correction term decreases as $S^{-1/2}$ and the scaling sets in more quickly than in the central rapidity region. In some processes the scaling in the variable $x_F$ is achieved in the fragmentation region, beginning with a rather low energy ($E \gtrsim$ 10–20 GeV) in the laboratory frame, in particular, in the case of inclusive $\pi^\pm$ and $K^+$ mesons produced in $pp$ collisions. In the case of secondary $p$, $K^-$, and especially $\bar{p}$, the scaling in $pp$ collisions is attained only at energies $S \sim 10^3$ GeV$^2$. This phenomenon is explained in terms of the muzzle hypothesis. If the quantum numbers of the $ab$, $ab\bar{c}$, $b\bar{c}$, and $a\bar{c}$ states are not exotic* and the $f$ and $\varphi$ trajectories cannot be exchanged, the scaling sets in at higher energies (late scaling). If one of the channels, $ab$, $ab\bar{c}$, or $a\bar{c}$, is exotic, then the scaling occurs frequently, beginning at low energies (early scaling).[52] These problems are, however, still far from being completely solved. It is useful here to mention an empirical fact that the use of the variable $x_E = E^*/E^*_{\max}$, instead of the Feynman variable $x_F = p^*_\parallel / |p^*_\parallel|_{\max}$, allows one to frequently reach a scaling description of the fragmentation region in a broader range of primary energies.

## Leading particles

Let us consider the range of values $1 - |x_F| \ll 1$ corresponding to the most energetic particles in the c.m. frame (the leading particles). These particles are produced as a result of a mechanism shown in the diagram in Fig. 3.12a, with a Reggeon exchange with suitable quantum numbers between the upper vertex $acR'$ and the lower vertex $bR'X$ ($X$ is the hadron jet). In the case under consideration, the inclusive hadron $c$ is emitted into the forward hemisphere. In a similar way, we consider the production of the leading hadrons which are emitted into the back hemisphere. Squaring the diagram in Fig. 3.12a, we obtain the diagram in Fig. 3.12b if the effective mass of the hadron jet $X$ is large. As a result, the invariant differential cross section for the production of inclusive particles $c$ is given by

*An exotic channel is a reaction channel in which there are no resonances.

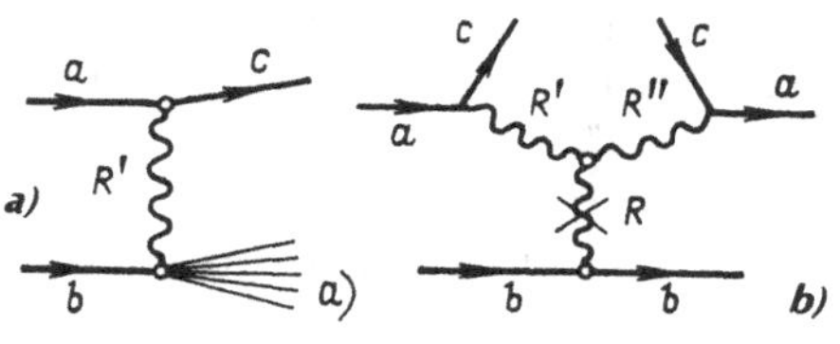


**Figure 3.12. (a) Reggeon diagram for the production of the leading particle and (b) the result of its squaring—three-Reggeon diagram.**

$$E_c \frac{d^3\sigma}{d^3p_c} = \sum_{R,R',R''} \Phi^{ab}_{RR'R''}(m_{c\perp})(1-x_F)^{1-\alpha_{R'}(t)-\alpha_{R''}(t)} \left(\frac{S_X}{S_0}\right)^{\alpha_R(0)-1}, \qquad (3.9)$$

where $S_X$ is the effective mass of the hadron jet:

$$S_X \approx S(1-x_F) + m_c^2 - 2m_{c\perp}^2/x_F.$$

In the limit $S\to\infty$, the principal contribution comes from the term with $R = P$ and the corrections due to the contributions of the Reggeons with $\alpha_R(0) \approx 0.5$ decrease as $S^{-1/2}$. Equation (3.9) is called a *three-Reggeon approximation*. This equation predicts rather well the shape of $x_F$ (or $x_E$) spectra, even if only the principal quantum-number-compatible trajectories $R'$ and $R''$ are taken into account in first approximation. In the reaction $p + p \to \pi^+ + X$, for example, the principal contribution to the inclusive spectrum is attributable to the Reggeon exchange with the neutron quantum numbers. Its trajectory is $\alpha_n(t) \approx 0.5 + \alpha_n'(0)(t - m_n^2)$. At $|t| \ll m_n^2$, using Eq. (3.9) as a guide, where $R = P$ with $\alpha_P(0) = 1$, we find $E_{\pi_+} d^3\sigma/d^3p_{\pi^+} \sim (1-|x_F|)^2$, in reasonably good agreement with experiment. Other inclusive spectra can also be easily parametrized and compared with the experimental data (as will be done below).

The three-Reggeon formula describes most effectively the spectrum of the leading protons in *pp* collision, where a sharp increase of the inclusive spectrum in accordance with $E_p d^3\sigma/d^3p_p \sim (1-|x_F|)^{-1}$, as $|x_F|\to 1$, is predicted. Such an increase is observed experimentally. The behavior of the spectrum in the region $0.8 \lesssim |x_F| \lesssim 0.95$ in this case can be described with allowance for the poles with the intercepts[58] $\alpha_P(0) = 1$ and $\alpha_{R',R'',R}(0) \simeq 0.5$.

The predictions concerning the power-law $x_F(x_E)$ dependence of the inclusive spectra obtained in the framework of the Regge model are very important when the experimental data are parametrized over a broad range of primary energies. In many cases the variable $x_E$ can be used to describe the observable spectra at $p_{c\perp} \lesssim 1$–$2$ GeV/$c$ down to very small values of $x_E$; i.e., the entire kinematic region can be described by a single formula which contains ordinary power functions such as $(1-x_E)^a$ (see the discussion below).

If the transition $a\to c$ cannot occur as a result of a single-Reggeon exchange because of the absence of Reggeons with appropriate quantum numbers (the exotic $a\bar{c}$ channel), the branch-point trajectories with the most $\alpha_c(0)$ crossings (the exchange of at least two Reggeons), instead of the $\alpha_{R'}(t)$ and $\alpha_{R''}(t)$ pole trajectories, should be substituted into Eq. (3.9). The position of the branch point which arises as a result of the exchange of $n$ Reggeons can be found from the relation[54] $\alpha_{cn}(t) = \max\left[\sum_{i=1}^{n} \alpha_i(t_i) - n + 1\right]$, where the maximum is calculated in the region of permissible values of the square of the 4-momentum, $t_i$, transferred by the Reggeons, with a fixed $t$ and under the condition $\sum_{i=1}^{n} \sqrt{-t_i} = \sqrt{-t}$, which reflects the law of conservation of transverse momentum at the vertex of the transition $a \to c$ with the emission of a few Reggeons.

### Hadron-nucleus interactions. The intranuclear-cascade model

In the classical version of this model,[48] the secondary particles, which are produced in the first collision of the incident hadron with the nucleon of the nucleus, interact with the rest of the nucleons in the same way as the observable particles do. The particles produced in the secondary collisions interact again with the nucleons, and so forth, i.e., the observable particles begin to cascade.[8] Barashenkov and Toneev[48] showed that this model can be used only at $E_0 < 5$–$10$ GeV, where the computer programs based on this model (see Chap. 6) apparently have no rival. At $E_0 \gtrsim 10$ GeV, the difference between theory and experiment is substantial. Attempts to modify the simple model—by introducing multiparticle interactions and reducing the nucleon density in the nucleus as the hadron avalanche is crossed[48]—have improved the agreement with the experimental data and permitted the upper energy limit to be raised to approximately 50 GeV in the calculations.

### The statistical hydrodynamic models

Among the statistical hydrodynamic models for the interaction of hadrons with nuclei the most common one is the Landau hydrodynamic model, which is based on an additional assumption concerning the interaction with the nuclear tube. The additional postulates are[8]:

(1) It is assumed that the incident hadron interacts with all nucleons of the nucleus which are situated along the hadron trajectory in nuclear matter. This interaction is of a collective hydrodynamic nature. The primary hadron, in effect, cuts out a tube of nuclear matter which is cylindrical in shape with a radius of about $m_\pi^{-1}$ and height equal to the size of the nucleus along the direction of motion of the hadron in the reference frame in which this collision is considered. In particular, the average number of nucleons in the nuclear tube is $\bar{\nu} = A\sigma_{\rm abs}^{N} / \sigma_{\rm abs}^{A}$ in the laboratory frame (the nucleus is at rest) [see Eq. (3.4)].

(2) The nuclear tube is assumed to be a single particle with a mass equal to the total mass of the nucleons in the tube; the nucleon structure of the tube is ignored.

(3) The collision of the hadron with the nuclear tube is considered in a system where the velocities of these entities are equal and opposite to each other (an equal-velocity system).

(4) In an equal-velocity system, the hadron and tube experience a Lorentz contraction in the longitudinal direction.

(5) At the time of collision, the hadron and nuclear tube form a single statistical hydrodynamic system, whose further evolution occurs according to the relativistic hydrodynamics and statistical physics laws which are supplemented by the equation of state (see the discussion above).

Points (1)–(5) supplement the standard Landau model. Analysis of the thermodynamics of the process and solution of the hydrodynamics equations lead to the following predictions [the velocity of sound in nuclear material is assumed to be $c_0 = 1/\sqrt{3}$ ($c_0^2 = 1/3$) (Ref. 8)]:

(1) In hadron-nucleus interactions the average multiplicity of secondary hadrons increases with the energy $E$ in the laboratory frame and with the increase in the number of nucleons $A$ in the nucleus in accordance with $\langle N \rangle \sim (1+\bar{\nu})E^{1/4}$.

(2) The rapidity spectrum of secondary hadrons is determined by the hydrodynamic dispersion laws and has an approximately Gaussian shape. This spectrum depends on the value of $\bar{\nu}$. If $\bar{\nu} < (1+c_0)/(1-c_0)$, we have

$$\frac{dN_s}{dy^*} = \frac{\langle N_s \rangle}{\sqrt{2\pi L}} \exp\left[ -\frac{(y^* + y_0)^2}{2L} \right], \tag{3.10}$$

where $y_0 = \operatorname{arctanh} v_0$, $L = 0.56 \ln \dfrac{E}{m_N} + 1.6 \ln \dfrac{2}{1+\bar{\nu}} + 1.6$,

$$v_0 = \tanh\left[ \frac{3(\bar{\nu}-1)}{1+\bar{\nu}} - \operatorname{arctanh} \frac{\bar{\nu}-1}{\bar{\nu}+1} \right],$$

$y^*$ is the rapidity in the c.m. frame, and $v_0$ is the c.m. velocity relative to the equal-velocity system.

The asymmetry of the rapidity distribution relative to the value $y^* = 0$ arises due to the asymmetry of the initial conditions in the c.m. system upon the collision of a hadron with the nuclear tube to the left and right of the collision plane. If $\bar{\nu} > (1+c_0)/(1-c_0)$, the hydrodynamics equations become more complex and their solution is found numerically for different values[8] of $\bar{\nu}$. The spectrum is approximately described by (3.10), where the parameter $y_0$ is the function $\bar{\nu}$ (and $c_0$).

(3) If $p_\perp \gg m_\pi$, the transverse-momentum secondary-hadron spectrum is approximately described by

$$dN_s/dp_\perp^2 \sim \exp(-m_\perp/T),$$

where $T \simeq m_\pi$. The transverse motion of hadrons in this case is primarily a thermal motion. At a moderately high energy, however, the lateral hydrodynamic motion becomes a factor, giving rise to a slight energy dependence of $\langle p_\perp \rangle$: $\langle p_\perp \rangle \sim E^\delta$, where $\delta \simeq \frac{1}{14} - \frac{1}{12}$. The exponential distribution indicated above refers to the range of values $p_\perp \lesssim 1$–$2$ GeV$/c$, in which the bulk of the particles are produced. Some of the particles may, however, be produced at the temperature of the hadronic fluid $T \gg m_\pi$. These fluctuational processes are said to be linked with the production of particles with $p_\perp \gtrsim 1\text{–}2$ GeV.

(4) The $A$ dependence of the average multiplicity of the shower particles predicted by the statistical hydrodynamic model has the form $\langle N_s \rangle = a + bA^{1/2}$. This dependence is approximated by the power function of $A$: $\langle N_s \rangle \sim A^\alpha$, where $\alpha \sim 0.15\text{–}0.20$ based on various calculations.[8]

(5) The relative yield of the shower hadrons of various types in hadron–nucleus collisions, which is the same as that in hadron–nucleon collisions, is determined by the Fermi or Bose distribution for the energy spectra of fermions or bosons produced in the rest frame of nuclear matter as it decays into real hadrons. If the mass of a hadron of type $i$ is large ($m_i \gg T \approx m_\pi$), the average-multiplicity ratio is[8]

$$\frac{\langle n_i \rangle}{\langle n_\pi \rangle} \approx \frac{q_i}{3} [F(1)]^{-1} \left(\frac{\pi}{2}\right)^{1/2} \left(\frac{m_i}{m_\pi}\right)^{3/2} \exp\left(-\frac{m_i}{m_\pi}\right),$$

where $q_i$ is the statistical weight of the hadron of type $i$ (the number of spin and isospin states is $q_\pi = 3$), and

$$F(1) = \int_0^\infty dx\, x^2 (\exp\sqrt{1+x^2} \pm 1)^{-1}.$$

Here " $\pm$ " refers to fermions (bosons). The factor $\exp(-m_i/m_\pi)$ suppresses the production of heavy particles.

The statistical hydrodynamic model under consideration predicts a weak $A$ dependence of the characteristics of the shower particles, consistent with experiment. This model also correctly predicts qualitatively the increase in the particle yield in the region $y^* < 0$ in comparison with the hadron–nucleon interactions [see Eq. (3.10)]. At high energy this model attributes the properties of the spectra of the gray and black tracks entirely to the properties of the excited nucleus. These properties must therefore be considered on the basis of purely nuclear models. The statistical hydrodynamic model can, on the whole, describe the multiple production of shower particles in hadron–nucleus collisions at the energies studied in a completely satisfactory manner.[8,49] It cannot, however, explain several important details such as the leading-particle spectrum.

### Multiperipheral Regge models with elastic and inelastic rescattering

Since for our purpose the MR model predictions concerning primarily the multiple production of hadrons by nuclei are of interest, we shall briefly discuss the principal predictions of these models.[8,50,51] In the MR models that have so far been developed, the physical picture of the hadron–nucleus interaction, put in a slightly simplified form, can be described as follows.

It is assumed that the principal process responsible for the multiple production of secondary hadrons is the creation of a single or several hadronic crests in the collision of a primary hadron (or of the components that constitute it) with the nucleons of the nucleus. It is assumed that at the currently available energies only a single hadronic crest can form per nucleon of the nucleus. The nucleons of a nucleus which have already participated in an interaction cannot form any crests.

A similar quasi-Glauber treatment, which was developed by Shabel'skiĭ[59] (see also Ref. 8), predicts that the cascade multiplication is suppressed as a result of secondary interaction of the hadrons, which are produced in the first event, with the nucleons of the nucleus in the part of the spectrum corresponding to relatively high energy. This prediction, in qualitative agreement with experiment, contradicts the simple cascade model[48] without any modifications indicated in the beginning of this section.

The MR model is supplemented with the hypothesis on the multiple elastic and inelastic rescattering of the primary hadron and of the hadronic states with a small effective mass (including the hadron resonances) which are formed on the nucleons of the nucleus by this hadron through diffraction. Its principal predictions are as follows[8,59]:

(1) In the high-energy part of the spectrum of inclusive secondary particles of type $c$, where $E_c \gtrsim E_a/2$, the ratio of the rapidity spectra in the $aA$ and $aN$ collisions is estimated to be $R_a(y) = \sigma_{\text{abs}}^{A(1)}/\sigma_{\text{abs}}^{A} \lesssim 1$, where $a$ is the incident hadron, $E_a$ is its energy, $A$ is the nucleus, $\sigma_{\text{abs}}^{A(1)}$ is the cross section for the production of a crest on the nucleus $A$, $\sigma_{\text{abs}}^{A}$ is the cross section for the absorption of hadron $a$ by the nucleus: $\sigma_{\text{abs}}^{A(1)} = \pi \int db^2 [\sigma_{\text{in}}^{N} T(b)] \exp[-\sigma_{\text{in}}^{N} T(b)]$; $b$ is the impact parameter, $T(b)$ is the optical thickness of the nucleus [see Eq. (3.6)], and $\sigma_{\text{in}}^{N}$ is the cross section for inelastic interaction of hadron $a$ with the nucleon. Here and below we are considering the events in which the inclusive particle has a small transverse momentum: $p_{c\perp} \ll E_c$.

(2) In the central part of the rapidity spectrum, where the inequalities $p' \ll p_c \ll p_a$ hold, the ratio is predicted to be

$$R_A(y) \simeq A\sigma_{\text{in}}^{N}/\sigma_{\text{abs}}^{A} \sim A^{1/3}. \tag{3.11}$$

Here $p'$ is a momentum chosen in such a way that it would not be necessary to go outside the limits of the central region of the variable rapidity ($p' \gtrsim 1$ GeV/$c$ is a crude estimate), $\sigma_{\text{in}}^{N} = \sigma_{\text{tot}}(aN) - \sigma_{\text{el}}(aN) - \sigma_D(aN)$, and $\sigma_D(aN)$ is the cross section for the diffractive production of hadron jets in the $aN$ collision.

In the model under consideration the invariant differential cross section for the inclusive process in the nucleus corresponds to the result of the impulse approximation (where all the nucleons in the nucleus act independently[59]):

$$E_c \frac{d^3\sigma}{d^3p_c}(aA \to cA') \simeq AE_c \frac{d^3\sigma}{d^3p_c}(aN \to cX), \tag{3.12}$$

where $A'$ are the undetectable secondary hadrons and nuclear decay products, and $X$ are the undetectable hadrons.

(3) A crude estimate of the rapidity interval in the laboratory frame, where relation (3.12) holds, yields

$$\ln\left(\frac{m_{c\perp}}{m_N}A^{1/3}\right)\lesssim y \ll \ln\left(\frac{2E_a}{m_{c\perp}A^{1/3}}\right).$$

In this region the average multiplicity of the shower particles increases logarithmically with the energy and the average multiplicity increases with the increase of $A$ in accordance with $\langle N_s\rangle \sim A^{1/3}$.

In the region $y \gtrsim \ln(2E_a/m_{c\perp}A^{1/3})$ the dependence of $N_s$ on $A$ decreases with increasing $y$ in comparison with the dependence on $A^{1/3}$ and when $y \gtrsim \ln(E_c/2m_{c\perp})$ it vanishes. At energies $E_a \lesssim 10^3$ GeV we should expect that $\langle N_s\rangle \sim A^{\alpha(E_a)}$, where $\alpha(E_a) \approx 0.15$–$0.20$. The condition $\langle N_s\rangle \sim A^{1/3}$ can be established only at very high energies ($E_a \gtrsim 10^4$–$10^5$ GeV).

**Quark-parton model for hadron-nucleus interactions**

The hypothesis which states that a relativistic hadron is a set of point objects—partons[9] proved to be extremely useful for a qualitative understanding of multiple production of hadrons by nucleons and nuclei.[51] The most common interpretation of the parton fluctuation of the hadron is the multiperipheral interpretation, which can be summarized by saying that as the hadron gradually decays into partons, it forms a parton crest (see Fig. 3.9). The high-energy partons interact hardly at all with the target hadron (nucleon) at rest; the interaction is accomplished by means of relatively slow partons with longitudinal momenta $p_{\parallel} \sim m$ (the parameter $m$ is usually on the order of 1 GeV). The time it takes a high-energy hadron to form such slow partons, according to the uncertainty principle, is[60]

$$\Delta t \sim (\Delta E)^{-1} \simeq E/\langle m_{\perp}^2\rangle, \tag{3.13}$$

where $E$ is the energy of the incident hadron in the laboratory frame, $m_{\perp}^2 = m^2 + \langle K_{\perp}^2\rangle$, and $K_{\perp}$ is the transverse parton momentum.

A detailed analysis of the process by which parton fluctuations of a hadron are generated shows[61] that the probability for the existence of fluctuations containing slow partons in a hadron is approximately equal to unity. These fluctuations appear in succession at short intervals of time, $\tau \sim m^{-1}$. At a high energy the formation of two parton fluctuations in succession with a lifetime (3.13) is unlikely. Such effects "die out" with increasing energy. Two fluctuations of this sort can, however, form simultaneously.[61] The slow partons of these fluctuations can interact with the target simultaneously. This effect manifests itself most clearly in the case of a compound target such as an atomic nucleus. If the incident hadron is a compound hadron, the parton fluctuations may very likely be generated both by one of the components of a hadron and simultaneously by several components. In particular, in an additive quark model such fluctuations are generated by the structural hadron quarks which determine the quark composition of the hadron if it is at rest. The parton fluctuations of the structural quarks apparently do not overlap each other spatially in the energy range studied.[62]

This situation assures a nearly independent interaction of each structural quark of an incident hadron with the target. The data on the total cross sections for the interaction of mesons with nucleons and of nucleons with nucleons provide a simple illustration of this property of the additivity of the structural quarks. At high energy the ratio $\sigma_{tot}(\pi N)/\sigma_{tot}(NN) = 2/3$ is valid within 10–15% for these quantities. The numerical value of this ratio corresponds to the number of pair collisions between the structural quarks in the meson–nucleon and nucleon–nucleon interactions.

In contrast with the model for sequential elastic and inelastic rescattering of the leading hadrons by the nucleons of the nucleus, the *additive quark model* takes into account the interaction of the structural (constituent) quarks with the nucleons of the nucleus.[62] The probability for the inelastic interaction $\nu$ of the structural quarks of the incident hadron from $n^*$ with the nucleons of the nucleus can be calculated from the equation

$$W_\nu = (\sigma_{\rm abs}^{A})^{-1} \int d^2b C_n^\nu \exp[-(n-\nu)\sigma_{\rm abs}^{qN} T(b)]\{1-\exp[-\sigma_{\rm abs}^{N} T(b)]\}^\nu, \tag{3.14}$$

where $\sigma_{\rm abs}^{qN}$ is the cross section for the absorption of a structural quark by a nucleon, $b$ is the impact parameter, $C_n^\nu$ is the binomial coefficient, and $T(b)$ is the optical thickness of the nucleus [Eq. (3.6)]. It is assumed that the cross sections of $qN$ and $\bar{q}N$ interactions at high energies are the same according to the Pomeranchuk theorem. The value of $\sigma_{\rm abs}^{qN}$ is determined from the experimental data on the cross sections of hadron–nucleon interactions by comparing it with the predictions of the additive quark model.[62]

The cross sections for the interaction of $u$ and $d$ quarks are usually assumed to be the same and the cross sections for the $s$ quarks and heavier quarks are assumed to be slightly smaller, consistent with the experimental data. In the derivation of Eq. (3.14) the nucleus was assumed to be a solid object (the optical model of the nucleus). This equation has a simple physical meaning: it takes into account the probability for the absorption, $\nu$, of the structural quarks under the condition that $n-\nu$ structural quarks do not interact inelastically (these quarks are called *spectator quarks*). The factor $C_n^\nu$ takes into account the number of possible combinations of $n$ quarks in $\nu$.

The quark–parton cascades in the nucleus are calculated[51,63] with allowance for the time it takes to generate a parton fluctuation of a structural quark. In Eq. (3.13) the quantity $E$ in this case is the energy transferred by the structural quark. The interaction of the slow parton with the nucleon of the nucleus disrupts the coherence of the given parton fluctuation. As a result, the partons convert into structural quarks in a time given by Eq. (3.13), where $E$ is the energy transferred by a given parton.

The structural quark is an object which has its own parton fluctuation

*The meson has two structural quarks (a quark and an antiquark), while a baryon has three structural quarks.

involving slow partons and which is capable of strong interaction. If the energy of the parton which belongs to the primary parton fluctuation satisfies the inequality $E/\langle m_{\perp}^2\rangle \gtrsim R$, where $R$ is the average size of the nucleus, then this parton will have no time to convert into a structural quark inside the nucleus and will become a "sterile" quark with respect to strong interactions. Such partons become structural quarks, while the latter become the observable hadrons which are no longer in the nucleus and which do not participate in the development of the intranuclear quark-parton cascade.

The foregoing remarks imply that at high energies the high-energy part of the secondary-hadron spectra is formed, according to the predictions of the model under consideration, due to the single inelastic collisions of the structural quarks in the primary hadron with the nuclear nucleons.

The development of a quark–hadron cascade in the nucleus is described quantitatively on the basis of the cascade equations of the type[51]

$$\frac{\partial N_q(\varepsilon,x)}{\partial \varepsilon} = \frac{dN(E\rightarrow\varepsilon)}{d\varepsilon}\,\theta\,[x - l(\varepsilon)]\exp\{-[x - l(\varepsilon)]\sigma_{\text{abs}}^{qN}\rho\}$$
$$+ \int_0^{x - l(\varepsilon)} d\tau \exp\{-[x - \tau - l(\varepsilon)]$$
$$\times\, \sigma_{\text{abs}}^{qN}\rho\} \int_\varepsilon^E d\omega\, \frac{\partial N_q(\omega,\tau)}{\partial\omega}\,\frac{dN(\omega\rightarrow\varepsilon)}{d\varepsilon}. \tag{3.15}$$

Here $\partial N_q(\varepsilon,x)/\partial\varepsilon$ is the energy spectrum of secondary structural quarks at a depth $x$ in nuclear matter, $dN(E\rightarrow\varepsilon)/d\varepsilon$ is the energy spectrum of the partons which are produced by the primary structural quark of energy $E$ in the first collision with the nucleon of the nucleus, $l(\varepsilon) = \varepsilon/\langle m_{\perp}^2\rangle$ is the mean free path of a parton in the nuclear matter before the secondary structural quark is formed, and $\rho$ is the density of the nucleons in the nucleus. The first term in this equation describes, as can easily be seen, the spectrum of secondary partons which are produced in the first $qN$ collision and which have no time to convert into structural quarks capable of strong interaction inside nuclear matter at a depth $x$. These partons form a spectrum of secondary structural quarks at a depth greater than $x$. The second term takes into account the conversion of partons into structural quarks inside the nuclear matter at a depth less than $x$ and the secondary interaction of the newly produced structural quarks in nuclear matter.

Equation (3.15) is written here in simplified form for the plane layer of nuclear matter. This equation can easily be generalized to describe a spherical nucleus[64] and to take into account the transverse-momentum distribution of partons and structural quarks,[63–65] the production of all types of quarks, the distribution of the primary energy of a hadron between its structural quarks,[63] the simultaneous production of one, two, or three (for primary baryons) parton crest by the structural quarks of the primary hadron with probabilities consistent with Eq. (3.14), and other features of the quark–parton cascade.

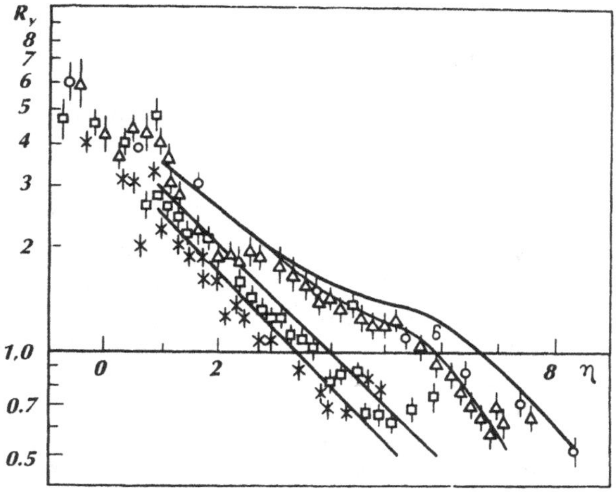


**Figure 3.13. Comparison of the predictions of the model of a quark–parton cascade in the nucleus for the ratio $R_y$ with the experimental data on the interaction of protons with energies $E_0 = 400$, 200, 50, and 24 GeV (the curves run from right to left) with emulsion nuclei (Ref. 63).**

The conversion of structural quarks into hadrons is resolved so far at the phenomenological level (the fragmentation, coalescence, and recombination models[66]). We wish to note here only that this problem is at present far from a complete solution. A comparison of the calculations carried out on the basis of the model under consideration with the experimental data shows that the spectra of secondary shower hadrons produced by nuclei are in fair agreement with the calculated spectra of secondary structural quarks.[51,53] Consequently, the model is apparently capable of describing the qualitative and quantitative features of the multiple hadron–nucleus processes in a completely adequate way (Fig. 3.13).

The additive quark model predicts for the secondary pion yield ratio in the case of the interaction of protons with nuclei and nucleons the following values[62]: for $x = E/E_0 \simeq \frac{2}{3}$ ($E_0$ is the energy of the primary protons, and $E$ is the energy of the pions)

$$R_x = W_1 . \tag{3.16}$$

In $\pi A$ collisions result (3.16) holds when $x \approx \frac{1}{2}$. In $pA$ reactions with $x = \frac{1}{3}$ it is expected that

$$R_x = W_1 + aW_2 , \tag{3.17}$$

where $R_x = [dN(pA \to \pi)/dx]/[dN(pp \to \pi)/dx]$; the probabilities $W_i$ are calculated from Eq. (3.14). The parameter $a$ depends on the model for the conversion of the structural quarks into pions, but from different estimates[50] $a \approx 1$. Results (3.16) and (3.17) follow from the assumption that in the corresponding regions of variation of $x$ the mesons are produced with the participation of only the spectator quarks. Predictions (3.16) and (3.17) are in agreement with experiment (see Ref. 62).

In the central region the model under consideration predicts the rapidity ratio of the pion spectra:

$$R_y \simeq N_q \sigma_{\text{abs}}^{qA} / \sigma_{\text{abs}}^{hA}, \tag{3.18}$$

where $N_q$ is the number of structural quarks in the incident hadron $h$. The cross sections for the absorption of quarks $q$ and hadrons $h$ by the nucleus $A$ are calculated from Eq. (3.5), where $\sigma_{\text{abs}}^{N}$ is understood to mean the cross section for inelastic interaction of $q$ or $h$ with the nucleon. Prediction (3.18) differs from the predictions of the model for multiple rescattering of the leading hadron (3.11). But both of these models predict that the spectra of the leading particles produced by the nuclei and nucleons are identical in shape.

The additive quark model with quark–parton cascades in the nucleus, which is described by equations such as (3.15), is apparently in better agreement, on the whole, with the experiments, at least in the case of shower particles, than the other models.[51] This model must, however, be used to carry out complex calculations of the intranuclear cascades, reducing the possibility of using it to calculate the passage of hadrons through condensed matter, which itself is complex. The semiphenomenological equations which describe the experimental data on the spectra of secondary hadrons produced in the hadron–nucleus interactions over a broad range of energies and value of $A$ can therefore be used more effectively as a reasonable compromise. The results of the theoretical models, which are in agreement with the experimental data, should be taken into account in choosing the analytic form of such representations (see the discussion below).

### The lepton-nucleus interactions

In the interaction of leptons (electrons, muons, neutrinos) with nucleons and nuclei the multiple production of hadrons occurs in the region of the so-called *deep inelasticity*, where the loss of lepton energy as a result of interaction is considerable, $\nu \gtrsim 2$–$3$ GeV and the square of the 4-momentum transfer to the secondary hadrons is large ($-t = Q^2 \gtrsim 1$–$2$ GeV$^2/c^2$). The electroproduction and muon and neutrino production of hadrons by nucleons and nuclei is described by the diagram of the type in Fig. 3.14, where the production of hadrons is accomplished through the interaction of a virtual photon (or $W^\pm, Z^0$ bosons—the weak-interaction quanta) with the nucleon (nucleus). The real high-energy photons can also generate hadrons by nucleons and nuclei. The existence of such processes suggest that a photon, real or virtual, has a virtual hadronic component, whose interaction with the target leads to the multiple production of secondary hadrons. The cross section for the absorption of real photons by nucleons is smaller by a factor $\alpha = \frac{1}{137}$ than the cross section for a hadron–nucleon inelastic interaction, since the probability for the transition of a photon to the hadronic state is proportional to the square of the electro-

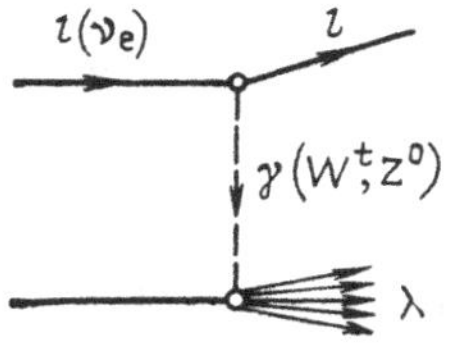


**Figure 3.14. Diagram of the deep inelastic lepton–nucleon interaction ($X$ is the hadron jet).**

magnetic coupling constant ($\alpha = e^2/\hbar c$, where $e$ is an elementary charge which is equal to the electron charge).

It has been shown experimentally that during the photoproduction in the central rapidity region, the shapes of secondary-hadron spectra are similar to the spectra generated in the hadron–nucleon collisions. In the fragmentation region of the photon the shapes of the spectra are similar to the spectra of the meson–nucleon collision, while the fragmentation of the target nucleon is the same as that in the hadron–nucleus collisions. The cross section of photoabsorption by a nucleus depends on the number of nuclear nucleons and is parametrized in the form[51] $\sigma_{\text{in}}(\gamma A) \simeq \sigma_{\text{in}}(\gamma N) A^{\beta}$, where $\beta \simeq 0.9$. If there is no hadronic component of the wave function of the real photon, we would expect that $\beta = 1$. Experiment shows that $\beta$ is significantly smaller than unity and hence the absorption of the hadronic component of the photon and the interaction of the secondary hadrons with the intranuclear nucleons occur inside the nucleus.

As the experimental data show, in the electroproduction and muon and neutrino production of hadrons by nuclei the dependence of the cross section for the absorption of virtual photons ($\gamma^*$) by a nucleus is different from that for the real photons. If the parametrization of the cross section is in the form $\sigma_{\text{in}}(\gamma^* A) \simeq \sigma_{\text{in}}(\gamma^* N) A^{\beta}$, the parameter $\beta$ in the slightly inelastic region is different from that in the deep inelastic region.[51] In the range of variation of the dimensionless variable $x = Q^2/(2m_N \nu + m_N^2)$ from $x = 0$ to $x \simeq 0.05$–$0.10$, the parameter $\beta$ increases from 0.9 to 1.0. In the region $x \gtrsim 0.05$–$0.10$ this parameter is approximately equal to 1.0 with some systematic deviations from this value in the direction of higher or lower values. The value $\beta = 1.0$ suggests that the nucleons of the nucleus produce an incoherent effect in the electroproduction processes in the deep inelastic region.

The data on the secondary-hadron multiplicity in deep inelastic processes show that $\langle N_s(\gamma^* A)\rangle / \langle N_s(\gamma^* N)\rangle > 1$ and $\langle N_s(\nu A)\rangle / \langle N_s(\nu N)\rangle > 1$. These results indicate that secondary interactions occur inside the nucleus after the completion of primary $\gamma^* N$ and $\nu N$ interactions.

Let us now discuss two of the most widely used theoretical models for the interaction of real and virtual photons of the electroweak interaction with nucleons and nuclei.

In the *vector dominance model* (see, e.g., Ref. 67) it is postulated that a photon ($\gamma$ or $\gamma^*$) can be represented as a superposition of neutral vector ($V$) mesons ($\rho^0$, $\omega$, and other mesons), which participate in the strong interaction with the target nucleon. The constant of the transition $\gamma(\gamma^*) \to V$ is determined in this case from the data on the radiative decay of $V$ mesons or on the production of $V$ mesons in the $e^+e^-$ annihilation process. The vector dominance model predicts, however, an incorrect dependence of $\sigma_{in}(\gamma A)$ and $\sigma_{in}(\gamma^* A)$ on $A$, since the photoabsorption cross sections in this model are proportional to the cross sections for the absorption of vector hadrons by the nucleus, $\sigma_{in}(VA) \sim A^{2/3}$.

The *quark–parton model* gives a qualitatively correct explanation of the $A$ dependence of the photoabsorption cross section. According to this model,[51] a photon first converts to a quark–antiquark pair ($q\bar{q}$) and then the components of this pair interact with the target nucleon. Since the cross section for a strong interaction of a quark parton with the target is large only when it is slow (when its momentum $p \sim m$), the $q\bar{q}$ pair is asymmetric with respect to energy. One of the components of the pair transfers almost all of the energy of the $\gamma(\gamma^*)$ ray and the other component is a relatively low-energy component. This slow quark has time to interact with the target during the lifetime $\tau$ of the quark fluctuation of the photon ($\tau \sim \nu/Q^2$). As a result, the spectator quark produces a high-energy secondary-hadron spectrum and the slow quark produces a low-energy part of this spectrum. In the quark–parton model, the photon is thus similar in its properties to one of the quarks (the virtual quark). This feature sharply sets the properties of this photon apart from those of the vector meson which consists of the $q\bar{q}$ pair which is symmetrical with respect to energy.

A detailed analysis of the lepton production of hadrons by nuclei on the basis of the quark parton model[68] shows that when the Bjorken dimensionless variable is $x = Q^2/2m_N\nu \lesssim A^{-1/3}\, m_\pi/m_N$, the distances at which the $\gamma^*$ ray exists as an asymmetric $q\bar{q}$ pair exceeds the size of the nucleus $R$ ($R \simeq m_\pi^{-1} A^{1/3}$, where $m_\pi$ is the pion mass). At these values of $x$ we should see the nuclear absorption of the hadronic component of the virtual photon, as has been observed experimentally. With $x \gtrsim m_\pi/m_N$ the $\gamma^*$ ray exists as an asymmetric $q\bar{q}$ pair at a distance no greater than the internucleon distance in the nucleus. In this case the nucleons in the nucleus act incoherently and the parameter $\beta = 1$. At $x \approx m_\pi/m_N$ we expect that $\beta > 1$. This effect is produced under the influence of the energy-momentum conservation law and the merging of the parton clouds (crests) of the nucleons in the nucleus in the reference frame where the nucleus moves at a relativistic energy.[68] At $x \lesssim m_\pi/m_N$ the value of $\beta$ decreases with decreasing $x$ and levels off at a value lower than unity when $x \lesssim A^{-1/3}\, m_\pi/m_N$.

The lepton-induced quark–parton cascade in the nucleus is calculated on the basis of a cascade equation in the same way as in the case of hadron-induced processes, with allowance for the qualitative effects considered above. The results of such calculations[51] are in agreement with the experimental data, although they are not described by simple analytical dependences. In the study of the high-energy part of the secondary-hadron spectra we can use as a guideline the data obtained from nucleons. The intranuclear cascades are important in the low-energy part of the spectrum.

**Production of hadrons with large $p_\perp$ by nuclei**

As was already mentioned in the discussion of the experimental data, the $p_\perp$ dependence of the invariant inclusive differential cross sections for the production of hadrons changes qualitatively at $p_\perp \gtrsim 1.5$–$2$ GeV/$c$. The replacement of the exponential (or Gaussian) dependence on $p_\perp$ by the power function is attributable to the hard quark–quark collisions which occur due to the exchange of gluons—neutral color strong-interaction photons.[69] These processes are described by the gauge theory of strong quark–gluon interaction called quantum chromodynamics.[70] The model for a single-gluon exchange between the quarks of the colliding hadrons predicts the following relation for large values of $p_\perp$:

$$Ed^3\sigma/d^3p \sim p_\perp^{-n}, \quad n=4 \tag{3.19}$$

at angles of emission of the secondary hadrons $\theta^* \sim 90°$ in the c.m. frame.

For experimentally accessible energies the invariant cross section depends on $p_\perp$ in a power-law fashion at $p_\perp \gtrsim 1.5$–$2$ GeV/$c$ (see the discussion of the experimental data in the beginning of this section), but the exponent $n$ in expression (3.19) is much larger than that predicted by quantum chromodynamics ($n \simeq 6$–$10$).[71]

Many studies have proposed various models which can approximate the experimentally observed dependence of $Ed^3\sigma/d^3p$ on $p_\perp$ at $p_\perp \gtrsim 1.5$–$2$ GeV/$c$. Without discussing these models in detail, we note that their principal results are described by (3.3). The most interesting dependence, which has not yet been explained conclusively, is the $A$ dependence of the invariant cross sections for the production of hadrons with large $p_\perp$ by the nuclear targets:

$$Ed^3\sigma/d^3p \sim A^{\alpha(p_\perp)}. \tag{3.20}$$

At $p_\perp \gtrsim 1.5$ GeV/$c$ the exponent $\alpha(p_\perp)$ in the proton–nucleus collisions is greater than unity and increases with increasing $p_\perp$, reaching the values of $\alpha(p_\perp \approx 5 \text{ GeV}/c) \approx 1.1$ for the secondary $\pi^\pm$ mesons, $\alpha(p_\perp \approx 6 \text{ GeV}/c) \simeq 1.3$ for protons, antiprotons, and $K^-$ mesons, and $\alpha(p_\perp \simeq 4 \text{ GeV}/c) \simeq 1.2$ for $K^+$ mesons.

The data on the production of the hadron jets with large $p_\perp$ in $pA$ and $\pi^\pm A$ collisions suggest that the parameter $\alpha(p_\perp)$ increases even more rapidly with increasing $p_\perp$. At the same time, the production of massive lepton pairs in $pA$ collisions was found to have a relation $d\sigma/dm_{\mu\mu} \sim A^{\alpha(m_{\mu\mu})}$, where $\alpha(m_{\mu\mu}) \approx 1$ at $m_{\mu\mu} \gtrsim 4$ GeV ($m_{\mu\mu}$ is the effective mass of the muon pair). This $A$ dependence is in good agreement with the model for the annihilation of a $q\bar{q}$ pair into a leptonic $\mu^+\mu^-$ pair through a virtual photon of the corresponding mass. In order for the massive pair to form, the quarks (antiquarks) must have a relatively high energy. The number of these quarks in a moving nucleus is proportional to the number of nucleons[51] $A$.

A generally accepted, satisfactory explanation of the $A$ dependence in (3.20) with $\alpha > 1$ so far has not been found, although models for multiple scattering of high-energy quarks by each other and by nuclear nucleons "operate" qualitatively in the necessary direction.[50]

Under the present circumstances the easiest way to calculate the spectra of hadrons with large $p_\perp$, which are produced as a result of the development of internuclear cascades in extended condensed matter, is to use equations such as (3.3) with an experimentally observable $A$ dependence in (3.20).

### Nuclear interaction at high energies

The very limited data on the multiple production of hadrons as a result of relativistic nucleus–nucleus collisions are described well by the model for multiple scattering of nucleons of one nucleus by nucleons of another nucleus.[72] The following are the principal results of this model.

(1) In the central rapidity region of secondary particles and at small $p_{c\perp}$, where $|x_F| \ll 1$ ($x_F = p^*_{c\parallel}/|p^*_{c\parallel\max}|$, and $p^*_{c\parallel}$ is measured in the c.m. frame of the $NN$ collision) the invariant differential spectrum of the inclusive particles $c$ (primarily the pions) is described by the equation

$$E_c \frac{d^3\sigma(A_1 + A_2 \to c + X)}{d^3p_c} = A_1A_2 \frac{E_c d^3\sigma(N + N \to c + X)}{d^3p_c}, \qquad (3.21)$$

where $A_1$ and $A_2$ are nuclei with the corresponding number of nucleons, and $X$ are the undetectable reaction products.

The ratio for the yield of secondary particles of species $c$ in $A_1A_2$ collisions $[E_c d^3\sigma(A_1 + A_2 \to c + X)/d^3p_c]/\sigma_{\text{in}}\ (A_1A_2) = F_{A_1A_2}$ and in $NN$ collisions $[E_c d^3\sigma(N + N \to c + X)/d^3p_c]\sigma_{\text{tot}}\ (NN) = F_{NN}$ for $|x_F| \ll 1$ is determined, according to (3.20), by the expression

$$F_{A1A2}/F_{NN} = A_1A_2\sigma_{\text{tot}}(NN)/\sigma_{\text{in}}(A_1A_2), \qquad (3.22)$$

where $\sigma_{\text{tot}}(NN)$ is the total cross section of the $NN$ interaction, and $\sigma_{\text{in}}(A_1A_2)$ is the cross section of the inelastic $A_1A_2$ interaction.

Equations (3.21) and (3.22) follow directly from the optical model for the collision between nuclei and are valid for moderately light $A_1$ and $A_2$ nuclei; Eq. (3.22) can also be written as

$$F_{A1A2}/F_{NN} = \langle\nu_{A1}\rangle\langle\nu_{A2}\rangle\Phi(A_1A_2),$$

where $\langle\nu_A\rangle = A\sigma_{\text{tot}}\ (NN)/\sigma_{\text{in}}\ (NA)$, $\sigma_{\text{in}}(NA)$ is the cross section of inelastic $NA$ interaction, and

$$\Phi(A_1A_2) = \sigma_{\text{in}}(NA_1)\sigma_{\text{in}}(NA_2)/\sigma_{\text{tot}}(NN)\sigma_{\text{in}}(A_1A_2).$$

The cross sections $\sigma_{\text{in}}(NA)$ are parametrized in the form $\sigma_{\text{in}}(NA) \simeq 39A^{0.72}$ mb and the cross sections $\sigma_{\text{in}}(A_1A_2)$ are parametrized in the form $\sigma_{\text{in}}(A_1A_2) \simeq \pi R_0^2(A_1^{1/3} + A_2^{1/3} - c)^2$, where $R_0 = 1.52\times10^{-13}$ cm, and $c \simeq 1.35$. Such parametrizations are consistent with the existing experimental data (a detailed comparison with experiment was carried out in Ref. 72).

(2) In the region where the nucleon of the nucleus $A_1$ fragments at $x_F \gtrsim 0.1$, the behavior of the spectrum as a function of $x_F$ is predicted to be

$$F_{A1A2}/F_{NN} \sim N_{A_1} A_2^{-\gamma(x_F)}, \qquad (3.23)$$

where $N_{A_1} = A_1\sigma_{\rm in}(NA_2)/\sigma_{\rm in}(A_1A_2)$ is the average number of nucleons in the nucleus $A_1$ which participate in the interaction. The behavior of the quantity $\gamma(x_F) > 0$ is determined from the data on $NA$ and $NN$ interactions. The parameter $\gamma(x_F)$ increases slowly with increasing $x_F$. At $x_F \simeq 0.5$ we have $\gamma(x_F) \simeq 0.2$.

(3) An equation similar to Eq. (3.23) applies in the region of fragmentation of the nucleons of the nucleus $A_2$. Relations (3.22) and (3.23) are valid when the energy of the nucleon of an incident nucleus $A_1$ is high ($E_{A_1}/A_1 \gtrsim 100$ GeV; Ref. 28). If, on the other hand, $A_1 \ll A_2$, these relations will apply even at energies on the order of several (tens) of GeV per nucleon of the nucleus.

At a moderately high energy, relations (3.22) and (3.23) generally do not apply, and either the phenomenological parametrizations of the experimental data or the intranuclear cascade model should be used.[48]

# 4. Phenomenological description of the inclusive spectra

The existing models for the description of the inclusive distributions of hadrons in $pp$ collisions at $E_0 > 1$ GeV are based on simplifying specific assumptions. These models therefore describe the process in a particular limited region of the kinematic variables. The complex-momentum method, which is generalized to the inclusive processes, for example, predicts the behavior of the distribution in the scaling variable $x_F$ separately for the fragmentation region and the pionization region, without yielding any information about the intermediate region. Since the transverse-momentum distribution is not determined in the complex-momentum method, it must be postulated by choosing a Gaussian or exponential shape and by fitting the parameters from a comparison with the experiment. Statistical and thermodynamic models also have an adequate number of adjustable parameters. Furthermore, such models give an incorrect dependence of the invariant cross section on the transverse momentum at $p_\perp \gtrsim 1$–$2$ GeV/$c$.

All these factors prevent a particular model from establishing a quantitative agreement between theory and experimental data over a broad kinematic interval without choosing free parameters. An urgent need to describe the inclusive spectra, especially in going up the energy scale, has initiated many attempts to construct relevant phenomenological equations.

The main features of the inclusive hadron distributions in the reaction $hp \to cX$ can be analyzed by using simple expressions of the type

$$Ed^3\sigma/d^3p = C_0(p_\perp)(1-x_F)^n; \quad n \neq n(p_\perp, p_0). \tag{3.24}$$

The values of the parameter $n$ for various reactions at primary hadron momenta $p_0 \sim 100$–$200$ GeV/$c$ in the laboratory frame at $p_\perp < 1.2$ GeV/$c$ and $0.2 < x_F < 1$ are presented[73] in Table 3.3.

**Table 3.3. Values of the parameter $n$ in Eq. (3.24) for various reactions.**

| Reaction | $\pi^+\to K^+$ | $\pi^-\to K^-$ | $K^+\to\pi^+$ | $K^-\to\pi^-$ | $\pi^+\to p$ | $\pi^-\to\bar{p}$ |
|---|---|---|---|---|---|---|
| $n$ | 1.28 | 1.40 | 2.28 | 2.50 | 1.78 | 2.13 |
| Reaction | $\pi^+\to\bar{p}$ | $\pi^-\to p$ | $K^+\to p$ | $K^-\to\bar{p}$ | $p\to\pi^+$ | $\bar{p}\to\pi^-$ |
| $n$ | 2.98 | 2.58 | 1.72 | 1.56 | 3.43 | 3.10 |
| Reaction | $p\to K^+$ | $p\to\pi^-$ | $\bar{p}\to\pi^+$ | $\pi^+\to\pi^-$ | $\pi^-\to\pi^+$ | $\pi^+\to K^-$ |
| $n$ | 2.87 | 4.53 | 3.71 | 3.32 | 2.94 | 2.30 |
| Reaction | $\pi^-\to K^+$ | $K^+\to\pi^-$ | $K^-\to\pi^+$ | $K^-\to p$ | $p\to K^-$ | $p\to\bar{p}$ |
| $n$ | 1.98 | 2.98 | 2.51 | 4.13 | 4.68 | 8.09 |

**Table 3.4. Average values of the inelasticity coefficient $\langle K_\gamma\rangle$.**

| Reaction | $E_0$, GeV | $CH_2$ | Al | Fe | Pb |
|---|---|---|---|---|---|
| $nA\to\pi^0X$ | 200–2000 | $0.17\pm0.01$ | $0.19\pm0.02$ | $0.19\pm0.02$ | $0.21\pm0.02$ |
| | | 0.17 | 0.19 | 0.21 | 0.23 |
| $\pi^\pm\to\pi^0X$ | 200–2000 | $0.33\pm0.02$ | $0.38\pm0.04$ | $0.37\pm0.05$ | $0.39\pm0.04$ |
| | | 0.31 | 0.31 | 0.31 | 0.30 |
| $pAS\to\pi^0X$ | $10^3$–$3\times10^4$ | — | — | — | $0.17\pm0.01$ |
| | | — | — | — | 0.19 |

The average inelasticity coefficient $\langle K\rangle$ is a useful concept in the analysis of an interaction. The calculated (lower numbers) and experimentally obtained (upper numbers) asymptotic values $\langle K_\gamma\rangle=\langle K_{\pi^0}\rangle$, taken from Ref. 74, are given in Table 3.4. At $E\sim20$ GeV in the $pN$ reaction $\langle K_\gamma\rangle\simeq0.15$, $\langle K_+\rangle\simeq0.46$, $\langle K_-\rangle\simeq0.13$, and $\langle K_0\rangle\simeq0.41$ (Ref. 74).

The expressions for effective invariant cross sections of the inclusive reactions

$$pA\to(\pi^\pm, K^\pm, \tilde{p}, \Sigma^\pm)+X_A;\quad pA\to(p,n)+X_A \tag{3.25}$$

for the proton and nuclear targets were proposed in Ref. 75. These equations have no more free parameters than the specific models and, most important, they have all the qualitative properties at small transverse momenta ($p_\perp\lesssim1.5$ GeV/$c$) and large transverse momenta ($p_\perp\gtrsim1.5$ GeV/$c$), describing the spectra in the entire region, $-1\leqslant x_F\leqslant1$.

As the kinematic variables one can use $x'=E^*/E^*_{\max}$ and $p_\perp$, where $E^*$ and $E^*_{\max}$ are the total energy of the inclusive particle in the c.m. frame and its maximum possible value. The variables $x'$ and $p_\perp$ are useful near the phase boundary and in the central region (see Sec. 1.3). In addition, these variables should reveal an early scaling.

Taking into account the qualitative predictions of the Regge and parton models, the following expression was obtained in Ref. 75 for the particle spectra of the first reaction in (3.25):

**Table 3.5. Values of the parameters in Eqs. (3.26) and (3.27).**

| Particle | m, GeV | ω, GeV | $\mu^2$ | $b_0$ | Γ, GeV | $C_1$ | $C_2$ | $C_3$ | $C_4$ | $C_5$ | $C_6$ |
|---|---|---|---|---|---|---|---|---|---|---|---|
| $p$ | 0.938 28 | 0.938 28 | 1.3 | 0.13 | 5.0 | 0.018 | 126 | 1.14 | — | — | — |
| $n$ | 0.939 57 | 1.077 85 | 1.3 | 0.14 | 5.0 | 0.018 | 126 | 1.14 | — | — | — |
| $\pi^+$ | 0.139 57 | 1.877 75 | 0.88 | 0.10 | 4.0 | 0.25 | 3.1 | 0.88 | 3.0 | 2 | 11.3 |
| $\pi^-$ | 0.139 57 | 2.016 13 | 0.86 | 0.11 | 3.5 | 0.20 | 4.0 | 1.18 | 3.0 | 2 | 9.8 |
| $K^+$ | 0.493 71 | 2.053 88 | 1.2 | 0.12 | 4.0 | 0.075 | 2.5 | 1.60 | 3.0 | 2 | 10.7 |
| $K^-$ | 0.493 71 | 2.370 27 | 1.2 | 0.12 | 4.0 | 0.078 | 6.1 | 2.46 | 3.5 | 2 | 8.8 |
| $\bar{p}$ | 0.938 28 | 2.814 84 | 1.1 | 0.12 | 5.0 | 0.080 | 8.6 | 2.30 | 4.2 | 2 | 10.5 |
| $\Sigma^+$ | 1.189 37 | 1.435 98 | 1.1 | 0.12 | 4.0 | 0.015 | 0.3 | 1.3 | 4.0 | 1 | 12.0 |
| $\Sigma^-$ | 1.197 35 | 1.575 55 | 1.1 | 0.12 | 4.0 | 0.014 | 0.4 | 1.6 | 3.8 | 1 | 12.0 |

$$\left.\begin{aligned}\frac{E}{\sigma_A}\frac{d\sigma_A}{d^3p} &= C_1 A^{b(p_\perp)}(1-x')^{C_2}\exp(-C_3 x')\Phi(p_\perp);\\ \Phi(p_\perp) &= \exp(-C_4 p_\perp^2) + C_5\frac{\exp(-C_6 x_\perp)}{(p_\perp^2+\mu^2)^4},\end{aligned}\right\} \qquad (3.26)$$

where

$$b(p_\perp) = \begin{cases} b_0 p_\perp, p_\perp \leqslant \Gamma;\\ b_0\Gamma, p_\perp > \Gamma, x_\perp \simeq 2p_\perp/\sqrt{S},\end{cases}$$

and the parameters $b_0$, $\Gamma$, $\mu^2$, and $C_i$, found by the method of least squares from the experimental data, are presented in Table 3.5. Also given in this table are the masses $m$ of the corresponding particles and the minimum masses $\omega$ of the $X_A$ system from (3.25) for the nucleon target.

The differential cross section of the second reaction in (3.25) was described in Ref. 75 by using the predictions of the Regge model for $|x_F|>0.7$, the two-cluster diffraction model for $|x_F|<0.7$, and the predictions of the parton model for large $p_\perp$. The three-Reggeon vertices $PPP$, $PPR$, $RRP$, and $RRR$ (see Sec. 3.1), which were parametrized in the form $G_{PPP(R)}(t) = G_0 \times \exp(at)$, were taken into account. Here $P$ is a pomeron with the trajectory $\alpha_P(t) = 1 + 0.15t$, and $R$ is the effective nonvacuum pole with the trajectory $\alpha_R(t) = 0.5 + 0.75t$ and $a = $ const. Finally,

$$\left.\begin{aligned}\frac{E}{\sigma_A}\frac{d^3\sigma_A}{d^3p} &= C_1\left[F_{PP} + C_2 A^{b(p_\perp)} x'(1-x')^{C_3 p_\perp^2}(p_\perp^2+\mu^2)^{-4.5}\right];\\ F_{PP} &= [1+S^{-1/2}(1-x')^{-1/2}](1-x')^{-1+0.3p_\perp^2}\exp(-3.5p_\perp^2),\end{aligned}\right\} \qquad (3.27)$$

where the function $b(p_\perp)$ is the same as that in Eq. (3.26) and the parameter values are given in Table 3.5. For the reaction $pA \to nX_A$ we have $F_{PP} = 0$.

Figures 3.15–3.21 compare the results of calculations based on Eqs. (3.26) and (3.27) with the experimental data of Ref. 75. These figures reveal a good agreement over a broad range of target nuclei ($p$, Be, Ti, and W) of the primary momenta $6 \leqslant p_0 \leqslant 1500$ GeV/$c$, transverse momenta $0.2 \leqslant p_\perp \leqslant 8$ GeV/$c$, and longitudinal momenta $0 \leqslant |p_\parallel^*| \leqslant p_{\max}^*$. This result is evidence that the sug-

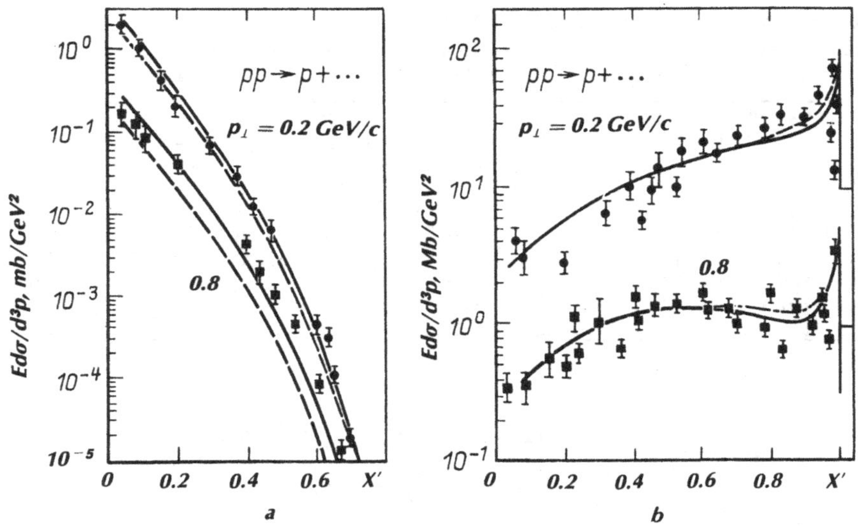


**Figure 3.15.** Invariant cross sections $Ed^3\sigma/d^3p$ for (a) the inclusive reaction $pp\to\bar{p}+\cdots$ [curves, calculation based on Eq. (3.26): —, $p_0 = 24$ GeV/$c$, —. $p_0 = 1500$ GeV/$c$; the experimental data for $p_0$ from 6 to 1500 GeV/$c$ were taken from Ref. 75: ●, $p_\perp = 0.2$ GeV/c; ■, $p_\perp = 0.8$ GeV/$c$, $x' = E^*E^*_{max}$; (b) the reaction $pp\to p+\cdots$ [curves, calculation based on Eq. (3.27)].

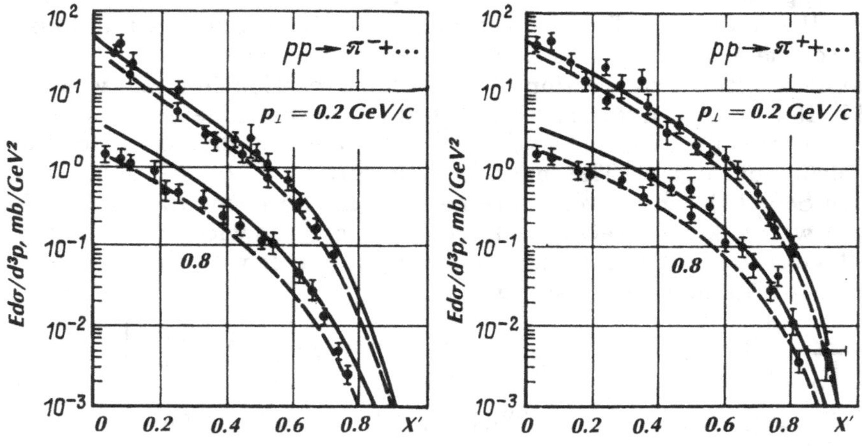


**Figure 3.16.** Invariant cross sections for (a) the reaction $pp\to\pi^-+\cdots$ and (b) the reaction $pp\to\pi^++\cdots$. (The notation is the same as in Fig. 4.15.)

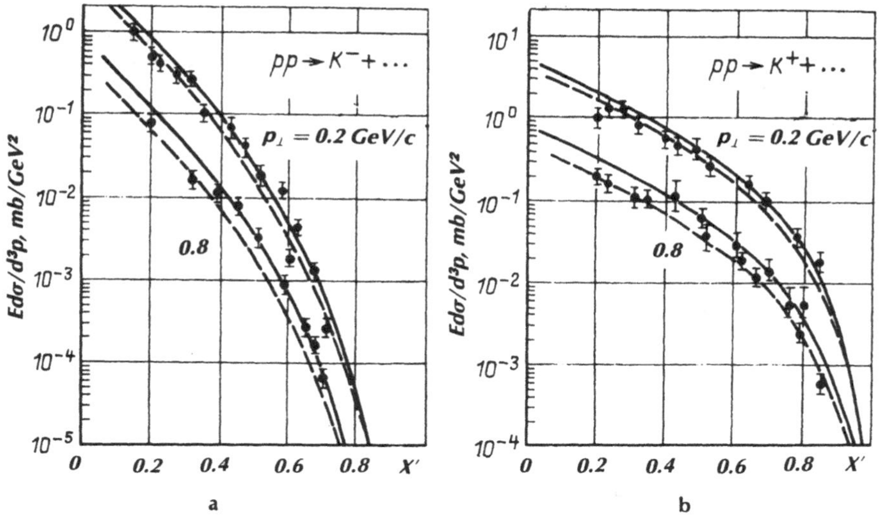


**Figure 3.17.** Invariant cross sections for (a) the reaction $pp \to K^- + \cdots$ and (b) the reaction $pp \to K^+ + \cdots$. (The notation is the same as in Fig. 3.15.)

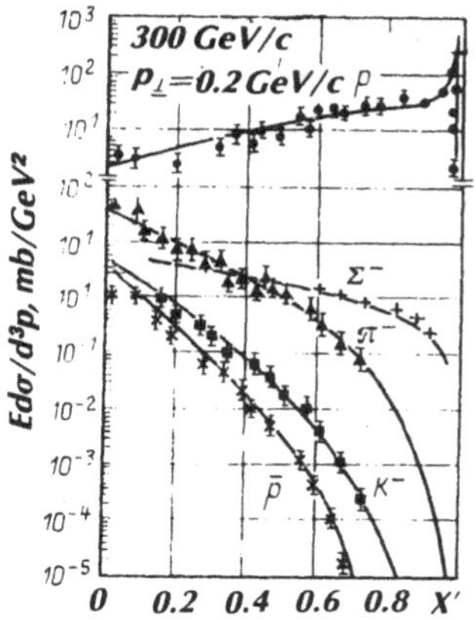


**Figure 3.18.** Invariant cross section for the production of inclusive hadrons in $pp$ collisions at $p_\perp = 0.2$ GeV/$c$ (curves, calculation from Ref. 75 for $p_0 = 300$ GeV/$c$; experimental points from Ref. 75 for $6 < p_0 < 1500$ GeV/$c$; $\Sigma^-$, experimental data (Ref. 75) on $pBe$ interaction for $p_0 = 29$ GeV/$c$ and $p_\perp = 0$; $x' = E^* E^*_{max}$).

gested equations are highly precise, which makes them a useful tool for the analysis of the inclusive distributions.

A slightly less accurate description of the spectra at small $p_\perp$ and of the particular features of the $A$ dependence are, however, the tradeoff for the universal applicability of Eqs. (3.26) and (3.27). The small transverse momenta play an important role in the calculation of electromagnetic cascades. In addition, information on the reactions of the type (3.25) must be available for all incident particles (over a broad range of the primary momenta $p_0 \gtrsim 0.1$ GeV/$c$) which participate in the development of the electromagnetic cascade.

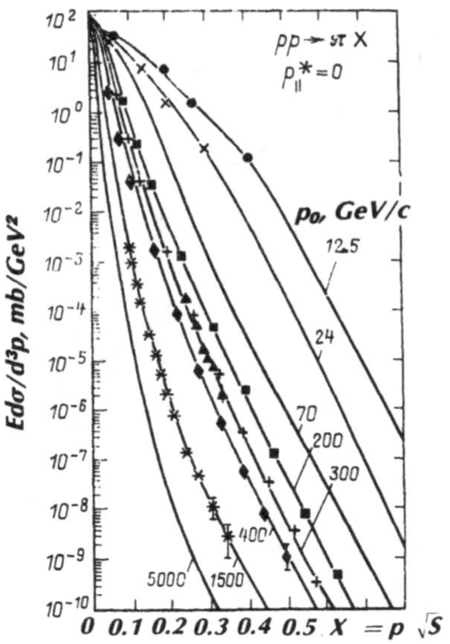


**Figure 3.19. Invariant cross section for the production of $\pi^0$ mesons in *pp* collision (the calculated curves and experimental points were taken from Ref. 75).**

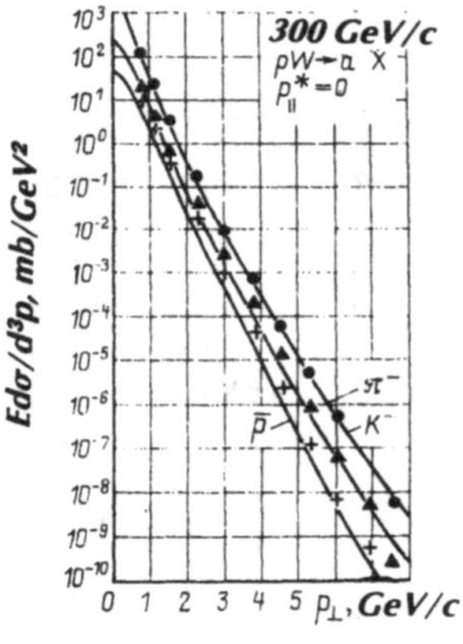


**Figure 3.20. Invariant cross section for the production of hadrons in *pW* interaction at $p = 300$ GeV/*c* (the calculated curves and experimental points were taken from Ref. 75).**

Analysis of various semiempirical equations published in the literature[76–83] shows that the following description of inclusive hadron distributions in *hA* interactions can now be used in the theoretical predictions of electromagnetic cascades.

### The region $E_0 > 5$ GeV

We assume that the incident particle is a nucleon. In the interval $0.8 \lesssim x_F \lesssim 1$ the leading-nucleon spectra can then be described by the *PPP* and *PPR* contributions of the three-Reggeon formalism. The average multiplicity of these

**Figure 3.21. Momentum distribution of protons and pions at an angle $\theta = 77$ mrad in the laboratory frame in $pT_i$ interaction at $p_0 = 300$ GeV/$c$. The solid curve was calculated from Eqs. (3.26) and (3.27), the dashed curve was plotted on the basis of the thermodynamic model, and the experimental data points were taken from Ref. 79.**

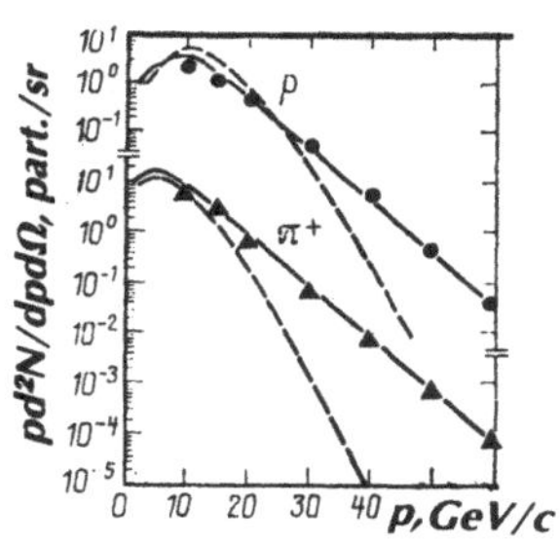


**Table 3.6. Values of the parameters $C_i$ in Eq. (3.28).**

| $A$ | $C_1$ | $C_2$ | $C_3$ | $C_4$ | $C_5$ | $C_6$ |
|---|---|---|---|---|---|---|
| $<40$ | 2.87 | $-2.99$ | 3.2 | 3.91 | 5.82 | 2.99 |
| $\geqslant 40$ | 2.80 | $-1.78$ | 0.3 | 5.38 | 5.80 | 2.80 |

nucleons is described by the empirical formula $\langle n\rangle \simeq 0.135(A+2)^{-0.355}$. The distributions of the other fast nucleons are found by means of the formulas[76,80] $(x = x_F)$

$$\frac{d^2N}{dxdp_\perp^2} = C_1 \frac{p^*_{\max}}{4\sqrt{S}}(1 + C_2|x| + C_3x^2) \times [\exp(-C_4 p_\perp^2) + C_5 \exp(-C_6 p_\perp)] , \tag{3.28}$$

where $x = x_F = p^*_\parallel / p^*_{\max}$, and the modified parameters $C_i$ which make a consistent description of the spectra possible, are given in Table 3.6.

The following equations can be used to describe the production of secondary $\pi$ and $K$ mesons[77–79]:

$$\left.\begin{aligned} \frac{d^2N}{dxdp_\perp^2} &= C_1 \frac{\pi}{x'}(1 - x_R)^{C_2} F \exp(-C_5 x_R)\Phi(p_\perp^2); \\ F &= 1 - \exp(-Tr^{*2}/C_3 - p_\perp^2/C_4); \\ \Phi(p_\perp^2) &= (1 - C_6)\exp(-C_7 p_\perp^2) + C_6 \exp(-C_8 p_\perp^2), \end{aligned}\right\} \tag{3.29}$$

where $x_R = p^*/p^*_{\max}$, $x' = E^*/E^*_{\max}$, and the parameters $C_i$ are given in Table 3.7. The quantities in the c.m. frame of the reaction are denoted by an asterisk.

We assume now that the incident particle is a $\pi^\pm$ meson. At $x_F \leqslant 0$ the nucleon spectra are described by Eq. (3.28) and at $x_F > 0$ they are described by an equation which describes the experimental data in an approximate way[84]:

$$\frac{d^2N}{dxdp_\perp^2} = B_1 \exp(-B_2 x)[\exp(-B_3 p_\perp^2) + B_4 \exp(-B_5 p_\perp)] . \tag{3.30}$$

Here $B_1 = 1.8$, $B_2 = 5.2$, $B_3 = 3.78$, $B_4 = 0.47$, and $B_5 = 3.6$.

**Table 3.7. Values of the parameters $C_i$ in Eq. (3.29).**

| Nucleus | Particle | $C_1$ | $C_2$ | $C_3$ | $C_4$ | $C_5$ | $C_6$ | $C_7$ | $C_8$ |
|---|---|---|---|---|---|---|---|---|---|
| $H_2$ | $\pi^+$ | 2.47 | 1.80 | 3.0 | 0.010 | 2.78 | 0.30 | 12.0 | 2.70 |
| | $\pi^-$ | 1.74 | 2.60 | 1.0 | 0.009 | 2.94 | 0.30 | 12.0 | 2.70 |
| | $K^+$ | 0.17 | 1.30 | 3.0 | 0.010 | 2.78 | 0.46 | 4.2 | 2.65 |
| | $K^-$ | 0.12 | 3.60 | 1.0 | 0.009 | 2.94 | 0.48 | 5.0 | 2.70 |
| Be | $\pi^+$ | 2.32 | 1.89 | 2.6 | 0.010 | 3.03 | 0.30 | 12.5 | 2.65 |
| | $\pi^-$ | 1.62 | 2.52 | 0.7 | 0.008 | 3.13 | 0.32 | 11.0 | 2.70 |
| | $K^+$ | 0.11 | 1.28 | 2.6 | 0.010 | 3.03 | 0.51 | 4.3 | 2.65 |
| | $K^-$ | 0.09 | 3.80 | 0.7 | 0.008 | 3.13 | 0.52 | 5.5 | 2.70 |
| Al | $\pi^+$ | 2.27 | 1.93 | 2.0 | 0.008 | 3.23 | 0.32 | 12.0 | 2.65 |
| | $\pi^-$ | 1.61 | 2.58 | 0.3 | 0.005 | 3.33 | 0.35 | 10.5 | 2.70 |
| | $K^+$ | 0.12 | 1.31 | 2.0 | 0.008 | 3.23 | 0.52 | 4.4 | 2.70 |
| | $K^-$ | 0.09 | 3.84 | 0.3 | 0.005 | 3.33 | 0.52 | 5.7 | 2.75 |
| Cu | $\pi^+$ | 2.25 | 1.89 | 1.8 | 0.006 | 3.57 | 0.35 | 11.2 | 2.65 |
| | $\pi^-$ | 1.59 | 2.53 | 0.2 | 0.002 | 3.70 | 0.37 | 10.0 | 2.70 |
| | $K^+$ | 0.13 | 1.35 | 1.8 | 0.006 | 3.57 | 0.54 | 4.6 | 2.75 |
| | $K^-$ | 0.10 | 3.90 | 0.2 | 0.002 | 3.70 | 0.55 | 6.0 | 2.80 |
| Pb | $\pi^+$ | 1.95 | 1.82 | 1.5 | 0.005 | 4.00 | 0.37 | 11.0 | 2.65 |
| | $\pi^-$ | 1.52 | 2.44 | 0.1 | 0.001 | 4.17 | 0.40 | 10.0 | 2.70 |
| | $K^+$ | 0.14 | 1.40 | 1.5 | 0.005 | 4.00 | 0.55 | 4.9 | 2.85 |
| | $K^-$ | 0.10 | 3.96 | 0.1 | 0.001 | 4.17 | 0.56 | 6.5 | 2.90 |

**Table 3.8. Values of the parameters $C_i$ in Eq. (3.31).**

| Reaction | $x$ | $C_0$ | $C_1$ | $C_2$ | $C_3$ | $C_4$ |
|---|---|---|---|---|---|---|
| $\pi^+\to\pi^+$ | $<0$ | 1.7 | 0.33 | 7 | 0.0001 | 5.7 |
| | $\geqslant 0$ | 1.7 | 0.22 | 5 | 0.115 | 5.7 |
| $\pi^-\to\pi^-$ | $<0$ | 1.7 | 0.31 | 13 | 0.0001 | 5.7 |
| | $\geqslant 0$ | 1.7 | 0.2 | 5 | 0.115 | 5.7 |
| $\pi^+\to\pi^-$ | $<0$ | 5.4 | $0.14-0.14/10\sqrt{S}$ | 13 | 0.0001 | 5.7 |
| $\pi^-\to\pi^+$ | $\geqslant 0$ | 5.4 | $0.14-0.14/10\sqrt{S}$ | 5.5 | 0.0001 | 5.7 |

A better description of the $A$ dependence of the spectra can be obtained by using Eqs. (3.28)–(3.31) for the hydrogen target in combination with the model-based predictions (3.16)–(3.18) and Fig. 3.13.

The inclusive distributions of secondary $\pi$ mesons are described approximately by an equation obtained in Ref. 80:

$$\frac{d^2N}{dxdp_\perp^2} = \frac{C_0}{2x'}\,[C_1 \exp(-C_2 x_F^2) + C_3]\,C_4^2 \exp(-C_4 p_\perp)\,. \tag{3.31}$$

The parameters $C_i$ are given in Table 3.8.

**Figure 3.22. Invariant cross section for the reaction $pAl \to \pi^\pm X$ at $E_0 = 100$ GeV and $p_\perp = 0.3$ GeV/$c$. Histograms, result of simulation using Eq. (3.29); points, experimental data taken from Ref. 84 (the calculated and experimental data for $\pi^-$ mesons were reduced by a factor of 2).**

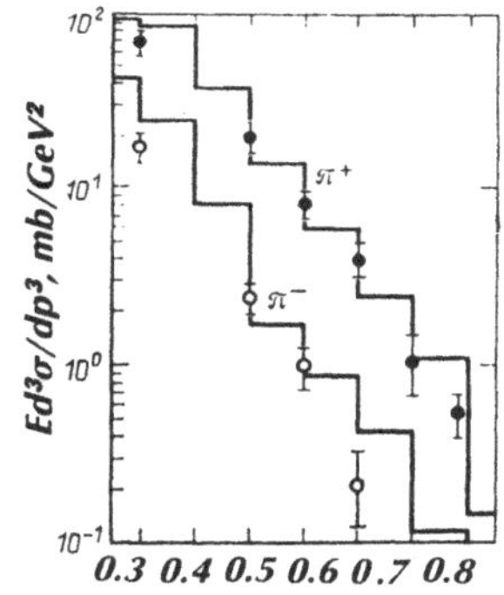


The distribution of slow cascade nucleons produced in $hA$ interaction at high energies in the laboratory frame can be described by equations obtained in Ref. 76:

$$\frac{d^2N}{dEd\Omega} = \left[\frac{n_{1i}\exp(-E/\alpha_{1i})}{\alpha_{1i}[1-\exp(-E_0/\alpha_{1i})]} + \frac{n_{2i}\exp(-E/\alpha_{2i})}{\alpha_{2i}[1-\exp(-E_0/\alpha_{2i})]}\right] g(E,\theta);$$

$$g(E,\theta) = \begin{cases} N_0\exp(-\theta^2/\lambda_0), & \theta < \pi/2; \\ N_0\exp(-\pi^2/\lambda_0), & \theta \geqslant \pi/2, \end{cases} \quad (3.32)$$

where the parameters have the following values at $E_0 \geqslant 5$ GeV: $n_{1p} = 0.21\sqrt{A}$, $n_{2p} = 0.0245\sqrt{A}$, $n_{1n} = 0.27\sqrt{A}$, $n_{2n} = 0.032\sqrt{A}$, $\alpha_{1p} = 0.027C$, $\alpha_{2p} = 0.16C$, $\alpha_{1n} = 0.023C$, $\alpha_{2n} = 0.15C$, $C = 1 - 0.001A$, $\lambda_0 = (0.12 + 0.00036A)/E$, and $N_0$ is a normalization factor such that $2\pi\int_0^\pi g(E,\theta)\sin\theta d\theta = 1$.

Figures 3.22–3.24 show some results of the inclusive distributions of protons and $\pi^\pm$ mesons in $pA$ and $\pi A$ collisions at energies $E_0 = 100$–$3000$ GeV, simulated by the Monte Carlo method (see Chap. 6). The Monte Carlo simulation was based on Eqs. (3.28)–(3.32).

## The region $E_0 < 5$ GeV

If the primary kinetic energy of the hadrons lies in the interval $0.1 < E_0 < 1$ GeV, we can use the equations from Ref. 83, which are particularly useful for analytic solutions of the intermediate-energy nucleon transport equation. The differential spectra of secondary nucleons in the laboratory frame can be represented as a sum of the quasifree and cascade components:

$$d^2N_{ij}/dE\,d\Omega = d^2N_{ij}^q/dE\,d\Omega + d^2N_{ij}^c/dE\,d\Omega\,, \quad (3.33)$$

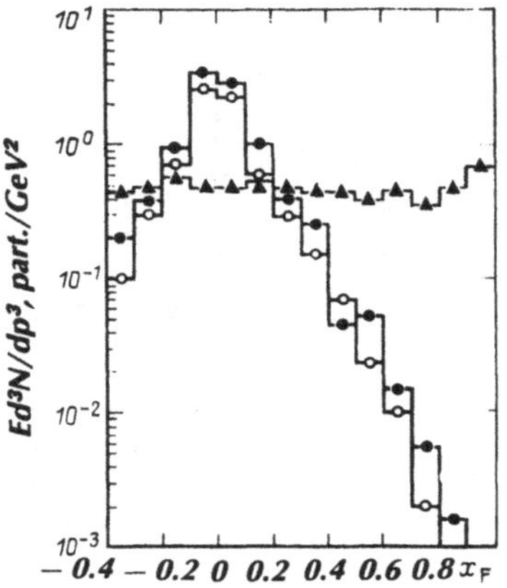


**Figure 3.23. Result of simulation of the inclusive hadron distributions in the reaction $pp \to cX$ at $E_0 = 3000$ GeV (the transverse momentum is in the range $0 < p_\perp \leqslant 0.2$ GeV/$c$; ▲, $p$; ●, $\pi^+$; 0, $\pi^-$).**

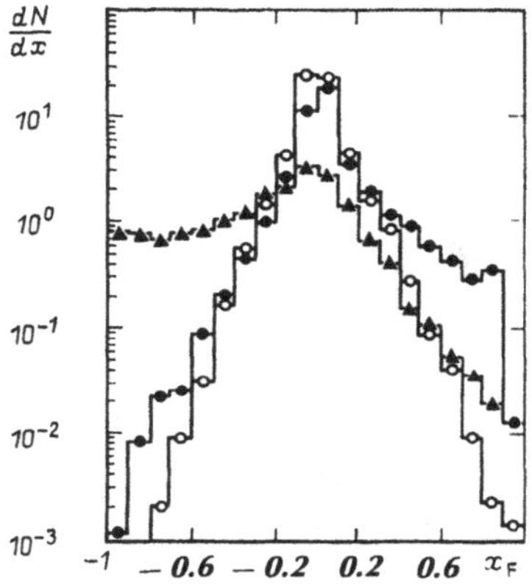


**Figure 3.24. The $p_\perp$-integrated inclusive distribution of protons (▲), $\pi^+$ mesons (●), and $\pi^-$ mesons (○) in the reaction $\pi^+ p \to cX$ at $E_0 = 3000$ GeV.**

where

$$\left.\begin{aligned} &\frac{d^2N^q_{ij}}{dEd\Omega}(E_0,E,\mu_s) = n_{ij}(E_0')\,\frac{\xi(E_0')+1}{2\pi E_0'}\left(\frac{E}{E_0'}\right)^{\xi(E_0')} \\ &\times \delta\left(\mu_s - \sqrt{\frac{E}{E_0}}\right); \\ &\frac{d^2N^c_{ij}}{dEd\Omega}(E_0,E,\mu_s) = \frac{\alpha}{4\pi}\left[\left(\frac{E}{E_0}\right)^{-\tau} - \left(\frac{E}{E_0}\right)^{\omega-\tau}\right](1+3\eta\mu_s); \end{aligned}\right\} \quad (3.34)$$

$\mu_s = \cos\theta$ is the cosine of the scattering angle in the laboratory frame. Here $n_{ij}$, $\xi$, $\alpha$, $\tau$, $\omega$, and $\eta$ are the parameters found by approximating the results of calculation of the nucleon–nucleus interactions on the basis of the intranuclear-cascade model.[83] Table 3.9 gives the values of the parameters $n_{nn}$ and $\xi$ for certain nuclei. For other reactions $n_{ij}$ are determined as follows:

**Table 3.9. Values of the parameters in Eq. (3.34).**

| | O | | Al | | Cu | | Pb | |
|---|---|---|---|---|---|---|---|---|
| $E_0$, GeV | $n_{nn}$ | $\xi$ | $n_{nn}$ | $\xi$ | $n_{nn}$ | $\xi$ | $n_{nn}$ | $\xi$ |
| 0.05 | 0.68 | 0.01 | 0.61 | 0.01 | 0.53 | 0.01 | 0.40 | 0.01 |
| 0.1 | 0.78 | 0.02 | 0.69 | 0.02 | 0.56 | 0.02 | 0.42 | 0.02 |
| 0.2 | 0.78 | 0.20 | 0.79 | 0.02 | 0.63 | 0.02 | 0.46 | 0.02 |
| 0.3 | 0.80 | 0.20 | 0.72 | 0.20 | 0.56 | 0.20 | 0.44 | 0.05 |
| 0.4 | 0.66 | 0.40 | 0.59 | 0.40 | 0.52 | 0.20 | 0.31 | 0.40 |
| 0.5 | 0.47 | 1.00 | 0.41 | 1.00 | 0.34 | 0.80 | 0.23 | 0.80 |
| 0.6 | 0.33 | 1.60 | 0.30 | 1.60 | 0.24 | 1.40 | 0.16 | 1.40 |

$$\left.\begin{aligned} n_{pn} &= [(A-Z)/(2A-Z)]n_{nn};\\ n_{np} &= [Z/(2A-Z)]n_{nn};\\ n_{pp} &= [(A+Z)/(2A-Z)]n_{nn},\end{aligned}\right\} \tag{3.35}$$

where $A$ and $Z$ are the atomic mass and atomic number, respectively.

The parameters of the second equation in (3.34) are virtually independent of $Z$ for $13 \lesssim Z \lesssim 82$ and incident-nucleon energies in the interval $0.1 \lesssim E_0 \lesssim 1$ GeV. These parameters are $\alpha = 0.3$ MeV$^1$, $\omega = 0.01$, $\tau \simeq 0.2$, and $\eta = 0.5$.

The inclusive hadron spectra in various reactions in the primary-hadron kinetic energy range $0.02 \lesssim E_0 \lesssim 5$ GeV can be described in greater detail by using the equation[81]

$$\frac{d^2N_{ij}}{dEd\Omega}(A, E_0, E, \theta) = F_q + F_L + F_c, \quad A > 1, \tag{3.36}$$

where the terms on the right side describe the angular and energy distributions of hadrons produced as a result of quasielastic scattering of the leading and cascade hadrons, respectively.

$$F_q = \frac{\eta^q_{ij} N_q(E_0, A)}{2\pi^{3/2}\delta_q(E_0, \theta)} f_q(\theta) \exp\left[-\frac{(E - E_q(E_0, \theta))^2}{\delta_q^2(E_0, \theta)}\right]; \tag{3.37}$$

$$F_L = \frac{\eta^L_{ij} N_L(E_0, A)}{2\pi} f_L(E, \theta) \frac{dN_L}{dE}(E_0, E); \tag{3.38}$$

$$F_c = \frac{\eta^c_{ij}}{2\pi} f^c_{ij}(E, \theta) \frac{dN_c}{dE}(E_0, E, A). \tag{3.39}$$

Here we present the parameters and functions contained in these expressions, making use, as in Ref. 81, of the kinetic energies $E_0$ and $E$ (MeV) and angles $\theta$ (rad) in the laboratory frame.

For Eq. (3.37):

$$\eta^q_{ij} = \begin{cases} 0.8, & \text{if } i=j (i,j=p,n); \\ 0.2, & \text{if } i \neq j; \end{cases}$$

$$N_q(E_0, A) = 1.17[1 - 0.8\exp(-g_1(E_0))]\{1 - \exp[-(E_0/350)^{1.5}]\}$$
$$\times \exp(-0.08\sqrt{A-1});$$

$$f_q(\theta) = a_q(E_0)\exp[-0.5a_q(E_0)\theta^2];$$

$$E_q(E_0,\theta) = E_0\cos^2\theta/[1 + E_0\sin^2\theta/(2m_N)] - 0.25;$$

$$g_1(E_0) = 4\times 10^{10}E_0^{-4} + 33.9E_0^{-0.65};$$

$$\delta_q(E_0,\theta) = 25(1 + 0.008E_0\theta);$$

$$a_q(E_0) = 4[0.5 + 0.01E_0^2/(2000 + E_0)](1 + E_0/2m_N),$$

where $m_N$ is the rest mass of the nucleon (MeV).

For Eq. (3.38):

$$\eta^L_{ij} = \begin{cases} \frac{2}{3} \text{ if } i=j; \\ \frac{1}{3} \text{ if the charge state of the secondary hadron differs from the charge state of the primary hadron by unity;} \\ 0 \text{ in the other cases;} \end{cases}$$

$$N_L(E_0, A) = 28.1\exp[-g_1(E_0)](A + 69)^{-1};$$

$$f_L(E,\theta) = g_2(u,k_4)\exp[-u(\theta + k_4\theta^2)];$$

$$\frac{dN_L}{dE}(E_0, E) = (1 - E/E_0)/E_0, \quad i=j; \quad 2/3E_0, \quad i \neq j;$$

$$g_2(u,k_4) = (1 + u^2)[1 + 5.2k_4(2 + u^2)^{-1}]/[1 + \exp(-u\pi)];$$

$$u = E/\tau^2; \quad \tau_2 = 200E_0/(E_0 + 2600);$$

$$k_4 = 1.21E_0[(2000 + E_0)\sqrt{1 + \ln A}]^{-1};$$

For Eq. (3.39):

$$f^c_{ij}(E,\theta) = q_c(E_0)\exp\left\{-\left[\frac{2000}{\tau_3 k_7}(\sqrt{1 + k_7E\theta\times 10^{-3}} - 1) + k_4u_3\theta^2\right]\right\};$$

$$\frac{dN_c}{dE}(E_0, E, A) = \frac{a_c(E_0,A)}{E + \delta_2}\left(1 - \frac{E}{\varepsilon_{ij\max}}\right)^{\gamma}\frac{E_0}{\varepsilon_{ij\max}};$$

$$a_c(E_0, A) = \frac{\varepsilon_c k_6}{E_0\{(\gamma + 1)^{-1} - \delta_2[E_0^{-1}\ln(1 + E_0/\delta_2(\gamma + 1))]\}};$$

$\varepsilon_c = \varepsilon - \varepsilon_q - \varepsilon_L$ is the kinetic energy carried off by the cascade particles; $\varepsilon_q$ and $\varepsilon_L$ are the kinetic energies carried off by the quasielastic-scattering nucleons and the leading particles, respectively; $\varepsilon_{ij\,\max}$ is the maximum allowable kinetic energy of the particles of type $j$; $\varepsilon = E_0 - E^*(E_0, A) - [N_N(E_0, A) - 1]E_c(A) - m_\pi N_\pi(E_0, A) - E_z(E_0, A)$; $E^*$, $E_c$, and $E_z$ are the average energies of excitation of the nucleus, of the detachment of a nucleon from the nucleus, and of the disintegration of the nucleus, respectively;

$$\gamma(E_0, A) = 3(0.001E_0)^{0.06}[1 - \exp(-k_2E_0)]; \quad k_2 = 5\times10^{-4}(1 + A^{1/3});$$

$$k_6 = [1 + \delta_2(\gamma + 1)/3E_0(\delta_2/E_0 + 1.75)]^{-1};$$

$$\delta_2 = 250A^{-1/2}[1 + 2.5E_0(1000 + E_0)^{-1}\exp(-0.02\,A)];$$

$$N_N(E_0, A) = (1 + \sqrt{A})\exp\{-\xi(A)\exp[-\lambda(A)u]\};$$

$$u = \ln(E_0/3.68); \quad \xi(A) = 0.087A^{2/3} + 4.15; \quad \lambda(A) = 0.72(1 + \ln A)^{-0.4};$$

$$N_\pi(E_0, A) = N_{\pi M}\exp(0.075\sqrt{A-1})g(E_0, A);$$

$$N_{\pi M} = 0.5335(E_0^2 + 2m_N E_0)^{0.125} - 2;$$

$$g(E_0, A) = \exp[-0.25\times10^6 E_0^{-2} - 0.7(A-1)^{1/4}/(1 + 10^{-3}E_0)];$$

$$q_c(E) = \frac{(1 + u_3^2)[1 + 5.2k_4u_3(2 + u_3^2)^{-1}]}{1 + \exp(-u_3\pi)};$$

$$u_3 = E/\tau_3; \quad \tau_3 = 200E_0(E_0 + 560)^{-1};$$

$$k_4 = \begin{cases} 1.21E_0(2000 + E_0)^{-1}(1 + \ln A)^{-0.5}, & j = n,p; \\ 0.3(E_0 - 1000)(E_0 + 1000)^{-1}, & j = \pi; \end{cases}$$

$$k_7 = 0.01\sqrt{E_0\times10^{-3}}(1 + 0.25\ln A).$$

The fraction $\eta_{ij}^c$ of the cascade particles [for the particular case of the primary neutrons in Eq. (3.39)] is determined from the relations

$$N_n/N_p = [1 + \exp(-E_0/2000)](A + 1 - Z)Z^{-1};$$

$$N_{\pi-} = \left(\frac{1}{3}N_{\pi M} + \frac{3 - a_n}{b_n}\right)\frac{N_\pi(E_0, A)}{N_{\pi M}};$$

$$N_{\pi^0} = \left(\frac{1}{3}N_{\pi M} + \frac{2a_n - 1}{b_n}\right)\frac{N_\pi(E_0, A)}{N_{\pi M}};$$

$$N_{\pi+} = \left(\frac{1}{3}N_{\pi M} - \frac{2 + a_n}{b_n}\right)\frac{N_\pi(E_0, A)}{N_{\pi M}}; \quad b_n = 6(1 + a_n);$$

$$a_n = Z(A - Z)^{-1}[1 - 0.5\exp(-10^{-9}E_0^3)].$$

The distribution of evaporative particles produced as a result of the decay of the excited states of the nuclei is isotropic in the angles. Taking into account the Coulomb potential barrier as a function of the temperature $T_0$ of the excited nucleus, the energy spectra of nucleons can be approximately described by

$$dN_j/dE = N_j(\widetilde{E}/T^2)\exp(-\widetilde{E}/T), \tag{3.40}$$

where $N_j$ is the yield of the evaporative particles of type $j$,

$$\tilde{E} = \begin{cases} E, & j = n; \\ E - V_0/(1 + 0.15T), & j = p, \end{cases}$$

$T = 0.4T_0$, $T_0 = (10E^*/A)^{1/2}$, $V_0 = 1.115A^{-1/3}$, and $E^*$ is the excitation energy of the residual nucleus (see Sec. 6.3).

The total and differential cross sections (mb) of the high-energy elastic and quasi-elastic scattering can be described adequately[85] by the equations for $A \geqslant 9$ and $E_0 \gtrsim 5$ GeV:

$$\left.\begin{aligned} \sigma_{el}(A) &= 6.38A^{1.04}; \\ \sigma_{el}(A)/\sigma_{abs}(A) &= 0.13A^{0.37}; \end{aligned}\right\} \tag{3.41}$$

$$\frac{d\sigma_{el}}{d\Omega} = 12.5A^{1.63}\exp(-14.5A^{0.66}|t|) + 17.5A^{0.33}\exp(-10|t|), \quad A < 62; \tag{3.42}$$

$$\frac{d\sigma_{el}}{d\Omega} = 50A^{1.33}\exp(-60A^{0.33}|t|) + 20A^{0.40}\exp(-10|t|), \quad A \geqslant 62. \tag{3.43}$$

Here $t = -2p^2(1 - \cos\theta) \approx -p^2\theta^2$, and $p$ is the particle momentum (GeV/$c$).

In conclusion, since the phase space $d^3p/E$ is invariant, we can write out the relationship between the invariant cross sections of inclusive processes for different choices of the independent variables:

$$E\frac{d\sigma}{d^3p} = \frac{x'}{\pi}\frac{d\sigma}{dx\,dp_\perp^2} = \frac{E}{p^2}\frac{d\sigma}{dp\,d\Omega} = \frac{1}{p}\frac{d\sigma}{dE\,d\Omega}.$$

# Chapter 4
# Electron-photon showers

## 1. Qualitative features of the development of electron-photon showers

In Chap. 2 we considered the principal elementary interactions of electrons (positrons) and photons with isolated atoms, with the atomic nucleus, and with atomic electrons. In this chapter we study the passage of high-energy electrons and photons through condensed matter (everywhere below, unless otherwise specified, we understand electron to also mean a positron). Such a process is comprised of many elementary processes. Let us first analyze qualitatively the particular case of the passage of a high-energy electron ($E_e \gg m_e c^2$) through a plane layer of matter of thickness measuring several radiation lengths. At low electron energies the principal mechanism of the energy loss of an electron is the ionization and excitation of the atoms of the medium (Fig. 4.1). As the energy of the electron is increased, its relative ionization energy loss decreases and the energy loss due to radiation begins to play an ever increasing role. At an electron energy equal to the critical energy $\epsilon_{\rm crit}$ the energy loss due to radiation is comparable to the ionization loss. At $E_e \gg \epsilon_{\rm crit}$ the energy loss due to radiation becomes dominant. The critical energy $\epsilon_{\rm crit}$ depends on the material (Table 4.1). To estimate $\epsilon_{\rm crit}$ (MeV), we can use an approximate equation[86] which is valid within 10% for materials with atomic number $13 \leqslant Z \leqslant 92$:

$$\epsilon_{\rm crit} \simeq 550/Z. \tag{4.1}$$

As a result of passage through a layer of material of thickness corresponding to one radiation length $t_r$, the electron energy decreases by a factor of "$e$", on the average, as a result of bremsstrahlung (see Sec. 2.5). The radiation lengths for some materials are given in Table 4.1. The approximate equation[86]

$$t_r \simeq 180A/Z^2, \tag{4.2}$$

which is valid within 20% for $13 \leqslant Z \leqslant 92$, can be used to estimate the value of $t_r$ (g/cm$^2$).

The range of an electron, before it interacts with atoms of the material as a result of which the electron emits a bremsstrahlung photon, can be estimated from Eq.(2.10) for the differential cross section of electron bremsstrahlung. This equation can be written in the form

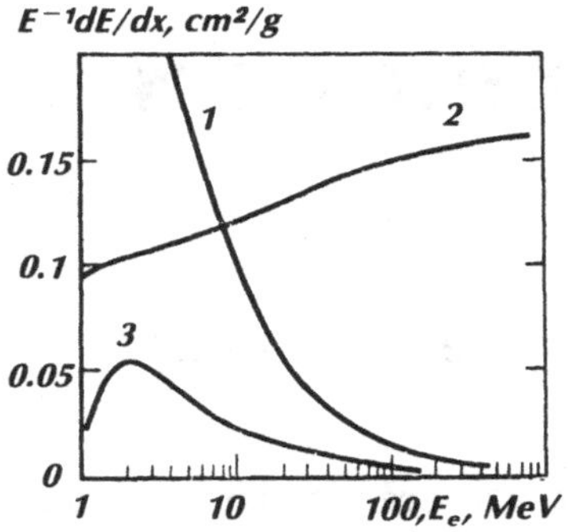


**Figure 4.1.** The relative energy loss of an electron per radiation length in lead (Ref. 31) due to the ionization and excitation of atoms (1), due to bremsstrahlung (2), and due to the production of δ electrons (3). [The inelastic scattering of electrons by atomic electrons is included in the ionization loss (1) if the energy loss in each event is no greater than 0.255 MeV; otherwise, this process is viewed as the Møller scattering (3)].

**Table 4.1.** Values of the radiative width $t_r$, of the critical energy $\epsilon_{crit}$, of the Møller unit of length $r_M$ (see the discussion below), and of the average ionization energy loss $dE/dx$ at the ionization minimum for certain materials.

| Material | $Z$ | $A$ | $\rho$, g/cm³ | $dE/dx$ MeV·cm²/g | $t_r$ g/cm³ | $t_r$ cm | $\epsilon_{crit}$, MeV | $r_M$, cm |
|---|---|---|---|---|---|---|---|---|
| C | 6 | 12.0 | 1.55 | 1.78 | 42.7 | 27.5 | 75.9 | 7.68 |
| Al | 13 | 27.0 | 2.70 | 1.62 | 24.0 | 8.9 | 39.3 | 4.76 |
| Ar | 18 | 40.0 | 1.40 | 1.51 | 19.55 | 14.0 | 29.8 | 9.96 |
| Fe | 26 | 55.9 | 7.87 | 1.48 | 13.84 | 1.76 | 20.5 | 1.82 |
| Cu | 29 | 63.5 | 8.96 | 1.44 | 12.86 | 1.43 | 18.7 | 1.62 |
| Sn | 50 | 118.7 | 7.31 | 1.28 | 8.82 | 1.21 | 11.4 | 2.25 |
| W | 74 | 183.9 | 19.32 | 1.17 | 6.76 | 0.35 | 7.9 | 0.93 |
| Pb | 82 | 207.2 | 11.35 | 1.13 | 6.37 | 0.56 | 7.2 | 1.65 |
| U | 92 | 238.0 | 18.95 | 1.09 | 6.00 | 0.32 | 6.6 | 1.03 |
| Water | | | 1.00 | 2.03 | 36.1 | 36.1 | 73.0 | 10.5 |
| Plastic scintillator | | | 1.032 | 1.97 | 43.8 | 42.9 | 87.1 | 10.4 |
| Nuclear emulsion | | | 3.815 | 1.44 | 11.02 | 2.94 | 16.4 | 3.8 |
| SF-5 lead glass | | | 4.08 | — | 10.4 | 2.54 | 15.8 | 3.41 |
| NaI | | | 3.67 | 1.32 | 9.49 | 2.59 | 12.5 | 4.39 |

$$d\Sigma_{re}(E,\omega) = (d\omega/\omega)\,(1 + w_0^2/u^2)^{-1}F_e(E,u), \tag{4.3}$$

where $d\Sigma_{re}$ is the macroscopic differential cross section expressed in units of radiation length; $u = \omega/E$; $\omega$ and $E$ are the energy of the bremsstrahlung photon and electron, respectively; $F_e$ $(E,u)$ is a function which depends only slightly on $E$ and $u$ (see Fig. 2.1), and $w_0 = \lambda_e\sqrt{n_e r_e/\pi}$ ($n_e$ is the electron density of the material, $r_e$ is the classical electron radius, and $\lambda_e$ is the Compton electron wavelength). The factor $F_p = (1 + w_0^2/u^2)^{-1}$ takes into account the polarization of the medium. Since $F_e \simeq 1$ with a good degree of accuracy, the total macroscopic bremsstrahlung cross section can easily be estimated:

$$\Sigma_{re}(E) \simeq \int_0^1 \frac{u\,du}{(u^2 + w_0^2)} \simeq \ln(w_0^{-1}), \quad w_0^2 \ll 1. \tag{4.4}$$

The total macroscopic bremsstrahlung cross section of a high-energy electron thus does not depend on its energy and is determined exclusively by the properties of the material. The mean-free-path of an electron before it emits a bremsstrahlung photon is determined (in units of radiation length) by the equation

$$\lambda_{re} \approx 1/\Sigma_{re} \simeq 1/\ln(w_0^{-1}). \tag{4.5}$$

For lead, for example, $\lambda_{re} \simeq t_r/9$. It follows from Eq. (4.5) that as a result of passage through a layer of the material of thickness corresponding to one radiation length, the electron emits $\ln(w_0^{-1})$ bremsstrahlung photons on the average. Since the electron energy in this case decreases by a factor of $e$, the average energy of each emitted photon is estimated to be

$$\langle \omega_\gamma \rangle \simeq 0.6 E_e / \ln(w_0^{-1}). \tag{4.6}$$

For lead $\langle \omega_\gamma \rangle \simeq E_e/15$. Since the bremsstrahlung photons are characterized by a very mildly sloping energy spectrum ($d\sigma/d\omega \sim \omega^{-1}$), the energy of the emitted photon is subjected to strong fluctuations and may differ appreciably from the mean value given by Eq. (4.6). During bremsstrahlung the probability for the emission of a high-energy photon (with an energy $\omega \simeq E_e/2$, for example) is relatively high. The ultrarelativistic electron that emits this photon, despite losing considerable energy, remains nonetheless a high-energy electron capable of freely emitting high-energy bremsstrahlung photons.

Interacting with the atoms of the medium, the emitted photon can either be absorbed as a result of the photoelectric effect or undergo a Compton scattering by the atomic electrons, or produce an electron–positron pair in the Coulomb field of the nucleus or atomic electrons (Fig. 4.2). The photoelectric effect plays an insignificant role at a photon energy $\omega_\gamma \gtrsim 1$ MeV. The lighter the substance, the smaller is the cross section of the photoelectric effect (the smaller is the dependence on the number of electrons in the atomic shell $Z^5$) in comparison with the Compton scattering cross sections ($Z$) and the cross section for the production of $e^+e^-$ pairs [$\sim Z(Z+1)$]. At a low photon energy the most probable interaction of the photon with matter is the Compton scattering by atomic electrons. The probability for the production of electron–positron pairs increases progressively with increasing photon energy and when the photon energy exceeds the value $2m_ec^2$ by several factors, this process becomes dominant. At photon energies $\omega_\gamma \gtrsim 1$ GeV the macroscopic cross section for the production of $e^+e^-$ pairs is virtually independent of the energy. At these energies the mean free path of a photon before the interaction, which produces the electron–positron pair, is

$$\lambda_{p\gamma} \simeq 9t_r/7. \tag{4.7}$$

Each produced electron and positron carries off one-half the photon energy on the average and each one can in turn emit, upon the interaction with the atoms of the medium, bremsstrahlung photons which subsequently produce

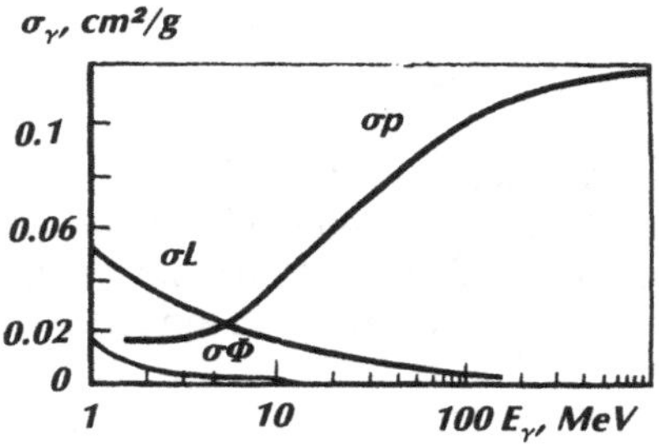


**Figure 4.2.** Macroscopic cross section for the interaction of photons with lead (Ref. 31).

electron–positron pairs, and so on. The multiplication of photons, electrons, and positrons occurs in an avalanchelike manner but not boundlessly. With an increase in the number of secondary particles, the average energy of each particle decreases. When the energy of the electron is lower than the critical energy, the electron begins to slow down, principally due to the ionization energy loss. The probability for Compton scattering increases with decreasing photon energy. The Compton electrons produced in this process, with an energy lower than the critical energy, do not participate in the multiplication process. Accordingly, when the energy of each of the secondary particles that forms an avalanche reaches a sufficiently low value (for lead, for example, $E_e \leqslant \epsilon_{\text{crit}} \simeq 7$ MeV and $\omega_\gamma \lesssim 5$ MeV), further particle multiplication in the avalanche ceases, and it begins to decay. The process which we described here is called *electron–photon shower*.

In summary, the longitudinal development of the electron–photon shower (i.e., the development of a cascade in the direction of motion of the primary particle) can be qualitatively characterized as follows. At first, the number of particles in the cascade increases rapidly with increasing depth of its penetration into the medium (section I of the cascade curve in Fig. 4.3). The average energy of the particle decreases rapidly in this case. In the region where the shower peaks (when the average number of particles in the shower reaches a maximum), the average energy of each charged particle is approximately equal to the critical energy for the given material. As the maximum is reached, the number of secondary particles in the shower begins to slowly decrease (section II of the cascade curve). This "tail" of the shower consists primarily of photons (and Compton electrons produced by these photons) with an energy which corresponds roughly to the total macroscopic cross section for the interaction of a photon with the material (for lead this energy is equal to several megaelectron volts). The Compton-scattered photons are thus the dominant photons when the shower is developed at a considerable depth.

In the case of a shallow penetration of the electron–photon shower into the material (up to the point at which the shower peaks) when there is a strong rapid multiplication of particles due to the production of $e^+e^-$ pairs by photons and due to the electron bremsstrahlung, the development of a shower in the transverse direction is attributable primarily to the multiple Coulomb scattering of electrons (positrons). In fact, the average angle of divergence

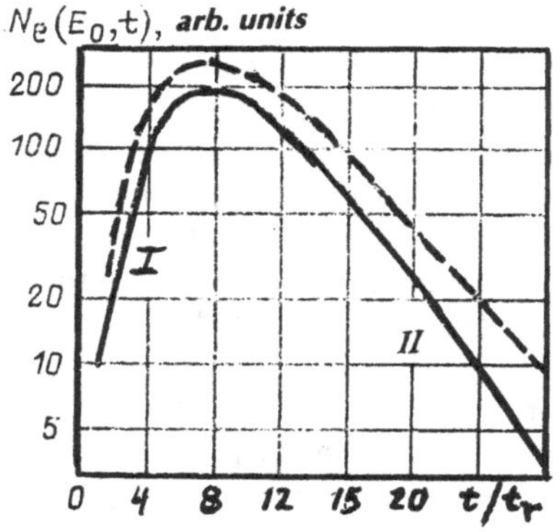


**Figure 4.3.** Experimental measurement of the average number of charged particles in the EPS initiated in lead by an electron with an energy $E_0 = 50$ GeV versus its penetration depth $t$ (Ref. 86). Solid curve, the level of possible fluctuations corresponding to one standard deviation from the mean value.

between the bremsstrahlung photon and the electron which has emitted this photon is[21]

$$\langle\theta_{e\gamma}\rangle_r \simeq (8m_ec^2/E_0)\ln(E_0/m_ec^2). \tag{4.8}$$

where $E_0$ is the energy of the primary electron. The average angle at which the electrons and positrons are emitted during the production of $e^-e^+$ pairs by a photon is[21]

$$\langle\theta_{e\gamma}\rangle_p \simeq (m_ec^2/E)\ln(E/m_ec^2),$$

where $E$ is the energy of the secondary electron (positron). The mean-square angle of the multiple Coulomb scattering of the relativistic electron in the material of thickness $t$ is given by Eq. (2.39)

$$\langle\theta_e\rangle_c = \frac{E_s}{E}\sqrt{t/t_r},$$

where $E_s \simeq 42m_ec^2$, and $E$ is the electron energy.

The mean-free path of the photon before the interaction, in which the electron–positron pair is produced, is $9t_r/7$ [see Eq. (4.7)]. The mean-free path of the electron before the interaction, in which a bremsstrahlung photon with an energy $\omega_\gamma > \omega_c$ is emitted, is described by the equation

$$\lambda_{re}(\omega > \omega_c) = 1/\Sigma_{re}(E,\omega > \omega_c),$$

where

$$\Sigma_{re}(E,\omega > \omega_c) \simeq \int_{u_c}^{1} \frac{u\,du}{(u^2 + w_0^2)}.$$

Here $E$, $u$, and $\omega_0$ are the same as in Eqs. (4.3) and (4.4), and $u_c = \omega_c/E$. For $u_c \gg \omega_0$ the mean-free path is $\lambda_{re}(\omega > \omega_c) = t_r/\ln(E/\omega_c)$.

The processes leading to the multiplication of particles are assumed to be those bremsstrahlung processes in which the emitted photon carries off an appreciable part of the energy of the primary electron ($\omega_c \gtrsim \langle\omega_\gamma\rangle \simeq E/15$, for example, for lead). Then $\lambda_{re}(\omega \gtrsim \langle\omega_\gamma\rangle) \simeq 0.4t_r$ and $\langle\theta_{e\gamma}\rangle_r/\langle\theta_e\rangle_c \simeq \ln(E/m_ec^2)/5$; $\langle\theta_{e\gamma}\rangle_p/\langle\theta_e\rangle_c \simeq \ln(E/m_ec^2)/30$ for lead. These estimates show that multiple Coulomb scattering plays a dominant role in the transverse development of a shower when the average energy of a single secondary charged shower particle is not particularly large. In the beginning of the electron–photon shower, on the other hand, when the energy of secondary particles is still large, the development of a shower in the transverse direction is very slight, since the absolute values of the angles $\langle\theta_{e\gamma}\rangle_r$ and $\langle\theta_{e\gamma}\rangle_p$ are very small ($\langle\theta_{e\gamma}\rangle_r \simeq 130$ and $\langle\theta_{e\gamma}\rangle \simeq 1/200$, for example, at $E = 1$ GeV). An electron with an initial energy $E_0 = 1$ GeV, after passing through a layer of the material of thickness corresponding to one radiation length, will have an energy $E = 0.3$ GeV, on the average, and the angle through which it scatters multiply along a path $t = 0.4t_r$ will be $\langle\theta_e\rangle_c \simeq \frac{1}{20}$. This example shows that multiple Coulomb scattering begins to play the dominant role in the transverse development of the electron–photon shower at a cascade-penetration depth of 1–2 radiation lengths, since the average energy of a single secondary shower particle decreases exponentially with increasing depth of its penetration.

In the case of deep penetration (at the shower tail) and at a considerable distance from the longitudinal electron–photon–shower axis the transverse development of the cascade is attributable primarily to the Compton scattering of low-energy photons.

The differential macroscopic cross sections and the total cross sections for the production of $e^-e^+$ pairs by a photon and for the electron bremsstrahlung depend only slightly on the energy of the primary particle (at $E \gtrsim 1$ GeV). Expressed in units of the radiation length [see Eqs. (2.8) and (2.14)], these cross sections are essentially the same for various substances, regardless of the atomic mass and atomic number of the substance. The penetration depth of the electron-photon shower can therefore be expressed in units of the radiation length $t_r$.

A rapid multiplication of particles in the electron–photon shower continues until the average energy of a single secondary particle becomes approximately equal to the critical energy $\epsilon_{\text{crit}}$. At a lower than critical energy the electrons slow down primarily as a result of the loss of energy due to ionization, and the multiplication of secondary particles ceases. Assuming, for simplicity, that during the electron bremsstrahlung the emitted photon carries off half the electron energy (the average energy of the bremsstrahlung photon is, in fact, much lower) and that during the production of the electron–positron pairs by the photon each secondary particle also carries off energy equal to half the photon energy, the number of generations of secondary particles in the cascade is estimated to be $K \sim \log_2(E_0/\epsilon_{\text{crit}})$, where $E_0$ is the energy of the primary particle. It follows from this estimate that the number of secondary particles $N$ in the shower is $N \sim 2^K$ or $N \sim E_0/\epsilon_{\text{crit}}$.

At the initial stage of development of the electron–photon shower (beginning at a depth $t \gtrsim 1\text{–}2t_r$), and when the shower peaks, the transverse development of the shower is determined by the multiple Coulomb scattering of secondary electrons. At the maximum of the electron–photon shower the average electron energy is approximately equal to the critical energy $\epsilon_{\text{crit}}$ for the given substance. The mean-square scattering angle of these electrons, after they have passed through a layer of the material of thickness corresponding to one radiation length is $\sqrt{\langle\theta_e^2\rangle_c} \sim E_s/\epsilon_{\text{crit}}$, and the mean-square displacement from the shower axis in order of magnitude is estimated to be $r_M \sim t_r E_s/\epsilon_{\text{crit}}$. The quantity $r_M$ is called the *Molière unit length*. For $13 \leqslant Z \leqslant 92$ the Molière unit within 10% is assumed to be[86]

$$r_M \simeq 7A/Z. \tag{4.9}$$

The values of $r_M$ for certain substances are given in Fig. 4.1. The Molière unit length physically characterizes the development of the electron–photon shower in the transverse direction. This shower development is nearly independent of the substance if the distance from the shower axis is expressed in Molière units $r_M$.

The exponential damping of the electron–photon shower at a large penetration depth into the substance [$\sim\exp(-t/\Lambda)$] is characterized by the *attenuation length* $\Lambda$, which is virtually independent of the primary-particle energy (see Sec. 4.5). The attenuation length $\Lambda$ is approximately equal to the range of the most penetrating photons before their interaction: $\lambda_{\max} = 1/\mu_{\min}$, where $\mu_{\min}$ is the minimum absorption coefficient for the absorption of photons in the material. The photon absorption coefficient is equal to the sum of the macroscopic cross sections of the fundamental interaction of the photon with the material (photoelectric effect, Compton scattering, pair production): $\mu = \Sigma_\Phi(\omega) + \Sigma_K(\omega) + \Sigma_p(\omega)$ (Ref. 87). Expressed in units of radiation length, the quantity $\Lambda$ depends only slightly on the material and is approximately equal to $2 - 3t_r$.

The development of the electron–photon shower in a substance is thus determined by three principal parameters: the radiation length $t_r$, the critical energy $\epsilon_{\text{crit}}$, and the Molière unit of length $r_M$. For definiteness, let us consider the development of an electron–photon shower in a light substance (with a small $Z$) and a heavy substance (with a large $Z$) which is initiated by a particle (an electron or a photon) with an energy $E_0(\omega_0)$. The critical energy of a light substance [see Eq. (4.1)] is higher than that of a heavy substance ($\epsilon_{\text{crit}}^{\text{Al}}/\epsilon_{\text{crit}}^{\text{Pb}} \simeq 6$, for example). Accordingly, in a shower which develops in a light substance the number of secondary particles is $\epsilon_{\text{crit}}^{l}/\epsilon_{\text{crit}}^{h}$ times smaller than in a heavy substance if the energy of the primary particles which initiate the electron–photon shower is the same. The spatial distributions of secondary particles in the cascades can also be distinguished in light and heavy substances. In light substances the electron–photon shower is more drawn out in the direction of motion of the primary particle than it is in heavy substances. The contours in Fig. 4.4 show the effective regions of development of the electron–photon shower in aluminum and lead at the same energy of the primary particle. Since the longitudinal development of the shower is determined by the radiation length of the substance, the electron–photon-shower length (in units of $t_r$) for light substances is approximately the same as that for heavy substances. The development of electron–phonon shower (EPS) in the transverse direction is characterized by the Molière unit $r_M$ which, being

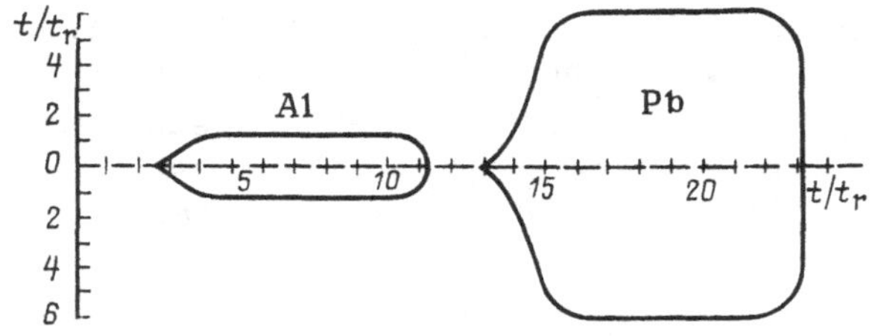


**Figure 4.4. Outlines of regions in which 90% of the energy of the EPS developed in aluminum and lead is released (Ref. 86).**

expressed in units of radiation length, is much smaller for light substances than for heavy substances. For aluminum, for example, $r_{\rm M}^{\rm Al} \simeq 0.5 t_r^{\rm Al}$ and for lead $r_{\rm M}^{\rm Pb} \simeq 3 t_r^{\rm Pb}$. The effective region of development of a shower in lead is therefore approximately six times broader than that in aluminum.

## 2. Principal quantitative characteristics of the electromagnetic cascade. Definitions

The study of the development of an electron–photon shower is of considerable practical interest. It is sufficient to mention such fields of application as the design and construction of electromagnetic calorimeters, which are generally used in experimental high-energy physics, and also the solution of a series of radiation problems with the use of high-energy electron and proton accelerators. In this section we consider the most important characteristics of the electron–photon shower which are of practical interest.

One of the important characteristics is the spatial distribution of secondary particles in the EPS. This distribution is described by the function $P_i^j\ (n, E_0, E_c, r, t)$, which is the probability density that $n$ particles of type $i$ with an energy $E$ higher than $E_c$ can be seen at a depth $t$ and at a distance $r$ from the shower axis in the EPS which is initiated by a particle of type $j$ with a primary energy $E_0 (i, j = \gamma, e^-, e^+)$. The superscript $j$ will be discarded below, keeping in mind that a particular type of particle which initiates a cascade is always considered in specific cases. If the distribution density $P_i$ is known, we can find the energy and spatial distribution density of the average number of particles of type $i$

$$N_{ir}(E_0, E_c, r, t) = \sum_{n=0}^{\infty} n P_i(n, E_0, E_c, r, t). \qquad (4.10)$$

In actual situations it is often important to know how a particular characteristic of the EPS fluctuates near its mean value. In the case of the spatial distribution of secondary particles in the EPS which we are considering, the problem of fluctuations can be solved if the function $P_i\ (n, E_0, E_c, r, t)$ is known. In some cases it is sufficient to know only the possible fluctuations which are characterized by the dispersion

$$\sigma_i^2 = \sum_{n=0}^{\infty} n^2 P_i(n) - N_i^2. \tag{4.11}$$

By analogy with (4.10), we can introduce the density function for particles of type $i$, whose distribution of motion forms an angle $\theta$ with the shower axis (with the direction of motion of the primary particle)

$$N_{i\theta}(E_0, E_c, \theta, t) = \sum_{n=0}^{\infty} nP_i(n, E_0, E_c, \theta, t).$$

The dispersion of this quantity, which is a measure of its fluctuations, is given by Eq. (4.11).

Integrating the function $N_{ir}$ $(E_0, E_c, r, t)$ [or the function $N_{i\theta}$ $(E_0, E_c, \theta, t)$] over the radius $r$ (or the function $N_{i\theta}$ over the angle $\theta$), we find the distribution of the average number of particles of type $i$ with respect to the depth $t$ (the so-called *cascade curve*)

$$N_i(E_0, E_c, t) = \int_0^{\infty} dr N_{ir}(E_0, E_c, r, t).$$

Such quantities as $N_i$ $(E_0, E_c, t)$ at the maximum $N_i^{\max}$ and the position of this maximum in the depth $t$ $(t_{\max})$ are used in actual situations.

The spatial distribution of the absorbed shower energy is an important characteristic of the EPS. The energy of the primary particle which initiates the EPS dissipates during the development of the shower through the transfer to the secondary particles and finally is absorbed by the substance as a result of ionization and excitation of the atoms of the medium. Frequently considered is the energy liberated by the medium, rather than that absorbed by it, keeping in mind that the energy of the primary particle, which is converted into the energy of the $\delta$ electrons and the excitation energy of the atoms of the medium as a result of the development of the cascade, does not disappear without a trace. This energy can be measured under certain conditions. To describe the energy dissipation, we introduce the function $w(E_0,E,\mathbf{r},t)$, which characterizes the probability for the liberation of energy $E$ at the point with the coordinates $t$ and $\mathbf{r} = (\rho,\varphi)$, where $(\rho,\varphi)$ are the polar coordinates of this point in the plane perpendicular to the shower axis. We can then use the following equation to describe the spatial distribution density of the released energy:

$$W(E_0,\mathbf{r},t) = \int_0^{E_0} Ew\,(E_0,E,\mathbf{r},t)\,dE.$$

Here $W(E_0,\mathbf{r},t)$ is the average energy released per unit volume near the point $(\mathbf{r},t)$. The fluctuations of the energy evolution in space are determined by the function $w(E_0,E,\mathbf{r},t)$ and are characterized by the dispersion

$$\sigma_E^2 = \left[\int_0^{E_0} E^2 w\,(E)\,dE - W^2\right] \Big/ \int_0^{E_0} w\,(E)\,dE.$$

Of particular importance from the practical point of view are such EPS characteristics as the average total length of the charged-particle tracks $\langle L\rangle$ and the average number of crossings $N_{cr}$ of a particular number of planes by charged particles (these planes are generally separated the same distance from each other and are oriented perpendicular to the shower axis). The average total length $\langle L\rangle$ of the charged-particle tracks in the EPS is determined by means of the function $P_{ch}(L)$, which is the probability for observing in the shower the total track length equal to $L$: $\langle L\rangle = \int_0^\infty LP_{ch}(L)\,dL$. We also introduce the function $P_{cr}(m,n)$ which is the probability for observing $n$ crossings of the plane with the index $m$ by secondary charged particles in the EPS. The average number of crossings of the plane $m$ by the charged particles is

$$N_{cr}(m) = \sum_{n=0}^{\infty} nP_{cr}(m,n).$$

The fluctuations of this quantity are characterized by the dispersion

$$\sigma_{cr}^2(m) = \sum_{n=0}^{\infty} n^2 P_{cr}(m,n) - N_{cr}^2(m).$$

The average total number of crossings of $K$ planes is

$$N_{cr} = \sum_{n=0}^{\infty} n\Big(\sum_{m=1}^{K} P_{cr}(m,n)\Big) = \sum_{m=1}^{K} N_{cr}(m).$$

The fluctuations of $N_{cr}$ are characterized by the dispersion

$$\sigma_{cr}^2 = \sum_{n=0}^{\infty} n^2\Big(\sum_{m=1}^{K} P_{cr}(m,n)\Big) - N_{cr}^2.$$

The relative fluctuations of the average number of crossings, $\delta = \sigma_{cr}/N_{cr}$, decrease with increasing number of planes $K$. The distance between the planes is $d = l/K$, where $l$ is the thickness of the flat layer of the substance, in which most of the EPS energy (95%, for example) is released. If the distance between the planes is less than the mean-free path of an electron before the interaction, $d \lesssim \lambda_{re} = t_r/\ln(1/w_0)$ [$\lambda_{re} \simeq 0.1t_r$ for lead; see Eqs. (4.4) and (4.5)], then there will be a strong correlation between the number of crossings of the charged-particle tracks with the adjacent planes. Because of this circumstance, the number of transverse planes where the crossings are calculated should be no greater than $K_0 \simeq l/\lambda_{re}$, since at $K > K_0$ the relative fluctuation of the number of crossings $N_{cr}$ remains nearly constant as the number of planes $K$ is increased.

## 3. Basic results of the analytic solutions of the problem concerning the development of electromagnetic shower

The well-known analytic methods for the solution of the problem of the electron–photon shower were developed in Refs. 34 and 88–92. We shall therefore briefly consider only the main results obtained from the analytic solutions of the given problem.

An analytic approach to the problem makes use of two principal approximations which were confirmed in various energy regions and by means of which an analytic solution of the problem concerning the development of the EPS can be obtained. In the so-called approximation *A* (Ref. 90), only the secondary particles with energies much higher than the critical energy ($E \gg \epsilon_{crit}$) are considered. In this case the following simplifications are fully justifiable: (1) the cross sections for the pair production and bremsstrahlung are assumed to be independent of the primary-particle energy; (2) the differential cross sections of these processes correspond to the case of the total screening of the nucleus by atomic electrons; (3) the ionization energy loss of electrons in comparison with the energy loss due to the bremsstrahlung is ignored; (4) the Compton scattering of photons as a process that leads to the photon "absorption" in comparison with the production of electron–positron pairs is ignored.

The kinetic equations which describe the development of the EPS, without regard for the Compton scattering of photons, ionization energy loss of electrons, and the angular divergence of secondary particles in the shower, have the form[91]

$$d\pi(E)/dt = 2\int_E^\infty \gamma(E')\Sigma_{p\gamma}(E',E)\,dE' + \int_E^\infty \pi(E')\Sigma_{re}(E',E'-E)\,dE' - \int_0^E \pi(E)\Sigma_{re}(E,E-E')\,dE'; \tag{4.12}$$

$$d\gamma(E)/dt = \int_E^\infty \pi(E')\Sigma_{re}(E',E)\,dE' - \int_0^E \gamma(E)\Sigma_{p\gamma}(E,E')\,dE'. \tag{4.13}$$

Here $\pi(E)dE = \pi(E_0,E,t)\,dE$ and $\gamma(E)\,dE = \gamma(E_0,E,t)\,dE$ are the mean numbers of electrons and photons with an energy in the interval from $E$ to $E + dE$ at the depth $t$ at which the shower develops ($t$ is expressed in units of radiation length $t_r$), $\Sigma_{p\gamma}(E,E')\,dE'$ is the probability per unit length along the EPS axis for the production by a photon with an energy $E$ of an electron (positron) with an energy in the interval from $E'$ to $E' + dE'$, and $\Sigma_{re}(E,E')\,dE'$ is the probability per unit length for the emission of a bremsstrahlung photon with an energy between $E'$ and $E' + dE'$ by an electron with an energy $E$.

If the EPS is initiated by a photon with an energy $E_0$, then the initial conditions will have the form $\pi(E)|_{t=0} = 0$; $\gamma(E)|_{t=0} = \delta(E - E_0)$. In the case of a primary electron we have $\gamma(E)|_{t=0} = 0$ and $\pi(E)|_{t=0} = \delta(E - E_0)$.

If the screening of the nucleus by atomic electrons is complete ($\xi \simeq 0$), the macroscopic differential cross sections for the production of electron–positron pairs $\Sigma_{p\gamma}(E,E')\,dE'$ and the photon bremsstrahlung $\Sigma_{re}(E,E')\,dE'$ will depend only on the ratio $E'/E$; i.e., $\Sigma(E,E')\,dE' = \Sigma(u)du$, where $u = E'/E$. Taking advantage of this circumstance and using the Mellin transform on the left and right sides of Eqs. (4.12) and (4.13), we obtain the system of equations[91]

$$d\pi(s)/dt = -A(s)\pi(s) + B(s)\gamma(s);$$
$$d\gamma(s)/dt = C(s)\pi(s) - D\gamma(s),$$

where $\pi(s) = \int_0^\infty E^s \pi(E)\,dE$; $\gamma(s) = \int_0^\infty E^s \gamma(E)\,dE$; $D = 7/9$; $A(s)$, $B(s)$, and $C(s)$ are well-known functions[91]:

$$A(s) = \frac{4}{3}\{\psi(s) + \eta\} - \frac{s(5s+7)}{6(s+1)(s+2)};\ B(s)$$
$$= \frac{2}{1+s} + \frac{8}{3(3+s)} - \frac{8}{s(2+s)};$$
$$C(s) = \frac{4}{3s} - \frac{4}{3(s+1)} + \frac{1}{(s+2)};\ \psi(s) = \frac{d}{ds}\ln(s!);$$

and $\eta = 0.577...$ is Euler's constant.

The average number of particles of a given type (electrons, for example) with an energy higher than $E_c$ is given by

$$N_e(E_0, E_c, t) = \int_{E_c}^{E_0} \pi(E_0, E, t)\,dE.$$

In the approximation $A$ of the cascade theory, this quantity is expressed in terms of the parameter $s$ in the following way[88]:

$$N_e(E_0, E_c, t) = H(s)\,(E_0/E_c)^s \exp\{\lambda_1(s)t\}/\sqrt{\varphi(s)}, \tag{4.14}$$

where

$$\varphi(s) = 2\pi s^a \sqrt{(d^2\lambda_1(s)/d^2s)\,t + a/s^2}.$$

In the case of a primary electron $H(s) = [D + \lambda_1(s)]/[\lambda_1(s) - \lambda_2(s)]$ and $a = 1$, while in the case of a primary photon $H(s) = C(s)/[\lambda_1(s) - \lambda_2(s)]$ and $a = 1/2$. The functions $\lambda_1(s)$ and $\lambda_2(s)$ are expressed in terms of the functions $A(s)$, $B(s)$, $C(s)$, and $D$ in the following way:

$$\lambda_1(s) = (1/2)\left\{[A(s) + D] - \sqrt{[A(s) - D]^2 + 4B(s)\,C(s)}\right\};$$
$$\lambda_2(s) = 1/2\left\{[A(s) + D] + \sqrt{[A(s) - D]^2 + 4B(s)\,C(s)}\right\}.$$

The parameter $s$ is related to $t$ and $x = E_0/E_c$ by the relation $t = -(\ln x - a/s)/[d\lambda_1(s)/ds]$.

As can be seen from Eq. (4.14), the quantity $N_e(E_0, E_c, t)$, which determines the cascade curve, depends only on $E_0/E_c$ and on the EPS penetration depth $t$.

The parameter $s$ which appears in the solution of the problem on the development of the EPS is called *shower age*. This name stems from the fact that the parameter $s$ increases with the development of the EPS. At the initial time $s = 0$, the maximum of the shower has $s = 1$, and in the region of damping of the EPS $s > 1$.

With the decrease in the energy of secondary particles (at $E \sim \epsilon_{\text{crit}}$), the ionization energy loss of electrons and the Compton scattering of photons can no longer be legitimately ignored. These processes are taken into account in the approximation $B$ of the cascade theory under the assumption that (a) the ionization loss is constant and does not depend on the electron energy, i.e., $dE/dt = \epsilon_i = \text{const}$, and (b) the Compton scattering is equivalent to the absorption of a photon with a given energy and the cross section of this scattering does not depend on the photon energy $\Sigma_C(E) = \text{const}$.

In the approximation $B$, the cascade-theory equations are obtained from Eqs. (4.12) and (4.13) by adding to their right sides the terms which take into account the change in the particle balance in the phase space $dtdE$ as a result of ionization energy loss $\epsilon_i(\partial\pi(E)/\partial E$ [in Eq. (4.12)] and as a result of the Compton scattering $\Sigma_C\gamma(E)$ [in Eq. (4.13)] (Ref. 88). In the approximation $B$, the integral electron energy spectrum is given by[89]

$$N_e(E_0, E_c, t) = D(s)\,H(s)P(s,\epsilon)\,(E_0/\epsilon_{\text{crit}})^s \exp\{\lambda_1(s)t\}/\sqrt{\varphi(s)}. \qquad (4.15)$$

Here $D(s) = \left[-\dfrac{d\lambda_1(s)}{ds}\dfrac{s}{H(s)}\right]^s/\Gamma(s+1)$, where $\Gamma(s+1)$ is the gamma function, and

$$P(s,\epsilon) = e^{\epsilon}\int_{\epsilon}^{\infty} e^{-y}(1-\epsilon/y)^s dy, \qquad (4.16)$$

where $\epsilon \simeq 2.3E_c/\epsilon_{\text{crit}}$, $\epsilon_{\text{crit}}$ is the critical energy, and the functions $H(s)$, $\lambda_1(s)$, and $\varphi(s)$ are the same functions as those in Eq. (4.14). As can be seen from Eqs. (4.15) and (4.16), the electron distribution density depends on the ratios $E_0/\epsilon_{\text{crit}}$ and $E_c/\epsilon_{\text{crit}}$, rather than on the energies $E_0$ and $E_c$:

$$N_e(E_0, E_c, t) = N_e(E_0/\epsilon_{\text{crit}}, E_c/\epsilon_{\text{crit}}, t).$$

Equation (4.15) is complex in form. If, however, we assume $E_c = 0$, we can obtain simple expressions for some parameters of the cascade curve, $N_e(E_0, 0, t)$. A simple expression can also be found for the total mean-free path of charged particles, $\langle L\rangle$; this expression, along with the expressions for $N_e^{\max}(E_0)$ and $t_{\max}$, is given in Table 4.2. The integrated electron spectrum near the shower maximum is described by the equation[88]

$$N_e^{\max}(E_0, E) = N_e^{\max}(E_0)\,f(\epsilon),$$

where $f(\epsilon) = 1 - \epsilon e^{\epsilon}\int_{\epsilon}^{\infty} e^{-y}dy/y$.

The three-dimensional problem on the EPS development can be solved analytically in the Landau approximation,[92] which can be used at small angles of divergence of the motion of secondary particles from the direction of the primary particle: $\sin\theta \simeq \theta \ll 1$. The spatial particle distribution density in

**Table 4.2. Some characteristics of the development of an electromagnetic shower in the *B* approximation (*L* and *t* are measured in units of radiation length, $x = E_0/\epsilon_{crit}$).**

| Parameter | Primary electron | Primary photon |
|---|---|---|
| $t_{max}$ | $1.01\,(\ln x - 1)$ | $1.01\,(\ln x - 0.5)$ |
| $N_e^{max}\,(E_0, E_c = 0)$ | $0.31x/\sqrt{\ln x - 0.37}$ | $0.31x/\sqrt{\ln x - 0.18}$ |
| $\langle L \rangle$ | $x$ | $x$ |

the shower in this case can be represented approximately in the form $F(r) = K(s)r^{s-2}(r+1)^{s-4.5}$. Here $r$ is the distance from the EPS axis in the Molière units $r_M$, $s$ is the shower age, and $K(s)$ is the normalization factor which can be calculated from the approximate formula (for $s < 1.6$): $K(s) \simeq 0.443s^2\,(1.9 - s)$. The results of a numerical calculation of the spatial distribution of electrons in the shower in the approximation $B$ are presented in Fig. 4.5.

Analytic solutions obtained on the basis of simplified models, which describe the EPS development, yield rather good quantitative results for light substances ($Z \lesssim 20$) and reasonably high secondary-particle energies ($E_c \gtrsim \epsilon_{crit}$). This feature greatly limits the application range of these solutions. In practice, the analytic methods cannot be used, on the other hand, to solve, for example, the problem on the fluctuation of the shower characteristics or shower development in the medium with a complex geometry. Such characteristics of the EPS development as the position and value of the shower maximum and the total length of the secondary charged-particle tracks, which were calculated on the basis of analytic equations can, however, be used to quantitatively estimate a broad range of problems which arise in actual situations.

## 4. Monte Carlo simulation of the electron-photon shower

A three-dimensional problem on the development of EPS in the layers of various substances, which is characterized by a complex geometry (specifically, these problems are generally of interest from the practical point of view) can be solved only by numerical methods, among which the principal one is the Monte Carlo method. Simulation of the development of a cascade by imitating (with a corresponding probability) the interaction of electrons and photons with matter makes it possible to (1) take into account all physical processes (which affect the development of a shower) of the interaction of electrons and photons with the atoms of the medium; (2) take into account all particular features of the energy behavior of the cross sections of these processes and the characteristic features of the energy and angular distributions of secondary particles; (3) solve the problem of the development of EPS in a

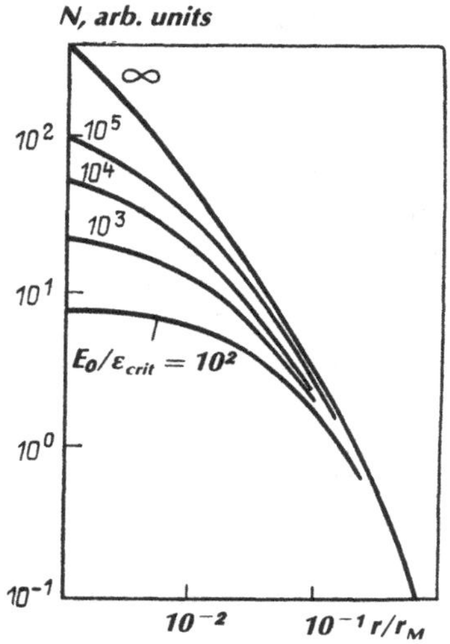


**Figure 4.5. The radial distributions of secondary electrons in the EPS at the shower maximum (Ref. 34) ($s = 1$). The primary-electron energy in units of the critical energy $\epsilon_{crit}$ is indicated on each curve.**

medium with a complex geometry and complex composition of the components; (4) solve in the most direct way the problem on the fluctuation of the shower characteristics.

An important factor in practical situations is the simplicity and clarity of the algorithm for the solution of the problem by the Monte Carlo method, allowing various functionals to be easily calculated and the external factors, which affect the development of a cascade (such as the magnetic field in a medium), to be taken into account. A drawback of the Monte Carlo method is its relatively slow convergence (the relative accuracy of the method is on the order of $N^{-1/2}$, where $N$ is the number of events that have been played out; see Sec. 1.5).

Many programs which allow the evolution of the EPS to be calculated by the Monte Carlo method have been developed.[21,42,93–95] These programs differ primarily in the ease with which they can be used and the degree of their versatility. These reasons and their accessibility account for the wide application of the SIMEX-1 program[93] in the USSR and the EGS-3 program[42] in other countries.

## A general scheme for the simulation of the development of an electromagnetic shower

The EPS simulation includes the simulation of the propagation of photons and electrons (positrons) in matter and their interaction with the atoms of the medium. Here the fate of each secondary particle of the EPS can be traced in detail. The data reflecting the charcteristics of all particles in the shower at a given stage of simulation of its development are stored in the appropriate section (data file) of the computer memory. This data file is usually set up in such a way that the characteristics of the particle with lowest energy are stored at its highest level. The fate of the particle from the highest level is traced first. The tracing stops if (a) there is an interaction, (b) the particle energy is lower than a previously specified cutoff threshold $E_c$, or (c) the particle enters a particular region of space. If there is a particle interaction, the highest level of the data file will contain the characteristics of the secondary particle with the minimum energy and the tracing is resumed but now with a new particle. Since the fate of the particle with the lowest energy is checked first, the depth (size) of the data file is maintained within $\log_2(E_0/E_c)$, where $E_0$ is the energy of the particle that initiates the cascade.

If the energy of the particle is lower than the threshold energy or if the particle enters a certain region of space (if it escapes from the substance, for example), it will no longer be monitored, and the fate of the particle from the next level, a lower level, of the data file, which at this time becomes the top level, will be traced. The simulation of a single cascade (its history) ceases when all secondary particles have been extracted from the data file and studied.

Such a tracing scheme is called a *lexicographic* scheme. The fate of each shower particle (the trajectory tree) is traced concomitantly with the accumulation [and (or) calculation] of information necessary for finding the required functionals: the spatial, angular, and energy distributions of particles of a particular species, fluctuation of their parameters, and other characteristics of the EPS development.

Before attempting to simulate specific processes, we shall consider the algorithm for the analysis of the evolution of the random value, which is a generalization of the rejection method (see Sec. 1.5) used extensively in the simulation of EPS. The distribution density function $f(x)$ for a random value of $x$ can be written in the form

$$f(x) = \sum_{i=1}^{n} \alpha_i \varphi_i(x)\, g_i(x), \tag{4.17}$$

where $\alpha_i > 0$, $0 \leqslant g_i(x) \leqslant 1$, and $\varphi_i(x) \geqslant 0$ for all $x$ from the region $(a \leqslant x \leqslant b)$ in which the function $\varphi_i(x)$ is determined, and $\int_a^b \varphi_i(x)\, dx = 1$. The algorithm for the analysis of the evolution of the random value of $x$, which is subject to distribution law (4.17), will then involve the following steps[21,41] ($\gamma$ and $\eta$ denote random numbers uniformly distributed over the interval from 0 to 1).

(1) Choose a random integer $j$ $(1 \leqslant j \leqslant n)$ with a probability proportional to $\alpha_i$; i.e., choose an integer which satisfies the condition

$$\sum_{i=1}^{j-1} \alpha_i \Big/ \sum_{i=1}^{n} \alpha_i < \gamma \leqslant \sum_{i=1}^{j} \alpha_i \Big/ \sum_{i=1}^{n} \alpha_i .$$

(2) Analyze (by the inverse-function method, for example; see Sec. 1.5) the evolution of the random value of $x$ from the distribution $\varphi_j(x)$.

(3) Single out such a random value of $x$ for which the function $g_j(x) \leqslant \eta$. Otherwise, reject the random value of $x$ and repeat the procedure, beginning with step (1).

This method can be used if the function $f(x)$ can be represented in a form such that $n$ will be a small value, $x$ can easily be analyzed from all distributions $\varphi_i(x)$, and the values of the functions $g_i(x)$ will differ from unity only slightly (i.e., the results of an analysis are seldom rejected). For an analysis based on this method the probability for a rejection is $W_r = 1 - 1/\sum_{i=1}^{n} \alpha_i$, and the number of attempts necessary for an analysis of the evolution of a single random quantity is $\overline{K} = \sum_{i=1}^{n} \alpha_i$.

**Simulation of the propagation of particles in the medium**

The mean-free path $\lambda$ of a particle ($\gamma$, $e^-$, $e^+$) before the interaction can be expressed in terms of its total cross section for the interaction with the atoms of the medium

$$\lambda = 1/(N_A \rho \sigma_{\text{tot}}/M) = 1/\Sigma_{\text{tot}},$$

where $N_A$ is Avogadro's number, $\rho$ is the density of the substance, $M$ is the molecular mass, $\sigma_{\text{tot}}$ is the total cross section for the interaction with the atom (molecule) of the medium, and $\Sigma_{\text{tot}}$ is the total macroscopic interaction cross section.

The probability for the interaction of a particle in the mean-free-path interval $(x, x + dx)$ is determined from the equation $dP(x) = dx/\lambda(x)$. Since the mean-free path of a particle can change during its propagation in the medium (if the particle, for example, loses energy or goes from one substance into another), $\lambda$ will generally depend on $x$; i.e., $\lambda = \lambda(x)$.

Let $dN$ denote the number of interactions in the interval from $x$ to $(x + dx)$. Clearly, $dN = N(x)\, dx/\lambda(x)$ or $dN/N = dx/\lambda(x)$, where $N(x)$ is the number of particles that arrive at the point $x$ without interaction. Integrating this equation, we find

$$N(x) = N_0 \exp\left\{ -\int_{x_0}^{x} dx/\lambda(x) \right\}, \quad N_0 = N(x_0).$$

The number of interactions is

$$n(x) = N_0 \left[ 1 - \exp\left\{ -\int_{x_0}^{x} dx/\lambda(x) \right\} \right]$$

and the probability for the interaction of the particle in the interval from $x_0$ to $x$ is

$$n/N_0 = 1 - \exp\left\{ -\int_{x_0}^{x} dx/\lambda(x) \right\}. \tag{4.18}$$

The integral

$$\int_{x_0}^{x} dx'/\lambda(x') = N_\lambda \tag{4.19}$$

is the number of mean-free paths that a particle travels as it moves from the point with the coordinate $x_0$ to the point with the coordinate $x$. At $\lambda(x') = \lambda = \text{const}$, $N_\lambda$ is simply equal to $(x - x_0)/\lambda$ in the interval from $x_0$ to $x$. The number of mean-free paths, $N_\lambda$, must satisfy, as is evident from Eq. (4.18), the distribution law

$$f(N_\lambda) = 1 - \exp(-N_\lambda). \tag{4.20}$$

A random value of $N_\lambda$ can be analyzed by the inverse-function method: $N_\lambda = -\ln\gamma$, where $\gamma$ is a random number which is uniformly distributed in the interval (0,1). Substituting this value on the right side of relation (4.19), we can determine the point at which the particle will interact.

The photon distribution is determined by three principal processes: the production of electron–positron pairs, Compton scattering, and the photoelectric effect. Since these processes have rather small cross sections, the passage of photons through matter can be simulated as a series of straight sections between the interactions (this chain may have "breaks" in the case of pair production, for example). If the medium in which the photon is moving is heterogeneous, i.e., if it consists of layers of different materials, and if the number of these layers is finite, the integral in expression (4.19) becomes

$$N_\lambda = \sum_{j=1}^{i-1} \left( \frac{x_j - x_{j-1}}{\lambda_j} \right) + \left( \frac{x - x_{i-1}}{\lambda_i} \right),$$

where $x \in (x_{i-1}, x_i)$, and $x_i$ $(i = 0,1...)$ are the coordinates of the layer boundaries of the material; in each layer the value of $\lambda$ is constant. In the case under consideration, the evolution of the mean-free path of the photon before its interaction can be analyzed according to the following scheme[41]:

(1) Analyze the evolution of the quantity $N_\lambda = -\ln\gamma$.

(2) Determine the mean-free path $\lambda$ for the instantaneous position of the photon with a given energy.

(3) Calculate a possible range of the photon, $l_1 = \lambda N_\lambda$.

(4) Determine the distance $d$ from the nearest layer boundary from the direction of motion of the photon.

(5) Trace the photon over a distance $l_2 = \min\{l_1, d\}$.

(6) If $(\lambda N_\lambda - l_2) = 0$, i.e., if $l_2 = l_1$, we can assume that the interaction has taken place, and the analysis of the evolution of the mean-free path may be assumed complete.

(7) If $(\lambda N_\lambda - l_2) > 0$, i.e., if $l_2 = d$, the photon has reached the layer boundary. If the layer boundary is chosen randomly (for example, to collect information on the shower development in this region) and if the next layer consists of the same material, then the procedure is repeated, beginning with step (3). If the layer material changes, the procedure is repeated, beginning with step (2).

The propagation of electrons and positrons in the material is determined by the $e^-(e^+)$ elastic scattering in a Coulomb field of the nuclei of the atoms of the medium, by the inelastic $e^-(e^+)$ scattering by atomic electrons, by radiation-induced $e^-(e^+)$ bremsstrahlung, and by positron annihilation. All these processes, with the exception of positron annihilation and bremsstrahlung, have a very large macroscopic cross section. As a result, the propagation of the charged component of the electromagnetic cascade as a discrete series of interactions with a linear motion between them, as in the case of photons,* essentially cannot be simulated.

The principal contribution to the cross sections for interaction of electrons with atoms (nuclei and atomic electrons) of the medium comes from the events with a small momentum transfer from the electron to the target. Such events have a very small effect on the fluctuations of the characteristics of the EPS. These processes can therefore be combined and their effect on the propagation of the electrons in the medium can be assumed to be continuous (averaged over many interaction events). The averaging is carried out over path segments considerably greater than the mean range of an electron relative to the interactions with a very small momentum transfer. On the other hand, the interaction of electrons with the atoms of the material, in which an appreciable part of the momentum is transferred from the electron to the target, cannot be averaged, since these events determine the fluctuations of the characteristics of the EPS.

The division of the processes into continuous (multiple) processes and discrete (multiple) processes is, in a certain sense, arbitrary but very useful from the practical point of view. Such a division allows one to avoid unfounded problems associated with the simulation of a very large number of interactions with a small momentum transfer. These interactions, while presented in an averaged form, can be described correctly in analytic form by using valid theoretical equations.

The inelastic scattering of electrons and positrons by atomic electrons is an example of a process which can easily be simulated by dividing the inelastic scattering events into two classes through the introduction of an energy threshold $T_b$. The inelastic scattering of an electron as a result of which a $\delta$ electron with a kinetic energy larger than $T_b$ is produced, is viewed as a discrete process. The mean range of an electron before such an interaction is described by an approximate equation, $\lambda_\delta \simeq 0.036 Z\beta^2 T_b t_r$ $(13 \leqslant Z \leqslant 92)$, which follows from expression (2.18) for the differential cross section of the Møller scattering. Here $\beta^2 = v^2/c^2 \simeq 1$ if the primary electron is relativistic; $T_b$ is the energy threshold (MeV). For lead $\lambda_\delta \simeq 0.6 t_r \beta^2$ for $T_b = 0.2$ MeV.

Collisions involving a small transfer of energy to the $\delta$ electrons $(T < T_b)$ are included in the continuous ionization energy loss.[21,42,93] Equation (2.24) which describes the average ionization energy loss by an electron $(dE/dx)_i$ takes into account the cases in which a large amount of energy is transferred to the $\delta$ electrons $(T > T_b)$. This equation must therefore be modified:

*The mean range of an electron before the interaction, as a result of which a bremsstrahlung photon is emitted, is determined, for example, from Eq. (4.4) and for lead is $\lambda_r \approx t_r/9$. The mean range of an electron in lead before the elastic Coulomb scattering, on the other hand, is $\lambda_c \approx 2.5 \times 10^{-5} t_r$ (see Sec. 2.6); i.e., $\lambda_r/\lambda_c \simeq 4 \times 10^3 \gg 1$. The mean range of a photon in lead, with an energy $\omega > 1$ GeV, before the interaction, is at least $9t_r/7$ [see Fig. 4.2 and Eq. (4.7)].

$$\left(\frac{dE}{dx}\right)_i^{\text{s.thr.}} = \left(\frac{dE}{dx}\right)_i - \int_{T_b}^{T_{\max}} T\frac{d\Sigma}{dT}\,dT, \tag{4.21}$$

where $T$ is the kinetic energy of the $\delta$ electron, and $d\Sigma/dT$ is the macroscopic differential cross section for inelastic scattering of an electron by an atomic electron [in fact, the Møller scattering; see Eq. (2.18)] or a positron by an atomic electron [the Bhabha scattering; see Eq. (2.19)]. The value of $T_b$ is assumed here to be large enough to regard the atomic electrons to be virtually free electrons.

The subthreshold energy loss of an electron $(dE/dx)^{\text{s.thr.}}$ can also include its radiation energy loss due to the emission of low-energy photons with energies $\omega < T_b$:

$$\left(\frac{dE}{dx}\right)^{\text{s.thr.}} = \left(\frac{dE}{dx}\right)_i^{\text{s.thr}} + \int_0^{T_b} \omega\,\frac{d\Sigma_r}{d\omega}\,d\omega,$$

where $\omega$ is the energy of the bremsstrahlung photon, and $d\Sigma_r/d\omega$ is the macroscopic differential bremsstrahlung cross section [see Eq. (2.8)].

The elastic scattering of electrons in a Coulomb field of the atomic nuclei of the medium can also be regarded as a continuous multiple Coulomb scattering. The angular distribution of the scattering of an electron after it passes through a layer of the material of thickness $t$ is described by the Møller equation [Eq. 2.31)].

The processes which continuously affect the propagation of electrons in the medium complicate the simulation of their passage through matter. The continuous energy loss by the electron causes the cross sections for the discrete interaction of the electron to change continuously. The electron, moreover, now moves from one interaction to another not rectilinearly but rather in a curved manner as a result of multiple Coulomb scattering.

## Simulation of the electron range before the discrete interaction

The range of an electron, $\lambda$, before a discrete interaction can be simulated by tracing the evolution of $N_\lambda$ from (4.20) and then solving Eq. (4.19) for $x$, assuming the range $\lambda$ to be $x - x_0$. This method of simulating the evolution of $N_\lambda$ may involve considerable technical difficulties associated with the solution of Eq. (4.19).

In practice, it is useful to simulate the electron range before a discrete interaction by introducing a fictitious macroscopic interaction cross section[42] $\Sigma_f$. The fictitious cross section $\Sigma_f$, which does not depend on the electron energy, is chosen in such a way that it would be larger than the actual total cross section of the discrete processes, $\Sigma_d(x)$, for any value of $x$ (or for any value of the electron energy $E$); i.e., $\Sigma_d(x)/\Sigma_f \leqslant 1$. The range of an electron before a discrete interaction can then be simulated as follows:

(1) The range $\lambda$ corresponding to a constant fictitious cross section $\Sigma_f$:$\lambda = -\ln\gamma/\Sigma_f$ is simulated.

(2) A random number $\eta$ is chosen and then compared with the quantity $\Sigma_d(x_0+\lambda)/\Sigma_f=\epsilon\leqslant 1$. If $\eta\leqslant\epsilon$, then a discrete interaction has taken place; otherwise, a new value is calculated, $x_0=x_0+\lambda$, and the procedure is repeated, beginning with step (1).

The propagation of electrons in a medium can also be simulated differently, without tracing explicitly the range of an electron before a discrete interaction.[93] The electron in this case is traced by partitioning its path into such predetermined small sections $\Delta x$ that the discrete interaction cross section $\Sigma_d$ can be assumed constant in each section: $\Sigma_d\simeq\Sigma_d(x)\simeq\Sigma_d(x+\Delta x)$. In each such section (step) a discrete interaction event is simulated according to the probability determined by the cross section $\Sigma_d$. The distortions of the distribution in the number of discrete interaction events [due to the approximation $\Sigma_d(x)=\text{const}$] and of the distribution in the interaction sites, which arise in this simulation method, can be reduced to a negligible level through a corresponding reduction of the step $\Delta x$. This procedure can be summarized as follows:

(1) The cross section $\Sigma_d=\Sigma_d(x)$ is calculated.

(2) The electron is traced in the step $\Delta x$: $x=x+\Delta x$.

(3) The new energy of this electron is calculated, $E=E-\Delta E$, where $\Delta E$ is the energy loss in the step $\Delta x$ due to the continuous processes, $\Delta E=\Delta x(dE/dx)^{\text{s.thr.}}$.

(4) A change in the direction of motion of the electron due to the multiple Coulomb scattering along with path $\Delta x$ is simulated and the corresponding corrections to the initial direction of motion of the electron are introduced.

(5) A random number $\gamma$, uniformly distributed over the interval (0,1), is chosen and compared with the quantity $\epsilon=1-\exp(-\Sigma_d\Delta x)$. If $\gamma<\epsilon$, then a discrete interaction has taken place. Otherwise, the procedure is repeated, beginning with step (1).

Let us further consider the combined method of simulating the propagation of electrons in a medium,[21] which is closely related to the simulation of the kinematic characteristics of the secondary particles produced as a result of discrete interactions. The fact that the cross section $\Sigma_d$ in this method is not used in the explicit form sets it apart from the other methods.

Let us assume that the differential cross section of the discrete process based on some kinematic variable can be described by (4.17).

$$d\Sigma_d(u)=\sum_{i=1}^{n}\alpha_i\varphi_i(u)\,g_i(u)\,du,$$

where $u=\omega/E$ in the case of bremsstrahlung, for example; $\omega$ and $E$ are the energy of the bremsstrahlung photon and the energy of the primary electron, respectively. According to the derivation of the functions $\varphi_i(u)$ and $g_i(u)$ [see Eq. (4.17)], the quantity $\Sigma_f=\sum_{i=1}^{n}\alpha_i$ may be regarded as a total fictitious cross section of the discrete process: $\Sigma_f\geqslant\int d\Sigma_d=\Sigma_d$. The range of the particle before a discrete interaction can then be simulated as follows:

(1) The range $\lambda$ is simulated in accordance with the fictitious cross section $\Sigma_f$.

(2) The length of the simulated range, $\lambda{:}x' = x + \lambda$, of the particle is traced.

(3) The kinematic characteristic $u$ from the corresponding distribution $\varphi_i(u)$ is simulated.

(4) If the simulated quantity of $u$ is rejected, the procedure is repeated, beginning with step (1). Otherwise, the interaction is assumed to be simulated completely.

It is easy to show that this procedure is equivalent to the simulation of a particle range on the basis of the actual interaction cross section $\Sigma_d$ and to the subsequent simulation of the kinematic characteristics on the basis of the differential cross section of the process. In fact, the probability that the range $\lambda$ simulated on the basis of a fictitious cross section $\Sigma_f$ will not be rejected is $\Sigma_d/\Sigma_f$. If $k-1$ simulated ranges are rejected before a "successful" simulation of a particle range is found, the path $l$ traversed by the particle under these conditions will be subject to the Poisson law (since each attempt to simulate the range is an independent attempt):

$$W_k(l) = \frac{(\Sigma_f l)^{k-1}}{(k-1)!}\,\Sigma_f \exp(-\Sigma_f l).$$

On the other hand, the probability for the occurrence of $k-1$ "unsuccessful" simulation attempts is

$$W_k = (\Sigma_d/\Sigma_f)(1-\Sigma_d/\Sigma_f)^{k-1}.$$

The ratio $\Sigma_d/\Sigma_f$ is the probability for a successful attempt to simulate a particle range and the second term in the equation for $W_k$ is the probability that $k-1$ successful simulation attempts will not precede this event. Summing over all possible unsuccessful attempts, we find a distribution law for a path traveled by the particle before the interaction,

$$W(l) = \sum_{k=1}^{\infty} W_k(l) W_k = \sum_{k=1}^{\infty} \frac{(\Sigma_f l)^{k-1}}{(k-1)!}\,\Sigma_f \exp(-\Sigma_f l)$$
$$\times \frac{\Sigma_d}{\Sigma_f}(l-\Sigma_d/\Sigma_f)^{k-1} = \Sigma_d \exp(-\Sigma_f l)\sum_{k=1}^{\infty}\frac{[l(\Sigma_f-\Sigma_d)]^{k-1}}{(k-1)!}$$
$$= \Sigma_d \exp(-\Sigma_f l)\exp(\Sigma_f l - \Sigma_d l) = \Sigma_d \exp(-\Sigma_d l).$$

The path $l$ of a particle before its interaction is distributed according to the law which is determined by the actual interaction cross section $\Sigma_d$, as was asserted above.

### Discrete-interaction simulation

After simulating the particle range before the discrete interaction and after determining the coordinates of the interaction point, it is necessary to determine the interaction type (if there are several competing processes). The interaction obeys the distribution law: $F(i) = \sum_{j=1}^{i} \sigma_j/\sigma_d$, where $\sigma_j$ is the total cross section of process $j$, and $\sigma_d = \sum_{j=1}^{n} \sigma_j$ is the total cross section of all possible discrete interactions. The interaction of the type $i$ is simulated by choosing a random number $\gamma$ and by determining the index $i$, which satisfies the condition $F(i-1) < \gamma = F(i)$.

After determining the interaction type, it is necessary to simulate the kinematic charcteristics of the interaction products such as the energy ($E$), the polar emission angle ($\theta$), and the azimuthal emission angle ($\varphi$). In general, the final state, produced as a result of the interaction, is characterized by several parameters $\lambda_i(E_i,\theta_i,\varphi_i)$:$\lambda = \{\lambda_1,\lambda_2,...,\lambda_n\}$, and the differential cross section of the process is $d^n\sigma = F(\lambda)d^n\lambda$. The total cross section of the process is given by $\sigma = \int F(\lambda) \times d^n\lambda$, and the function $f(\lambda) = F(\lambda)/\sigma$ is the density function of the parameters $\{\lambda_i\}$, from which the appropriate sampling should be made. If a combined method (i.e., a generalized rejection method) is used to simulate the parameters $\{\lambda_i\}$, the function $f(\lambda)$ can be normalized arbitrarily.

### Simulation of bremsstrahlung by an electron (positron)

The differential cross section for the electron (positron) bremsstrahlung is given by Eq. (2.10). Since the angles of emission of a secondary electron and photon are small [see Eq. (4.8)], we can assume in first approximation that during the bremsstrahlung the secondary products are moving in the direction of the primary electron.

It follows from expression (2.10) for the bremsstrahlung cross section, with allowance for the Landau–Pomeranchuk effect, that for parameter values $s > 1$ cross section (2.10) becomes the cross section given by Eqs. (2.1) and (2.2). At $s < 1$ the secondary-photon energy should therefore be simulated, according to Ref. 21, on the basis of expression (2.10) and at $s > 1$ it should be simulated on the basis of the differential cross section given by Eqs. (2.1) and (2.2), with allowance for the polarization of the medium which is taken into account by the factor in Eq. (2.9). The macroscopic differential bremsstrahlung cross section (in radiation-length units) can be represented in the form[21]

$$d\Sigma_{re}/du = \alpha_1 f_1(u)\, g_1(u) + \alpha_2 f_2(u)\, g_2(u), \tag{4.22}$$

where $u = \omega/E_0$, $\omega$ is the energy of the bremsstrahlung photon, $E_0$ is the energy of the primary electron, and

$$\alpha_1 = [1 - f_c(Z)/\ln(183Z^{-1/3})]\left\{\frac{4}{3} + \frac{1}{9[\ln(183Z^{-1/3}) - f_c(Z)]}\right\}$$

$$\times \{(1/2)[\ln(1+w_0^2) - \ln(w_0^2)]\};$$

$$\alpha_2 = (1/2)[1 - f_c(Z)/\ln(183Z^{-1/3})].$$

The function $f_1(u)$ is given by the equation

$$f_1(u) = \frac{2}{[\ln(1+w_0^2) - \ln(w_0^2)]} \frac{u}{(u^2+w_0^2)}.$$

The density function $f_2(u)$ of the second term in expression (4.22) is $f_2(u) = 2u^3/(u^2 + w_0^2)$. The polarization of the medium (which is appreciable at $u \lesssim w_0$) can be ignored in this term without introducing a significant error, since it is taken into account in the first term, where it plays an important role. It can thus be assumed that $f_2(u) \simeq 2u$.

A simulation of the value of $u$ from the density function $f_1(u)$ can be carried out by the inverse-function method (see Sec. 1.5): $u = w_0\sqrt{\exp(\gamma) - 1}$. A simulation of the value of $u$ from the function $f_2(u)$ can be carried out either by the inverse-function method, which yields $u = \sqrt{\gamma}$, or by choosing the maximum value of the two uniformly distributed random values: $u = \max\{\lambda_1, \lambda_2\}$.

The simulated value of $u$ is rejected if $u > 1 - m_e c^2/E_0$, a situation which occurs rather infrequently and which corresponds to the case in which the energy of the emitted photon is higher than the kinetic energy of the primary electron, which is forbidden kinematically.

The functions $g_1(u)$ and $g_2(u)$, on the basis of which the simulated quantity $u$ is rejected, are determined as follows. If the correction to the Landau–Pomeranchuk effect is negligible ($s > 1$), we have

$$g_1(u) = \begin{cases} (1-u)[1 - A(3.89\xi - 0.9805\xi^2)], \xi \leqslant 1; \\ (1-u)[1 - AH(\xi)], \xi > 1; \end{cases} \tag{4.23}$$

$$g_2(u) = \begin{cases} 1 - B(3.242\xi - 0.625\xi^2), \xi \leqslant 1; \\ 1 - BH(\xi), \xi > 1, \end{cases} \tag{4.24}$$

where $\xi$ is the screening length: $\xi = (136/Z^{1/3}) u m_e c^2 / E_0(1-u)$, $A = \{4\,[\ln(183Z^{-1/3}) - f_c(Z)] + 1/3\}^{-1}$, $B = \{4[\ln(183Z^{-1/3}) - f_c(Z)]\}^{-1}$, $H(\xi) = 4.184\ln(\xi + 0.952) - 0.280$, and $f_c(Z)$ is the Coulomb correction [see Eq. (2.2)].

If the Landau–Pomeranchuk effect ($s < 1$) must be taken into account, the functions $g_1(u)$ and $g_2(u)$ are

$$g_1(u) = (1/3)\,(1-u)\,[2\varphi(s) + h(s)]\,b(s); \tag{4.25}$$

$$g_2(u) = \varphi(s)\,b(s). \tag{4.26}$$

The parameter $s$, which is used to choose from (4.23)–(4.26) the correct expressions for the functions $g_1(u)$ and $g_2(u)$, is determined by Eq. (2.11). In practice, this choice can be made by comparing to unity not the parameter $s$ but the parameter value $s'$ approximately equal to it:

$$s' = 1.37 \times 10^3 [(m_e c^2/E_0)\, u t_r/(1-u)]^{1/2},$$

where $t_r$ is the radiation length, $g/cm^2$.

Expression (2.2) for the function $F_e(E_0, u)$, which determines the behavior of the differential bremsstrahlung cross section at $s > 1$ (except at very small values of $u$, $u \sim w_0$), is valid only at moderately large values of the screening length $\xi$, because in the limit $\xi \to \infty$, where $f_1(\xi) \simeq f_2(\xi) \to 0$, the function

$F_e(E_0,u)$ becomes a negative function, in contradiction of the notion of the differential cross section. In practice, the negative values of the cross sections can be avoided by assuming $F_e(E_0, u) = 0$ for $\xi \geqslant \xi_{\max}(Z, E_0)$, where

$$\xi_{\max}(Z, E_0) = \exp\left\{\left[21.12 - \frac{4}{3}\ln Z - 4f_c(Z)\right]/4.184\right\} - 0.952.$$

**Simulation of the inelastic scattering of electrons and positrons by atomic electrons**

The differential inelastic electron–electron scattering cross section is given by the Møller equation [Eq. (2.18)]. The differential macroscopic cross section of this process, expressed in units of the radiation length, can be described in the form[21] $d\Sigma_M/du = af(u)g(u)$, where $u = T/T_0$; $T$ and $T_0$ are the kinetic energy of the secondary $\delta$ electron and the kinetic energy of the primary electron, respectively, and

$$a = 2\pi r_e^2 t_r \frac{\rho Z N_A}{A}\frac{3}{T_b}; f(u) = \frac{u_0}{1-2u_0}\frac{1}{u^2}$$

$$\left(u_0 < u < \frac{1}{2}\right) g(u) = \frac{1-2u_0}{\beta_0^2}\left[1 + \frac{u^2}{(1-u)^2}\right.$$

$$\left. + \frac{T_0^2}{E_0^2}u^2 - \frac{2T_0+1}{E_0}\frac{u}{(1-u)}\right]; u_0 = T_b/T_0.$$

Here $T_b$ is the threshold energy. Beyond this energy the inelastic electron–electron scattering is treated as a discrete process [see Eq. (4.21)].

A simulation of the variable $u$ from the distribution $f(u)$ is carried out by the inverse-function method: $u = u_0/[1 - \gamma(1 - 2u_0)]$.

By analogy with the Møller scattering, the cross section for the scattering of a positron by an electron [Eq. (2.19)] can be represented in the form[21] $d\Sigma_B/du = af(u)g(u)$. Here $d\Sigma_B/du$ is the differential macroscopic cross section of the process expressed in units of radiation length, $u = T/T_0$, $T$ is the kinetic energy of the $\delta$ electron, $T_0$ is the kinetic energy of the primary positron,

$$a = 2\pi r_e^2 t_r(\rho Z N_A/A)\,3/T_b,$$

$$f(u) = \frac{u_0}{1-u_0}\frac{1}{u^2}(u_0 < u < 1),$$

$$g(u) = \frac{1-u_0}{\beta_0^2}\{1 - [(2-y^2)u - (3 - 6y + y^2 - 2y^3)u^2$$

$$+ (2 - 10y + 16y^2 - 8y^3)u^3 - (1 - 6y + 12y^2 - 8y^3)u^4]\},$$

$$y = 1/(1 + T_0/m_e c^2),$$

and $T_b$ is the threshold energy.

A simulation of the quantity $u$ is carried out on the basis of the equation $u = u_0/[1 - \gamma(1 - u_0)]$.

**Simulation of the production of electron–positron pairs.**
The differential macroscopic cross section for the production of electron-positron pairs by a photon, by analogy with the bremsstrahlung, can be represented in the form[21]

$$d\Sigma_{p\gamma}/du = \alpha_1 f_1(u)\, g_1(u) + \alpha_2 f_2(u)\, g_2(u).$$

The quantity $d\Sigma_{p\gamma}$ is expressed in units of radiation length; $u = E/\omega$ is the ratio of the energy of the secondary electron (positron) to the energy of the primary photon;

$$\alpha_1 = [1 - f_c(Z)/\ln(183Z^{-1/3})\{2/3 + (36[\ln(183Z^{-1/3}) - f_c(Z)])^{-1}\};$$

$$\alpha_2 = [1 - f_c(Z)/\ln(183Z^{-1/3})]\frac{1}{12}\left\{\frac{4}{3} + \frac{1}{9[\ln(183Z^{-1/3}) - f_c(Z)]}\right\};$$

$$f_1(u) = 1; f_2(u) = 12(u - 1/2)^2.$$

The simulation of the quantity $u$ from the density function $f_1(u)$ is carried out by simply assuming that $u = \gamma$. The variable $u$ from the function $f_2(u)$ can be simulated either by the inverse-function method: $u = [(\gamma - 1/2)/4]^{1/3}$ or by choosing a value from the three random values, $\gamma_1, \gamma_2, \gamma_3$, which differs from $\frac{1}{2}$ to the maximum extent.

To take the Landau–Pomeranchuk effect into account, by analogy with the bremsstrahlung, it is necessary to introduce the parameter $s' = 1.37 \times 10^3 \times [m_e c^2 t_r/\omega u(1-u)]^{1/2}$ in order to characterize the production of $e^+e^-$ pairs. If $s' > 1$, the Landau–Pomeranchuk effect is unimportant, and the differential cross section for the pair production is described by Eqs. (2.12) and (2.13). In this case,

$$g_1(u) = \begin{cases} 1 - C(2.914\xi - 0.447\xi^2), \xi \leqslant 1; \\ 1 - CH(\xi), \xi > 1; \end{cases}$$

$$g_2(u) = \begin{cases} 1 - A(3.898 - 0.9805\xi^2), \xi \leqslant 1; \\ 1 - AH(\xi), \xi > 1; \end{cases}$$

where the screening parameter $\xi = 136 m_e c^2 Z^{-1/3}/\omega u(1-u)$, $C = 4[\ln(183Z^{-1/3}) - f_c(Z)] - \frac{1}{6}$, and $H(\xi)$ and $A$ are the same quantities as those in Eqs. (4.23) and (4.24).

At $s' < 1$ the Landau–Pomeranchuk effect must be taken into account: $g_1(u) = (1/2)[\varphi(s) + h(s)]\, b(s)$; $g_2(u) = \varphi(s) b(s)$, where $\varphi(s)$, $h(s)$, and $b(s)$ are determined from Eqs. (2.10) and (2.11).

**Simulation of the Compton scattering of a photon by an electron**
The differential cross section for the Compton scattering of a photon by a free electron is described by Eq. (2.16). The macroscopic cross section of this process, expressed in units of radiation length, can be represented in the form[21]

$d\Sigma_c/du = \alpha f(u)\,g(u)$, where $u = \omega/\omega_0$ is the ratio of the energy of the secondary $\gamma$ ray to the energy of the primary $\gamma$ ray,

$$\alpha = 2\pi r_e^2 t_r(\rho Z N_A/A)\ln(1+2k_0)/k_0,\ k_0 = \omega_0/m_e c^2,$$
$$f(u) = [\ln(1+2k_0)]^{-1}(1/u),$$
$$(1+2k_0)^{-1} < u < 1,$$

and

$$g(u) = (1/2uk_0)[1 + (k_0^2 - 2k_0 - 2)u + (1+2k_0)\ u^2 + k_0^2 u^3].$$

A simulation of the quantity $u$ from the function $f(u)$ is carried out by the inverse-function method: $u = (1+2k_0)^{-\gamma}$.

## Simulation of the multiple Coulomb scattering

The density function of scattered electrons in the polar angle $\theta$ is given by Eq. (2.31). For practical purpose, we shall restrict the discussion to the first three terms in expansion (2.31)., At $v \lesssim 2$ the term $f^{(0)}(v) = 2\exp(-v^2)$, which is the dominant term, describes the angular distribution of particles that undergo multiple scattering at small angles. For large values of $v$ ($v \gtrsim 2$) the term $f^{(1)}(v)$, which describes a single scattering at a large angle, is the dominant term. The term $f^{(2)}(v)$ may be regarded as the correction to the sum of the first two terms at $v \approx 2$.

The multiple Coulomb scattering in thin layers of the material can thus be simulated by simulating the quantity $v$ from the distribution

$$F(v) = v[f^{0}(v) + (1B)\,f^{(1)}(v) + (1/B^2)\,f^{(2)}(v)].$$

The polar angle $\theta$ is related to $v$ by $\theta = \chi_c\sqrt{Bv}$. The parameters $B$ and $\chi_c$ are determined by Eqs. (2.33) and (2.34), respectively.

Following Ref. 42, we represent the function $F(v)$ in the form

$$F(v) = \sum_{i=1}^{3} \alpha_i f_i(v) g_i(v),$$

where $\alpha_1 = 1 - \lambda/B$, $f_1(v) = 2v\exp(-v^2)$, $v\in(0,\infty)$, $g_1(v) = 1$, $\alpha_2 = \mu G_2/B$, $f_2(v) = 1/\mu$, $v\in(0,\mu)$, $g_2(v) = (v/G_2)\,(\lambda f^{(0)}(v) + f^{(1)}(v) = f^{(2)}(v)/B)$, $\alpha_3 = G_3/2\mu^2 B$, $f_3(v) = 2\mu^2 v^{-3}$, $v\in(\mu,\infty)$, and $g_3(v) = (v^4/G_3)\,(\lambda f^{(0)}(v) + f^{(1)}(v) + f^{(2)}(v)/B)$.

The recommended numerical values of these parameters, which follow from Ref. 42, are $\lambda = 2$, $\mu = 1$, $G_2 = 1.8$, and $G_3 = 4.05$.

The quantity $v$ from the distribution $f_1(v)$ is simulated by using the inverse-function method: $v = \sqrt{-\ln\gamma}$.

The quantity $v$ from the distribution $f_2(v)$ is simulated on the basis of the equation $v = \mu\gamma$.

The simulation of the quantity $v$ from the function $f_3(v)$ can be carried out by first considering the function in the variable $\eta = 1v$. The density function for $\eta$ can then be written

$$f_\eta(\eta) = 2\mu^2\eta,\ 0 < \eta < 1/\mu,$$

and the "rejection" function is given by

$$g_\eta(\eta) = (\eta^{-4}/G_3)(\lambda f^{(0)}(\eta^{-1}) + f^{(1)}(\eta^{-1}) + f^{(2)}(\eta^{-1})/B).$$

In this approach, a random value of $\eta$ is simulated by choosing the maximum value of $\eta$ of the two uniformly distributed random values of $\eta$, $\eta = \max\{\gamma_1, \gamma_2\}/\mu$ or by using a direct method (i.e., the inverse-function method), $\eta = \sqrt{\gamma}/\mu$.

The functions $g_2(v)$ and $g_3(v)$ are linear combinations of the functions $f^{(0)}(v)$, $f^{(1)}(v)$, and $f^{(2)}(v)$. In practice, the functions $f^{(1)}(v)$ and $f^{(2)}(v)$ are calculated (tabulated) in advance on the basis of Eq. (2.32). The asymptotic behavior of these functions is used for large values of $v (v \gtrsim 10)$: ${}^{(fn)}(v) \sim v^{-2n-2}$.

As an example, Fig. 4.6 compares the results of the calculations on the basis of the scheme considered above with the data of an experiment on the scattering of electrons in a thin gold foil. The theoretical predictions are found to be in good agreement with the experimental data up to very large scattering angles.

## 5. The results of the experimental and theoretical studies of electromagnetic showers

The development of an electron–photon shower in condensed matter has been studied adequately, both theoretically[21,88,89,94,96] and experimentally[86,97,98] (see also the original studies which are cited in the indicated reviews). The development of EPS in various substances is nonetheless continuing to be studied extensively at various energies, both theoretically (calculation methods) and experimentally. These studies have been motivated primarily by the urgent need demonstrated by the experimental high-energy physics and cosmic-ray physics. A comprehensive understanding of the systematic features of an electron–photon shower and a quantitative description of its development are necessary, for example, in designing electromagnetic calorimeters—one of the principal pieces of equipment in the arsenal of present-day experimental high-energy physics, which is used to produce electron beams in high-energy proton accelerators, to solve radiation problems at accelerators, etc.

### Development of an electromagnetic shower in the longitudinal direction

The experimentally measured curves for the charged particles (electrons and positrons with energies $E \gtrsim E_c \sim 0.5$ MeV) in EPS are plotted in Fig. 4.7 as functions of the penetration depth $t$ of the shower. At the initial stage of development of the EPS, the dominant processes, which lead to a rapid multiplication of secondary particles, are the electron and positron bremsstrahlung and the production of electron–positron pairs. With an increase in the penetration depth $t$ of the shower, the particle multiplication slows down. If the secondary-particle energies are on the order of the critical energy $\epsilon_{crit}$, the

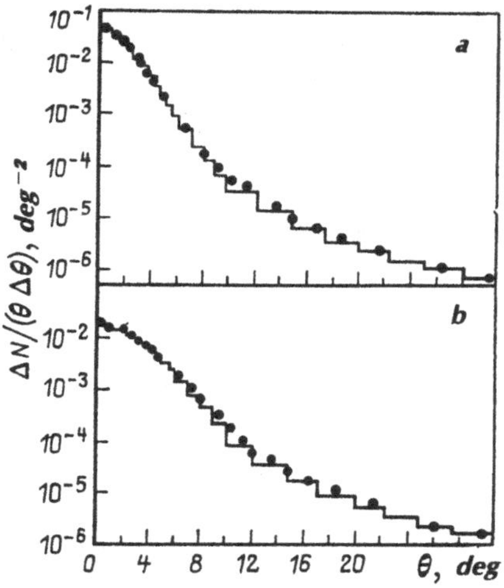


**Figure 4.6.** Distribution through a scattering angle θ of electrons with a kinetic energy of 15.7 MeV scattered in a gold foil of thickness (a) $18.66\times10^{-3}$ and (b) $37.28\times10^{-3}$ g/cm² (Ref. 42). Histogram, Monte Carlo calculation. ●, experimental data.

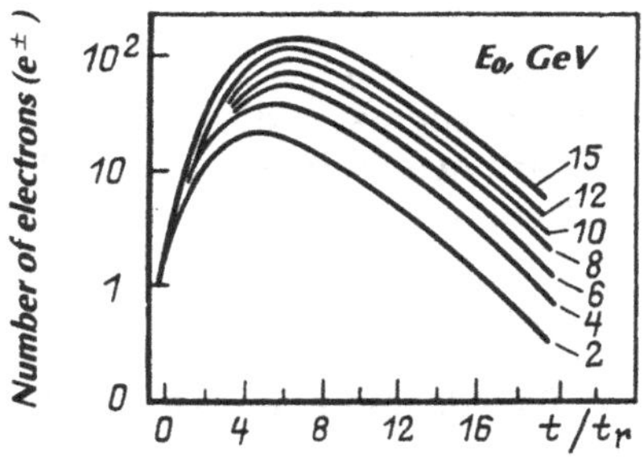


**Figure 4.7.** Cascade curves for showers initiated by electrons in lead (Ref. 98) (the primary-electron energies are indicated on the curves).

cascade-particle multiplication ceases and the number of particles in the EPS reaches a maximum. The position of this maximum in the variable $t$ is nearly independent of the primary energy $E_0$ of the particle (the electron in this case) which initiates the cascade. The approximate position of this peak in the variable $t$ is given by[96]

$$t_{\max} = \ln(E_0/\varepsilon_{\text{crit}}) - 1,$$

where $t_{\max}$ is measured in radiation lengths. The number of charged particles at the shower maximum can be accurately determined by an approximate equation[96]

$$N_e^{\max} = 10.71 E_0^{0.935},$$

where $E_0$ is the energy (in GeV).

After the EPS reaches a maximum number of secondary particles, it begins to decay gradually. The decay of the shower is characterized by the attenuation length Λ (see Sec. 4.1). The attenuation lengths of the electromagnetic showers measured experimentally in various materials are given in Table 4.3. Also given in this table are the approximate values of the photon

**Table 4.3. Attenuation length of electromagnetic showers $\Lambda$ initiated by electrons with an energy $E_0$ (Ref. 96) and the approximate values of the absorption lengths and energies of the most deeply penetrating photons (Ref. 87).**

| Substance | $E_0$, GeV | $\Lambda$ g/cm² | $\Lambda$ Radiation length | $\lambda_\gamma$, g/cm² | $E_\gamma$, MeV |
|---|---|---|---|---|---|
| Al | 6 | 64.3 | 2.7 | 46 | 20 |
| | 0.6–1 | 62.5 | 2.6 | | |
| | 6 | 39.0 | 3.0 | | |
| Cu | 1 | 35.7 | 2.8 | 33 | 8 |
| | 0.6 | 34.5 | 2.7 | | |
| Sn | 0.9 | 30.3 | 3.4 | 29 | 5 |
| W | 5–15 | 28.0 | 4.1 | 27 | 4 |
| | 6 | 24.7 | 3.9 | | |
| Pb | 1 | 21.7 | 3.4 | 24 | 3.5 |
| | 0.6 | 21.3 | 3.3 | | |

absorption lengths, $\lambda_\gamma = 1/\mu_{\text{min}}$, for minimum absorption, and the photon energies corresponding to them. It can be seen from these data that the values of $\Lambda$ are nearly independent of the primary energy of the cascade-initiating particle (an electron in this case) and that these values are approximately equal to the values of $\lambda_\gamma$ for the corresponding materials. These results confirm that the photons with the maximum penetrability play the key role in the development of a cascade in the bulk of the material.

The development of a photon-initiated cascade in the longitudinal direction is similar to the development of an electron-initiated cascade. The cascade curve for a primary photon can be approximated in the following way[97]:

$$N_e(E_0, E_c, t) = At^\alpha \exp(-bt). \tag{4.27}$$

Here the penetration depth $t$ is measured in units of the radiation length. The values of the parameters $A$, $\alpha$, and $b$ for the cascades initiated by photons in lead glass are given in Table 4.4. The results of Monte Carlo calculations of the longitudinal distribution of the EPS are in good agreement with the experimental data (Fig. 4.8).

In practice, it is frequently convenient to use such a characteristic of the longitudinal development of a shower as $t_{\text{med}}$, which is defined as the penetration depth of the shower and which upon being reached results in the absorption in the material of exactly half of the shower energy. The fraction of the extracted energy is virtually independent of the total energy of the cascade if the longitudinal dimension of the energy-evolution region is measured in units of $t_{\text{med}}$ (Fig. 4.9). It follows from the data shown in Figs. 4.8 and 4.9 that approximately 98% of the energy of the electromagnetic shower is released in the bulk of the absorber. This energy amounts to[96] $t\,(98\%) \simeq 3t_{\text{med}}$. The quantity $t\,(98\%)$ can also be expressed in terms of $t_{\text{max}}$ and $\Lambda$ (Ref. 96): $t\,(98\%) \simeq t_{\text{max}} + 4\Lambda$.

**Table 4.4. The parameter values in Eq. (4.27) (Ref. 96).**

| $E_{\gamma 0}$, GeV | 0.1 | 0.3 | 0.5 | 0.7 | 1.0 | 5.0 |
|---|---|---|---|---|---|---|
| $A$ | 4.54 | 7.18 | 8.24 | 8.32 | 8.58 | 10.88 |
| $\alpha$ | 1.00 | 1.45 | 1.65 | 1.84 | 2.03 | 2.74 |
| $b$ | 0.515 | 0.493 | 0.476 | 0.470 | 0.468 | 0.454 |

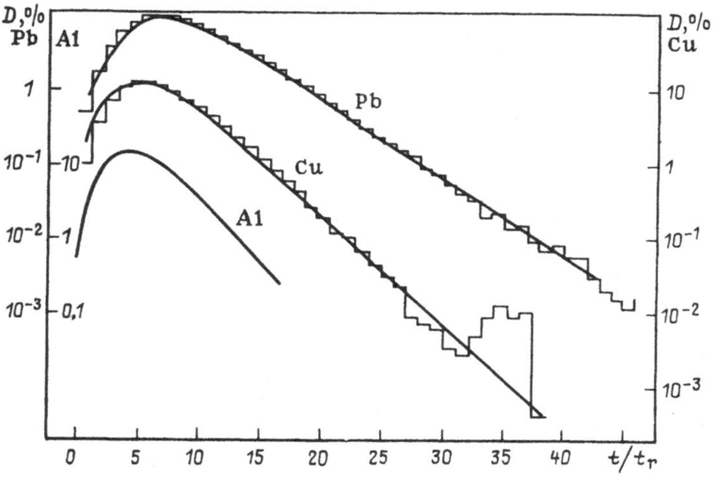


**Figure 4.8. The part of the energy released per unit radiation length vs the penetration depth of the cascades initiated by 6-GeV electrons in various materials (Ref. 96) (histogram, results of the Monte Carlo calculations).**

In such light absorbers as water and aluminum the results of theoretical calculations are also found to be in good agreement with experiment over a broad range of the penetration depths $t$ and distances from the shower axis $r$ (Fig. 4.10).

The agreement between the results of theoretical calculations and experiment should be treated with a certain caution in several cases, because it is frequently impossible to accurately determine the energy threshold $E_c$ for the detection of secondary particles in the experimental studies of the systematic features of the development of the EPS, which were carried out on the basis of an electronic (counter) technique. Since the principal characteristics of the cascade are very sensitive to the energy detection threshold, a comparison of the calculations with the experiment may turn out to be incorrect in several cases. In these situations there is a danger in obtaining an "excellent" agreement with the experiment by choosing an appropriate detection threshold or other parameters, whose exact values are either not known or cannot be extracted from the description of the experiment.

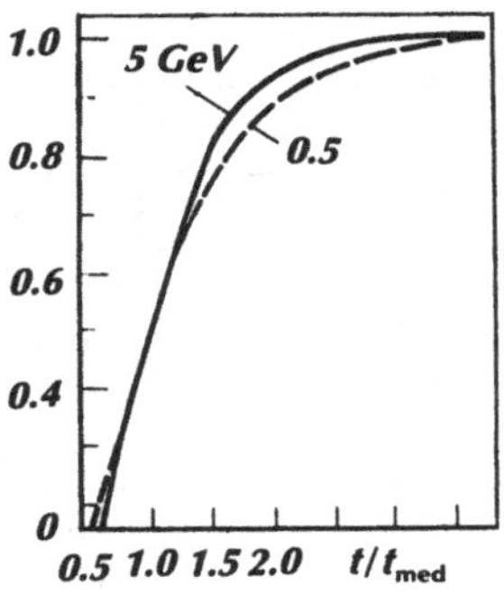


Figure 4.9. The energy released before reaching the depth $t/t_{med}$ in a photon-initiated cascade (Ref. 96).

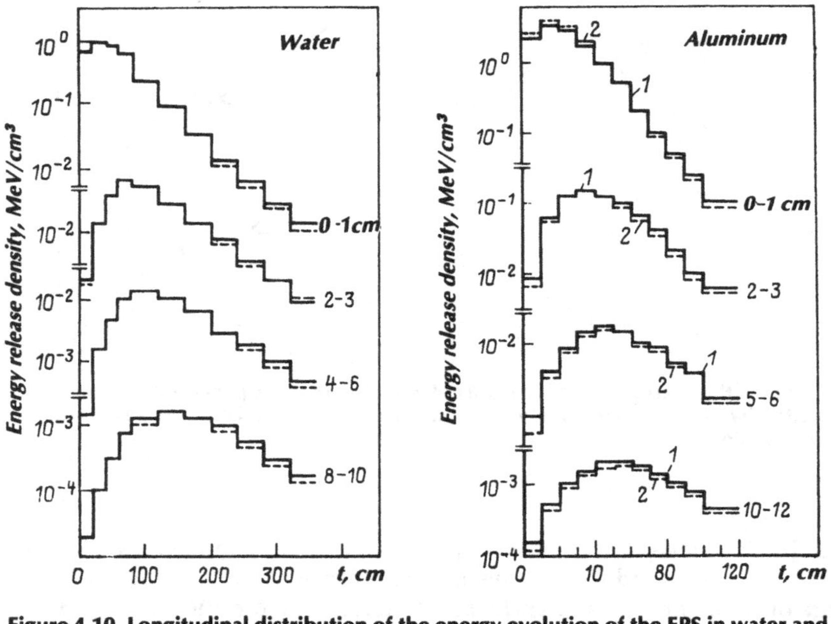


Figure 4.10. Longitudinal distribution of the energy evolution of the EPS in water and aluminum initiated by a 1-GeV electron (Ref. 42). 1, The experimental results; 2, the results of Monte Carlo calculations (appearing next to each curve, which describes the energy evolution density of the shower, is a radially directed interval for which this curve was obtained.)

Figure 4.11 shows the results of an experimental study of the development of a positron-initiated electromagnetic shower. The experiment was carried out with use of a streamer chamber placed in a magnetic field. This method made it possible to accurately determine the energy cutoff threshold and to vary it at will. Also shown in this figure are the results of calculations of the development of an electromagnetic shower under conditions corresponding to those of the experiment. At secondary-electron energies $E \geqslant 5$ MeV the results of calculations are found to be in good agreement with the

**Figure 4.11. The average number of electrons and positrons in a shower initiated by a 1-GeV positron versus the penetration depth of the shower in lead (Ref. 42) at various secondary-electron (positron) energies *E*. 1, Monte Carlo calculations with allowance for the magnetic field; 2, Monte Carlo calculations without consideration of the magnetic field (data points, experimental results).**

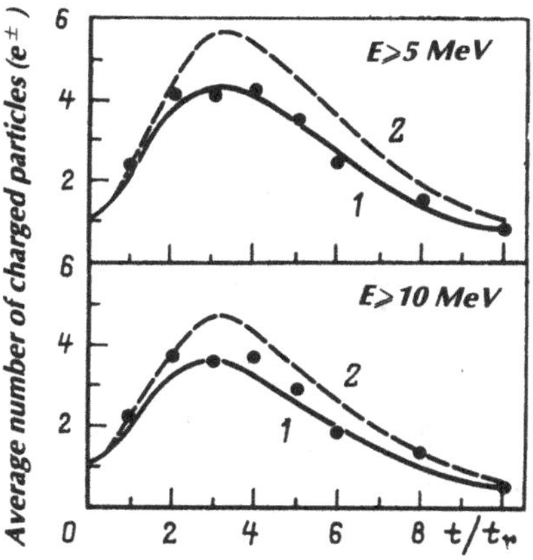


**Figure 4.12. The universal cascade curve (Ref. 97).**

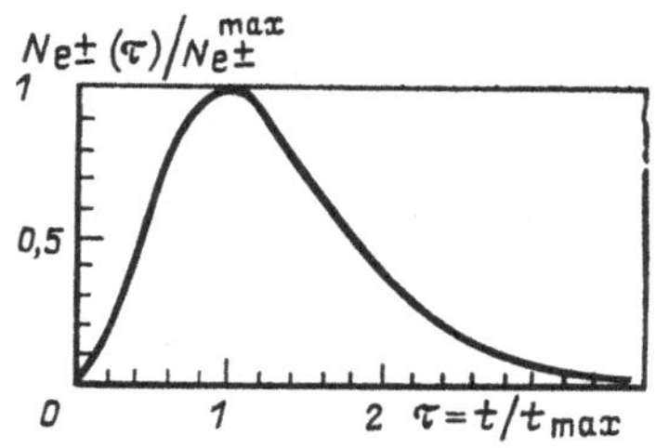


experiment and at $E \geqslant 10$ MeV they are in satisfactory agreement with the experiment. At $E \geqslant 25$ MeV (not shown in the figure) the agreement between the calculation and experiment is slightly worse.

Let us consider one more result which is important for practical applications: the theoretical cascade curve (Fig. 4.12), which is largely universal and which gives the dependence of the relative number of charged particles in the electromagnetic shower, $N_e/N_e^{\max}$, on its penetration depth $\tau$, expressed in units $\tau = t/t_{\max}$ ($t_{\max}$ is the maximum penetration depth of the EPS). $N_e^{\max}$ and $t_{\max}$ are determined from the approximate equations[98]: $N_e^{\max} = 10E_0^{0.9}$ and $t_{\max} = 1.08 \ln(E_0/0.05)$, where $E_0$ is the total energy of the shower (GeV), and $t_{\max}$ is given in radiation lengths. This universal cascade curve describes the behavior of the EPS, regardless of the energy of the particle that initiates the cascade and regardless of the type of material in which the shower develops.

The average total length of the tracks of charged particles in an electromagnetic cascade, a quantity which is of practical importance, may also be regarded as a characteristic of the longitudinal development of the EPS shower. Monte Carlo calculations show that the total length of the electron and positron tracks is proportional to the energy that initiates the EPS[97]:

$$\langle L \rangle = a\,(E_0/\varepsilon_{\text{crit}})t_r. \tag{4.28}$$

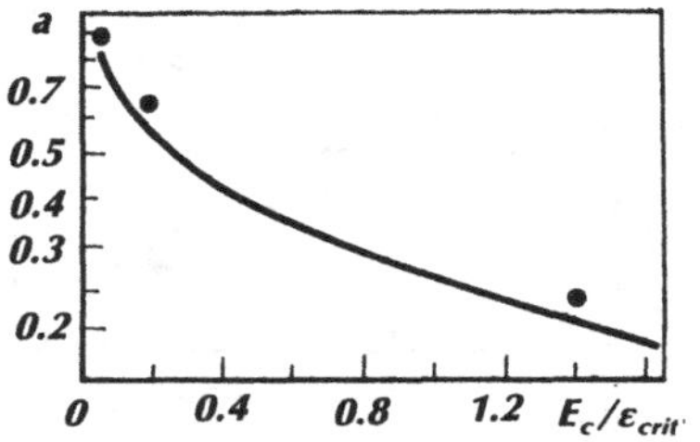


**Figure 4.13. The coefficient $a = FE_c/\epsilon_{crit}$ in Eq. (4.28) vs the ratio $E_c/\epsilon_{crit}$ (Ref. 86). Curve, calculations based on Eq. (4.29) in approximation $B$ of the cascade theory; data points, Monte Carlo calculations.**

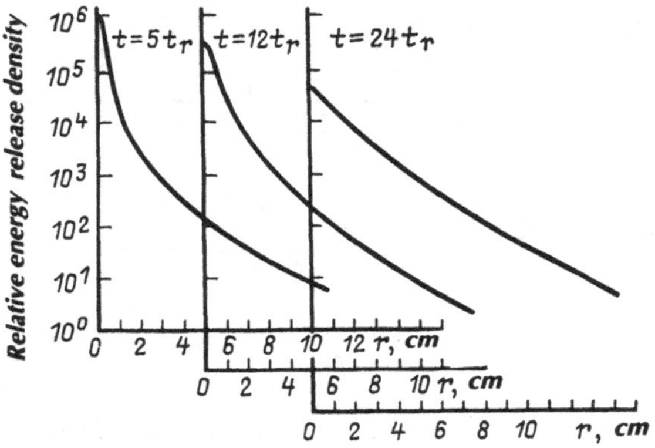


**Figure 4.14. The experimental dependence of the energy-evolution density on the distance $r$ from the shower initiated by a 6-GeV electron in lead (Ref. 96).**

The coefficient $a$, a function of the detection energy threshold $E_c$, increases as this threshold is lowered and in the limit $E_c \to 0, a \to 1$, in complete agreement with the result which follows from the analytic solution of the problem in approximation $B$ (see Table 4.2). In approximation $B$ the coefficient $a$ is given as a function of the detection energy threshold by[86]

$$a = F(\eta) = 1 + \eta e^{\eta} Ei(-\eta) \simeq e^{\eta}[1 + \eta \ln(\eta/1.526)], \tag{4.29}$$

where $\eta = 2.29 E_c/\varepsilon_{crit}$. The approximate expression for the function $F(\eta)$ (Fig. 4.13) is characterized by an error no greater than 10% for $\eta \leqslant 0.3$. Expression (4.29) describes well the behavior of the factor $a$ as a function of $\eta$ for light substances. It was shown in Ref. 86 that expression (4.29) describes well the data obtained for the values of the factor $a$ for light and heavy substances if the parameter $\eta$ is defined as $\eta = 4.58(Z/A)(E_c/\varepsilon_{crit})$.

### Development of an electromagnetic shower in the transverse direction

The transverse distribution of particles in the EPS has two characteristic regions: a narrow central region and a broad peripheral region (Fig. 4.14). The

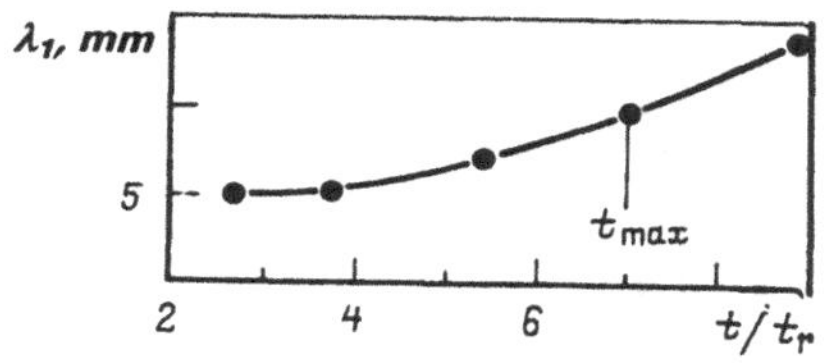


**Figure 4.15. Experimental dependence of the width of the central region of the transverse profile of the shower initiated by a 32-GeV electron in lead (Ref. 96).**

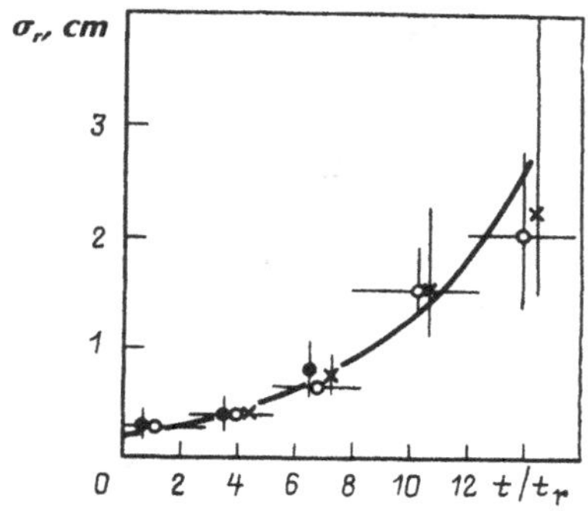


**Figure 4.16. The experimentally measured width of the shower initiated in lead by an electron with energy of 1 GeV (*x*), 2 GeV (○), and 4 GeV (●) (Ref. 96).**

shower profile extending in the transverse direction can be approximated with good accuracy by an approximate expression[97]

$$F(r,t) = C_1 \exp(-r/\lambda_1(t)) + C_2 \exp(-r/\lambda_2(t)).$$

In this expression the first term describes the central region and the second term describes the peripheral region. The curve $\lambda_1(t)$ is shown in Fig. 4.15. As the penetration depth $t$ of the shower is increased, the width of its central region increases (see Figs. 4.14 and 4.15).

In the transverse distribution of particles in the EPS the central region is formed as a result of multiple Coulomb scattering of electrons (and positrons) in the material. This part of the transverse distribution virtually ceases to depend on the atomic number of the material if the distance $r$ from the shower axis is expressed in the Molière units [we recall that the Molière unit of length is $r_M = (E_s/\varepsilon_{crit})t_r$, where $E_s = 21.2$ MeV, and $\varepsilon_{crit}$ is the critical energy of a given material]. Since the average energy per secondary particle in the shower decreases with the penetration depth, the scattering of charged particles increases. Passing through the bulk of the material, moreover, the charged particles are scattered through large angles. All these factors account for the broadening of the central region of the transverse distribution of particles in the EPS with increasing penetration depth.

At a large distance from the shower axis and at a large penetration depth ($t \gtrsim 2t_{max}$) the transverse distribution of secondary particles is determined by photons with the maximum penetrating power. In the case of the peripheral region, when $t \gtrsim t_{max}$, $\lambda_2 \simeq \Lambda/\sqrt{3}$, where $\Lambda$ is the attenuation length of the EPS

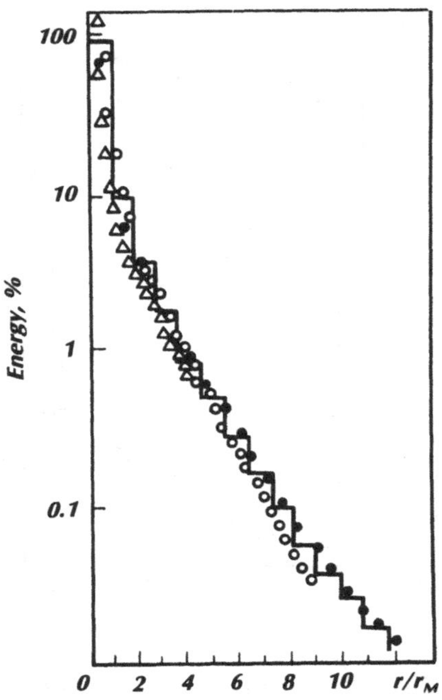


**Figure 4.17. The part of the energy of the EPS released outside the cylinder of radius *r* (Ref. 96) in experiments with an aluminum absorber (△), a copper absorber (●), and a lead absorber (○) for EPS initiated by a 6-GeV electron (histogram, the results of Monte Carlo calculations for a copper absorber).**

in the longitudinal direction. Taking the central and peripheral regions into account, we see that the shower width increases with increasing depth $t$ and that it does not depend on the primary-electron energy (Fig. 4.16).

An important characteristic of the development of the EPS is that part of the energy which is released in the region bounded by the cylindrical surface with a radius $r$, whose axis coincides with the shower axis. It is also possible to introduce a characteristic such as the energy release outside of this cylindrical region (Fig. 4.17). For comparison, Fig. 4.17 shows the results of a Monte Carlo calculation. The theoretical predictions are in good agreement with the experiment.

The results of the measurements and of the theoretical calculations show that the size of the cylindrical region, in which approximately 95% of the shower energy is released, is $r\,(95\%) \simeq 2r_{\mathrm{M}} \approx 2t_r(E_s/\varepsilon_{\mathrm{crit}})$.

A comparison of the results of theoretical calculations of various characteristics of electromagnetic showers with many experimental data shows that the theory is in good agreement with the experiment. This circumstance, in turn, allows the calculation methods to be used extensively for the solution of a broad range of practical problems involving the development of the EPS. The advantage of this approach is its universal applicability, and its reliability is evident from the results presented above, which show the Monte Carlo calculations to be in good agreement with the experimental data on all principal characteristics of an electromagnetic cascade in condensed matter.

# Chapter 5 Solution of equations for nucleon–meson cascades

## 1. Approximate methods of cascade calculations

A theory of the passage of radiation through matter requires a thorough knowledge of the characteristics of interaction of the beam particles with the atoms (electrons, nuclei) of the medium. A clear picture of the interaction of particles of various species with matter makes it possible to identify in many specific cases the main features of the processes and hence to simplify the analytic or numerical solution of the transport problems. The relationship among the various physical processes and their systematic features, analyzed in Chaps. 1–3, can be used to describe the phenomena observed in a complex process such as an electromagnetic cascade in condensed matter.

Most of the theoretical results obtained in the solution of the problems involving the passage of high-energy particles through matter are based on the use of the Monte Carlo method. This method can be used most effectively to describe a large variety of physical phenomena accompanying the development of an electromagnetic cascade for nearly any geometry of the medium with complex nonuniformities. The limited capabilities of the Monte Carlo method as applied to large extended systems, the need for high-speed variational calculations, and an attempt to identify and study the systematic features of radiation transport (admittedly with a limited number of phase-space variables) have created the need, however, for the development of original approximate (analytical and numerical) methods. Historically, these were the first methods to be developed and they are still the principal methods in some applications.

In this chapter we describe the most common approximate methods of solving the problem of radiation transport in a semi-infinite homogeneous layer of the material on which a broad unidirectional beam of high-energy hadrons is incident along the normal. We consider the propagation at high energies of three principal components of the hadronic cascade: protons, neutrons, and $\pi^{\pm}$ mesons (see Sec. 1.1), although the described methods can also be used for other particles. Unless otherwise stipulated, the kinetic energies of primary-beam particles, $E_0$, and those of the propagating particles, $E$, are measured in gigaelectron-volts, while the layer depth $x$ is measured in g/cm$^2$. The systematic features of the development of high-energy electromagnetic cascades in matter are discussed in the next chapter.

## 2. Solution of the cosmic-radiation transport equations

Cosmic-ray physics has been dealing for some time now with the problem of the passage of high-energy particles through matter. Since specific methods and specific approximations are customarily used in this field of physics, we shall consider them separately.

The earliest theoretical calculations of the cascade process in the atmosphere were carried out by Rozental'.[99] The principal assumptions usually used in cosmic-ray physics in solving the kinetic equations are the following[100–102]:

(1) The primary cosmic radiation incident on the earth's atmosphere consists of the flux of protons and $\alpha$ particles with a purely power-law energy spectrum: $P_j(E) = C_j E^{-\gamma-1}$; $\gamma = 1.7 \pm 1$.

(2) The $\pi$ mesons do not generate secondary nucleons.

(3) The hadron–nucleus interaction cross sections are independent of the energy. The macroscopic cross sections are $\Sigma_p = \Sigma_n = \lambda_N^{-1}$; $\lambda_N$ is the mean-free path of a nucleon in the atmosphere before the nuclear interaction.

(4) The electromagnetic interactions of protons and $\pi^{\pm}$ mesons are ignored; i.e., there is no fourth term in Eq. (1.28).

(5) The forward approximation (1.34) is used for the angular distribution of relativistic secondary hadrons.

(6) The inelastic coefficients are independent of the energy. This assumption implies that the scattering indicatrices are homogeneous (the scaling in the variable $u$): $K(E', E)dE = K(u)du$; $u = E/E'$.

The use of these approximations makes it possible to greatly simplify Eqs. (1.26) and (1.27) and to find their solutions.

The equation for the spatial-energy spectrum of nucleons, for example, becomes

$$\partial\Phi_j(x,E)/\partial x = -(1/\lambda_j)(1-G_{jj})\Phi_j(x,E) + \sum_{i\neq j}\frac{1}{\lambda_i}G_{ij}\Phi_i(x,E);\ j=p,n;\ i=p,n,\alpha, \tag{5.1}$$

where

$$\left.\begin{aligned}\Phi_p(0,E) = P(E) = C_p E^{-\gamma-1};\quad \Phi_n(0,E) = 0;\\ \Phi_\alpha(x,E) = \alpha(E)\exp(-x/\lambda_\alpha);\quad G_{ij}\int_0^1 u^\gamma K_{ij}(u)du.\end{aligned}\right\} \tag{5.2}$$

Equation (5.1) for protons can be solved by means of the function

$$\Phi_p(x,E) = \frac{P(E)}{2}\left[\exp\left(-\frac{x}{\lambda}\right) + \exp\left(-\frac{x}{\tilde{\lambda}}\right)\right] + \frac{\alpha(E)}{2}\frac{\lambda}{\lambda-\lambda_\alpha}G_{\alpha N}\left[\exp\left(-\frac{x}{\lambda}\right) - \exp\left(-\frac{x}{\lambda_\alpha}\right)\right], \tag{5.3}$$

where the parameter values for the earth's atmosphere are[101] $G_{pp} = G_{nn} = 0.2$, $G_{pn} = G_{np} = 0.1$, $G_{NN} = G_{pp} + G_{pn} = 0.3$, $G_{\alpha N} = 2.6$, $\lambda_p = \lambda_n = \lambda_N = 95$ g/cm$^2m$ $\lambda_\alpha = 46.5$ g/cm$^2$, $\lambda = \lambda_N(1 - G_{NN})^{-1} = 135$ g/cm$^2$, $\tilde{\lambda} = \lambda_N(1 - G_{pp} + G_{np})^{-1} = 105$ g/cm$^2$, $P(E) = 2.09E^{-\gamma-1}$, and $\alpha(E) = 0.11E^{-\gamma-1}$.

The depth of the atmosphere $x$ is linearly related to the density $\rho$ and temperature $T$ of the air by $x = \text{const}\,\rho T/\cos\theta_0$, where $\theta_0$ is the zenith angle. As $x \to \infty$, we have $\Phi_n/\Phi_p \to 1$.

Using approximations (1)–(6) and solving (5.3), we find the $\pi$-meson distribution[101]

$$\Phi_\pi(x, E, \theta_0) = \sum_{k=1}^{\infty} \pi_k(\lambda, E, \theta_0) x^k \exp(-x/\lambda). \tag{5.4}$$

Here

$$\pi_1(\lambda, E, \theta_0) = \frac{G_{N\pi}}{\lambda_N}\left[P(E) + \frac{\lambda}{\lambda - \lambda_\alpha}\alpha(E)\, G_{\alpha N}\right] \times\left(1 + \frac{B_\pi}{E\cos\theta_0}\right)^{-1}; \tag{5.5}$$

$$\pi_k(\lambda, E, \theta_0) = \pi_{k-1}(\lambda, E, \theta_0)\left[\frac{1}{\lambda} - \frac{1}{\lambda_{k-1}}\right]\left(k + \frac{B_\pi}{E\cos\theta_0}\right)^{-1}, \quad k \geqslant 2, \tag{5.6}$$

where the parameters of the earth's atmosphere have the following values: $G_{N\pi} = G_{p\pi}^{+} + G_{p\pi^-} = G_{n\pi}^{+} + G_{n\pi}^{-} = 0.08$; $B_\pi = 113$ GeV; $\lambda_{k-1} = \lambda_\pi [1 - G_{\pi\pi}(k-1, \lambda, E, \theta_0)]^{-1}$, $k \geqslant 2$; $\lambda_\pi = 115$ g/cm$^{-2}$; $G_{\pi\pi}(k-1, \lambda, E, \theta_0) = \pi_{k-1}^{-1}(\lambda, E, \theta_0)\int_E^\infty \pi_{k-1}(\lambda, E', \theta_0) \times g_{\pi\pi}\left(\frac{E}{E'}\right)\frac{dE}{E'}$; and $g_{\pi\pi}(u) = 20\exp(-7u)$. The rest of the notation in Eqs. (5.4)–(5.6) is the same as that in expression (5.3).

In the limit $E \to \infty$ the solution of the distribution function is

$$\Phi_\pi(x, E) \approx f(E) x \exp(-x/\lambda_0),$$

and

$$f(E) = \frac{G_{N\pi}}{\lambda_N}\left[P(E) + \frac{\lambda}{\lambda - \lambda_\alpha}\alpha(E) G_{\alpha N}\right],$$

and $\lambda_0$ is the limit $\lambda_{k-1_{E\to\infty}} \to \lambda_0$.

Having the solution of (5.4), we can calculate the densities of the sources (1.30) of muons and muonic neutrinos from the decays $\pi^\pm \to \mu^\pm + \nu_\mu(\bar{\nu}_\mu)$, and we can solve the corresponding muon-transport problem. The function (5.4) can be used to calculate the source of photons from the decays $\pi^0 \to \gamma\gamma$ and to solve the equations for the electron–photon shower (see Chap. 4).*

*At extremely high energies the photons of the muon bremsstrahlung and the direct production of electron-positron pairs must also be taken into account.

## 3. Analytic methods of calculating high-energy radiation transport

Let us consider other approximate methods for the solution of the transport problem involving the transport of hadrons through matter, which were developed (like all subsequent models) in the context of high-energy physics for use with proton accelerators.

In the early studies[103,104] and in cosmic-ray physics a direct approximation was used for the angular distributions of secondary hadrons. In this approximation the solution for the unscattered component of radiation can easily be written. For neutrons we have

$$\Phi_{0n}(x, E) = \Phi_{0n}(0, E)\exp[-\Sigma_n(E)x]. \tag{5.7}$$

For protons and $\pi^{\pm}$ mesons we have

$$\Phi_{0j}(x, E) = \Phi_{0j}(0,E_j)\frac{\beta_j(E_j)}{\beta_j(E)}\,exp\left[-\int_E^{E_j}\frac{\Sigma_j(E') + \Sigma_{jD}(E')}{\beta_j(E')}\,dE'\right], \tag{5.8}$$

where $E_j(x, E)$ are determined from the equation $x = \int_E^{(E_{jx}, E)} dE'/\beta_j(E')$.

For the monochromatic beam with an energy $E_0$ we have

$$\Phi_{0j}(0,E_j) = N_{0j}\delta[E_0 - E_j(x,E)]. \tag{5.9}$$

To find a solution for the scattered component of $\Phi_j(x, E)$, it is necessary to assume that at sufficiently high energies the dependence of the cross sections $\Sigma_j(E) \simeq \Sigma_j$, the electromagnetic interaction of protons and $\pi^{\pm}$ mesons $[\beta_j(E) \simeq 0]$, and the decay of $\pi^{\pm}$ mesons $[\Sigma_{\pi D}(E) \simeq 0]$ can be ignored. In the one-dimensional case the differential operator in (1.28) will then become

$$\hat{L} = \partial/\partial x + \Sigma_j. \tag{5.10}$$

An analytic solution of the boundary-value problem (1.26), (1.27) can be found by choosing a special form of the scattering indicatrix (1.34). In Ref. 104 the scattering indicatrix was expressed in the form

$$K_{ij}(E',E) = N_j E'^{l}/E^{l+1}, \tag{5.11}$$

where the numerical values of the parameters $N_j$ are proportional to the asymptotic values of the inelasticity coefficients of particles of species $j$; $N = \sum_j N_j = 0.64$ and $l = 0.216$. Substituting expression (5.11) into relation (1.29) and using (5.10), we find

$$\Phi_j(x, E) = N_j\Sigma_j\,\exp(-\Sigma_j x)\,(E_0^l/E^{l+1}) \times\left[\frac{x}{B(E_0, E)}\right]^{1/2} I_1[2\sqrt{xB(E_0, E)}]\,, \tag{5.12}$$

where $B(E_0, E) = N\ln(E_0/E)$, $I_1(y)$ is a modified Bessel function of the first kind, and $E_0$ is the energy of primary protons in a broad unidirectional beam.

The simple solution (5.12) is in satisfactory agreement with the experiment of Ref. 104.

It was shown in Ref. 103 that in the approximations indicated above the solution of the transport problem (1.26), (1.27) for nucleons and $\pi$ mesons with the operator $\hat{L}$ in the form (5.10) can be obtained by quadrature if the scattering indicatrix (1.34) is represented in the general form

$$K_{ij}(E',E)=\frac{g_i(E')}{g_j(E)}\,[F_j(E',E)+\eta(E')h(E)\delta_{ij}], \tag{5.13}$$

under the condition $\sum_j F_j(E',E)+\eta(E')h(E)=M(E')H(E)$, where $g_j(E)$, $M(E')$, and $H(E)$ are arbitrary functions.

The expressions for the differential pion and nucleon flux densities can be found in explicit form by choosing among the inclusive cross sections (see Chap. 3) those cross sections that satisfy relation (5.13).

If, for example,

$$K_{ij}(E',E)=\frac{g_i(E')}{g_j(E)}\,[\alpha_j(E')+\delta_{ij}\eta(E')],$$

$$M(E')=\alpha(E')+\eta(E');\Sigma=\Sigma_j;C(E_0,E)=\Sigma\int_E^{E_0}M(E')dE';$$

$$B(E_0,E)=\Sigma\int_E^{E_0}\eta(E')dE,$$

then after Laplace transformations with respect to $x$, the transport equations can be solved, if condition (5.9) is satisfied, in the form[103]

$$\begin{aligned}\Phi_j(x,E)=&\frac{\exp(-\Sigma x)}{g_j(E)}\,\Sigma N_{0j}\Big\{[\eta(E_0)\delta_{pj}+\alpha_j(E_0)]\sqrt{\frac{x}{B(E_0,E)}}\\&\times I_1[2\sqrt{xB(E_0,E)}]+M(E_0)\int_E^{E_0}dE'\,\frac{\alpha_j(E')\Sigma x}{C(E_0,E')+B(E',E)}\\&\times I_2[2\sqrt{x[C(E_0,E')+B(E',E)]}]\Big\}\\&-\frac{\exp(-\Sigma x)}{g_j(E)}\int_E^{E_0}dE'\int_0^x dx'\Big\{I_0[2\sqrt{B(E',E)(x-x')}]\\&\times\frac{\partial}{\partial E'}\,[g_j(E')\tilde{G}_j(x',E')\exp(\Sigma x')]\\&+\Big[\frac{\partial}{\partial E'}\sum_i g_i(E')\tilde{G}_i(x',E')\exp(\Sigma x')\Big]\int_E^{E'}dE''\,\Sigma\alpha_j(E'')\\&\times\Big[\frac{(x-x')}{C(E',E'')+B(E'',E)}\Big]^{1/2}I_1[2\sqrt{(x-x')[C(E',E'')+B(E'',E)]}]\Big\},\end{aligned}$$

where $\tilde{G}_j(x, E)$ is the density of external sources [Eq. (1.30)], $I_0$, $I_1$, and $I_2$ are the modified Bessel functions of appropriate orders, and $i, j = p, n, \pi^{\pm}$.

Examples in which other scattering indicatrices were used and the description of the transformation techniques can be found in Ref. 103.

The effects which were ignored above in the solution of the boundary-value problem may be viewed as small perturbations and may be included in the density of the external sources:

$$\tilde{G}_j(x, E) = G_j(x, E) + \left[\frac{\partial}{\partial E}\beta_j(E) + \Sigma_j - \Sigma_j(E) - \Sigma_{jD}(\Sigma)\right]\Phi_j(x, E). \quad (5.14)$$

At high energies the perturbation-theory methods can also be used effectively to find the angular dependence of the differential flux density. Instead of (1.34), we can write the scattering indicatrix in the form[105]

$$\mathscr{P}_{ij}(E', E, \mu_s) = K_{ij}(E', E)\frac{\delta(1-\mu_s)}{2\pi} + F_{ij}(E', E, \mu_s), \quad (5.15)$$

where $F_{ij}$ is the correction term which makes Eq. (5.15) correct (or approximately correct). A strong angular anisotropy of secondary hadrons, which are formed in high-energy hadron–nucleus interactions, accounts for the fact that the second term in Eq. (5.15) is much smaller than the first. (The use of a simple approximation at high energies has some advantage in this case.) The iteration method of solving the system of cascade equations can thus be used. One of the methods described above can be used as a zeroth approximation to solve the problem in the simple approximation

$$\Phi_j^{(0)}(x, E, \mu) = \Phi_j(x, E), \mu = \cos\theta. \quad (5.16)$$

In the first approximation the system of equations for the differential flux density in a semi-infinite plane layer of thickness $H$ is

$$\left.\begin{aligned} \hat{L}_j\Phi_j^{(1)} &= Q_j^{(1)} + G_j^{(1)};\\ \Phi_j^{(1)}(0, E, \mu) &= 0, 0 < \mu \leqslant 1;\\ \Phi_j^{(1)}(H, E, \mu) &= 0, -1 \leqslant \mu < 0, \end{aligned}\right\} \quad (5.17)$$

where the operator $\hat{L}_j$ is the same as that in (1.28), with allowance for Eq. (1.33)

$$Q_j^{(1)}(x, E, \mu) = \sum_i \int_E^{E_0} K_{ij}(E', E)\Sigma_j(E')\Phi_j^{(1)}(x, E', \mu)dE'; \quad (5.18)$$

$$G_j^{(1)}(x, E, \mu) = G_j(x, E, \mu) + \sum_i \int_E^{E_0} F_{ij}(E', E, \mu)$$

$$\times \Sigma_i(E')\Phi_j^{(0)}(x, E', \mu)\, dE', \quad (5.19)$$

Note that the quantity $\mu = \cos\theta$ appears in Eqs (5.17)–(5.19) as a parameter rather than as a variable. This means that the same algorithms are used to solve Eqs. (5.17) and the equations in the simple approximation described above. Substituting the solution of Eqs. (5.17) into Eq. (5.19) and repeating the iterations the necessary number of times, we find the solution of the angular-flux problem.

In the *small-angle-approximation*, two methods of solving the boundary-value problem (1.26), (1.27) for the hadronic cascade in a plane semi-infinite layer were developed in Refs. 106–108.

Keeping in mind that the iteration procedure will further be used [Eq. (5.14)], we assume that at $E > 1$ GeV the following relations are valid:

$$\left.\begin{aligned}&\Sigma_j(E) = E_0;\ \Sigma_j^{\text{ion}}(E) + \Sigma_{jD}(E) = \Sigma_{Ej}(E);\\ &\mathscr{P}_{ij}(E', E, \mu_s) = \frac{g(E')}{g(E)} K_j(E', E, \mu_s),\end{aligned}\right\} \tag{5.20}$$

where $\Sigma_0 = \text{const}$, $\Sigma_j^{\text{ion}}(E)$ is the macroscopic cross section for taking the particle of $i$ th species out of a given energy interval as a result of ionization loss, and $g(E)$ is a function which smooths out the energy dependence of the scattering indicatrix.

The quantity $\Sigma_j^{\text{ion}}$ can be found from the relation

$$\Sigma_j^{\text{ion}}(E)\Phi_j(x, E, \mu) = -\frac{\partial}{\partial E}[\beta_j(E)\Phi_j(x, E, \mu)]. \tag{5.21}$$

Integrating expression (5.21) over all values of $x$ and $\mu$ and taking into account that at high energies $\beta_j'(E) \approx 0$, we find

$$\Sigma_j^{\text{ion}}(E) \approx -\beta_j(E)\xi_j'(E)/\xi_j(E),$$

where

$$\xi_j(E) = \int_{-1}^{1} d\mu \int_0^\infty dx \Phi_j(x, E, \mu). \tag{5.22}$$

The function (5.22) can be found by using the iteration procedure or the solution which was found by the approximate methods that were described above. In the rough approximation[106]

$$\Sigma_j^{\text{ion}}(E) \approx C(A)\beta_j(E)/E, \quad E > 1 \text{ GeV},$$

where $C(A)$ is a constant which depends exlusively on the atomic mass of the atoms of the medium.

The function $g(E)$ can be parametrized in the form $g(E) = E^{-n}$, where $n \approx 2$. After introducing the notation[103]

$$\left.\begin{aligned}&\Psi_j(x, E, \mu) = g(E)\Phi_j(x, E, \mu)\exp(E_0, x),\\ &\Psi = \sum_j \Psi_j;\ K_0(E', E, \mu_s) = \sum_j K_j(E', E, \mu_s),\end{aligned}\right\} \tag{5.23}$$

and then expanding in Legendre polynomials, using relation (5.20), and summing the equations, we can transform system (1.26) and reduce it to a single integrodifferential equation for the function $\Psi_l(x, E)$ and to $j$ differential equations for the functions $\Psi_{jl}(x, E)$.[106,108] Here $\Psi_l$, $\Psi_{jl}$, $K_{0l}$, and $K_{jl}$ are the $l$ th harmonics of the expansion of the functions (5.23) in Legendre polynomials.

At small angles the terms with large $l$ are the dominant terms in these expansions. Since $P_l(\cos\theta) \simeq J_0(l\theta)$, where $J_0(y)$ is the zeroth-order Bessel function, we have

$$\Psi_j(x, E, \cos\theta) = \int_0^\infty J_0(l\theta)\Psi_{lj}(x, E)l dl. \tag{5.24}$$

At large values of $l$ the systems of equations infinite in $l$ reduce approximately to the equations

$$\frac{\partial \Psi_l(x, E)}{\partial x} + \Sigma_E(E)\Psi_l(x, E) = \Sigma_0 \int_E^{E_0} dE' K_{0\,l}(E', E)\Psi_l(x, E') + G_l(x, E); \tag{5.25}$$

$$\frac{\partial \Psi_{j\,l}(x, E)}{\partial x} + \Sigma_{Ej}(E)\Psi_{j\,l}(x, E) = \Sigma_0 \int_E^{E_0} dE' K_{j\,l}(E', E)\Psi_{j\,l}(x, E) + G_{j\,l}(x, E). \tag{5.26}$$

At high energies we have $\Psi_l(0, E) = \Psi_{j\,l}(0, E) = 0$.

In Eqs. (5.25) and (5.26) $\Sigma_E = \sum_j \Sigma_{Ej}$, $G_l = \sum_j G_{j\,l}$, and $G_{j\,l}(x, E)$ are the harmonics of the expansions of the external-source density in Legendre polynomials.

Applying to Eqs. (5.25) and (5.26) the Laplace transform with respect to $x$ and differentiating these equations with respect to energy, we find an explicit expression for the $l$ th harmonic of the differential flux density of hadrons of species $j$.

$$\Phi_{j\,l}(x, E) = \frac{\Sigma_0}{2\pi i} \exp(-\Sigma_j(E)\,x) \frac{1}{g(E)} \times \int_{c - i\infty}^{c + i\infty} \exp(\lambda, x) \frac{d\lambda}{\lambda + \Sigma_{Ej}(E)} \int_E^{E_0} dE' K_{j\,l}(E', E) \times [\,f_l(\lambda, E') + f_{0\,l}(\lambda, E')], \tag{5.27}$$

where $f_l(\lambda, E) = \int_0^\infty \exp(-\lambda x)\Psi_l(x, E)\,dx$, and $f_{0l}(\lambda, E)$ is a Laplace transform for functions of the type (5.7) and (5.8) in the problem involving normal incidence of a broad unidirectional hadron beam on a barrier.

This problem can be solved on the basis of Eq. (5.27) by evaluating an integral of the type (5.24). Expression (5.27) with $l = 0$ yields a direct estimate of the energy spectrum.

The solution found for the primary-beam particle energy between 20 GeV and 70 GeV is in satisfactory agreement with the experimental data.[106,108]

The spatial and angular distributions of cascade particles in a planar layer, bombarded by unidirectional monoenergetic beam of hadrons of species $i_0$ with an energy $E_0 \gg 1$ GeV, can be determined by using a method developed in Ref. 107. According to this method, the differential flux density in the collision integral should be expanded in a series in the angular variable up to the second-order terms, inclusively. We integrate the equations which we obtained over the energy within the limits of certain groups or, if $E_{\min} \gg 1$ GeV, we integrate over the entire energy spectrum. Here we take advantage of the fact that at high energies $\Sigma_j(E) + \Sigma_{Ej}(E) \approx \Sigma_j$ [see Eq. (5.20)]. The spatial and angular distribution function $\Phi_j(x, \mu)$ will then satisfy the following system of differential equations:

$$\mu \frac{\partial \Phi_j(x,\mu)}{\partial x} + \Sigma_{0j}\Phi_j(x,\mu) = \sum_i \{\alpha_{ij}\Phi_i(x,\mu) + \beta_{ij}L[\Phi_i(x,\mu)]\} + \eta_{ij}\exp(-\Sigma_i x); \tag{5.28}$$

$$\Phi_j(0, \mu > 0) = 0; \quad \Phi_j(H, \mu < 0) = 0,$$

where

$$L = \frac{\partial}{\partial \mu}\left[(1-\mu^2)\frac{\partial}{\partial \mu}\right];$$

$$\left.\begin{aligned} \alpha_{ij} &= \int_{E_{\min}}^{E_{\max}} \frac{dE}{E_{\max} - E} \int_E^{E_{\max}} dE' \int_{-1}^{1} d\mu_s \Sigma_i(E) \mathscr{P}_{ij}(E', E, \mu_s); \\ \beta_{ij} &= \frac{1}{2} \int_{E_{\min}}^{E_{\max}} \frac{dE}{E_{\max} - E} \int_E^{E_{\max}} dE' \int_{-1}^{1} d\mu_s (1-\mu_s) \\ &\quad \times \Sigma_i(E) \mathscr{P}_{ij}(E', E, \mu_s); \\ \eta_{ij} &= \int_{E_{\min}}^{E_{\max}} \frac{dE}{E_{\max} - E} \int_E^{E_{\max}} dE' \Sigma_i(E) \mathscr{P}_{ij}(E', E, \mu_s = 1). \end{aligned}\right\} \tag{5.29}$$

Strictly speaking, a weighting function $\Phi_j(x, E, \mu)$ should be included in taking an average of (5.29). If, however, the width of the group $E_{\max} - E_{\min}$ is small, expressions (5.29) are approximately correct.

At $i = j$ the fundamental solution of the system of equations (5.28) can easily be found

$$f_j(x,\mu) = \frac{\exp(-\alpha_{ij}x)}{2\sqrt{\beta_{ij}x}}\left[\frac{1}{\sqrt{\pi}}\exp\left(-\frac{1-\mu}{4\beta_{ij}x}\right) + \frac{1}{\sqrt{\beta_{ij}x}}\exp\left(-\frac{1-\mu}{2\beta_{ij}x}\right)\right], \quad i = j.$$

The function which we are seeking can then be determined by an iteration method:

$$\Phi_j(x,\mu) = \sum_i \sum_{\nu=0}^{N} f_{if}^{(\nu)}(x,\mu). \tag{5.30}$$

Here $\nu$ is the iteration number, and

$$f_{ij}^{(\nu)}(x,\mu) = f_{ij}^{(0)}(x,\mu) + \sum_{\nu'=1}^{\nu} f_{ij}^{(\nu')}(1 - \delta_{ij}),$$

where

$$f_{ij}^{(0)}(x,\mu) = \int_0^x dx' \int_{-1}^{1} d\mu' f_j(x,x',\mu,\mu')\eta_{ij} \exp(-\Sigma_i x),$$

and

$$f_{ij}^{(\nu)}(x,\mu) = \int_0^x dx' \int_{-1}^{1} d\mu' f_j(x,x',\mu,\mu')[\alpha_{ij} + \beta_{ij} L] f_{ij}^{(\nu-1)}.$$

The sum (5.30) calculated for $N = 1$ is nearly identical to the results of a numerical solution of Eqs. (5.28) for moderately high energies and moderately large $x$ and $\theta$. This solution is, moreover, in agreement with the results of Monte Carlo calculations and experimental data for $E_0 = 10$ GeV and 70 GeV (Ref. 107).

## 4. Numerical solutions of cascade equations

At high energies of the primary hadrons, $E_0 > 1$ GeV, the following are the best known numerical solutions of the cascade equations: (1) the multigroup method of discrete ordinates[109]; (2) the multilevel-iteration method[78]; (3) the successive-collision method.[110] A one-dimensional problem, in which a uniform plane layer of thickness $H$ is bombarded by a broad unidirectional hadron beam, is considered in all the cases. No simplifying assumptions concerning the functional dependence of the scattering indicatrix are made in this case.

In the first method of solution the system of kinetic equations for the flux density of nucleons and $\pi^{\pm}$ mesons is integrated over the energy within the limits of each of the $q$ energy groups, into which the entire energy interval under study is divided. In this case the concept of the macroscopic extraction cross section $\Sigma_j^{\text{ion}}(E)$ is used. In addition, the standard multigroup methods are used. The group cross sections and group indicatrices at the 42 points of a nonuniformly divided angular interval $(0,\pi)$ are calculated in the 20-group approximation.

The method of successive approximations is used to solve a system of three equations ($j = p, n, \pi^{\pm}$). The solution of the problem for the $n$th approximation can be broken up into $q_{\max}$ cycles (according to the number of groups), each of which is an iteration process which makes use of the nonlinear acceleration in the calculation of the angular profile of the solution (the mean-flux method). The iterated solution is found by integrating the equations over the characteristics. A multigroup system of equations can be solved numerically by using the well-known POZ-5 reactor programs.[111]

The results of calculations based on the described procedure are in fair agreement with the experimental data at $E_0 \sim 20$ GeV. The information on the spatial, angular, and energy distributions has made it possible to identify some of the systematic features. At energies $E_0 \gtrsim 50$–$100$ GeV the scattering anisotropy cannot, however, be described adequately, because of the limited number of nodes of the angular grid of the discrete-ordinate method. A more detailed analysis showed that this method is not stable enough for processes with a large secondary-particle multiplicity. The averaging of group constants is, moreover, a nontrivial problem. This method can apparently be used only at energies $E_0 \lesssim 10$ GeV in the region where the approximate analytic methods of solution have not yet been fully refined.

The second approach to the solution of the problem of the transport of high-energy particles was developed in Ref. 78. We shall single out in the differential flux density function that part $\Phi_{0\,j}(x, E, \mu)$ which corresponds to the unscattered flux, Eqs. (5.7) and (5.8). The system of equations for the scattered radiation flux can be represented in the form

$$\left.\begin{array}{l} \hat{L}\Phi_j = \hat{S}\Phi_j + G_j; j = p, n, \pi^+, \pi^-, K^+, K^-; \\ \Phi_j(0, E, \mu) = 0, \ 0 < \mu \leqslant 1; \\ \Phi_j(H, E, \mu) = 0, \ -1 \leqslant \mu < 0, \end{array}\right\} \tag{5.31}$$

where

$$\left.\begin{array}{l} \hat{L} = \mu \dfrac{\partial}{\partial x} + \Sigma_{j\,\mathrm{tot}}(E); \Sigma_{j\,\mathrm{tot}}(E) = \Sigma_j(E) + \Sigma_{j\,D}(E); \\ \hat{S}\Phi_j = \displaystyle\int_0^{2\pi} d\varphi \int_{-1}^{1} d\mu' \int_E^{E_0} dE' \Sigma_j(E') \mathscr{P}_{jj}(E', E, \mu_s) \Phi_j(x, E', \mu'); \end{array}\right\} \tag{5.32}$$

$$\left.\begin{array}{l} G_j(x, E, \mu) = G_{0\,j}(x, E, \mu) + \dfrac{\partial}{\partial E} [\beta_j(E) \Phi_j(x, E, \mu)] + \displaystyle\sum_{i \neq j} \hat{S}\Phi_i; \\ G_{0\,j}(x, E, \mu) = 2\pi \displaystyle\sum_i \int_E^{E_0} dE' \Sigma_i(E') \mathscr{P}_{ij}(E', E, \mu) \Phi_{0\,i}(x, E'). \end{array}\right\} \tag{5.33}$$

The functions $\Phi_{0\,i}(x, E)$ are determined by Eqs. (5.7) and (5.8).

Using the Green's function $\Gamma_j(x', x, E, \mu)$ of the operator $\hat{L}$, we find the solution of Eqs. (5.31) for the range of angles $0 < \mu \leqslant 1$

$$\Phi_j(x, E, \mu) = \int_0^x \Gamma_j(x', x, E, \mu)(G_j + \hat{S}\Phi_j)\, dx'. \tag{5.34}$$

In the range of angles $-1 \leqslant \mu < 0$ the limits of integration 0 and $x$ in Eq. (5.34) must be replaced by the limits of integration $x$ and $H$, respectively.

The Green's function of the operator $\hat{L}$ is given by

$$\Gamma_j(x', x, E, \mu) = \exp\left[-\frac{x - x'}{\mu} \Sigma_{j\,\mathrm{tot}}(E)\right].$$

The third term on the right side of Eq. (5.33) can be taken into account iteratively. Introducing the notation

$$f_j(x, E, \mu) = \int_0^x \Gamma_j(x', x, E, \mu) G_j(x', E, \mu)\, dx', \tag{5.35}$$

we can then write Eq. (5.34) for a running iteration as a Volterra integral equation of the second kind:

$$\Phi_j(x, E, \mu) - \int_0^x \Gamma_j(x', x, E, \mu)\hat{S}\Phi_j(x', E, \mu)\, dx' = f_j(x, E, \mu). \tag{5.36}$$

We shall break up the energy interval of the cascade particles into two regions: (1) 0.4 GeV$\leqslant E \leqslant E_0$ and (2) 15 MeV$\leqslant E \leqslant$400 MeV. (At $E \lesssim$ 400 MeV the ionization range is shorter than the range before the inelastic nuclear interaction, and the angular distributions become considerably less anisotropic.) By analogy with relations (5.15)–(5.19), in the first region we use the perturbation theory methods in the angles for the forward hemisphere $0 < \mu \leqslant 1$. Representing the scattering indicatrix in the form (5.15), we find an expression of the type (5.18), instead of Eq. (5.32). Accordingly, the third term in expression (5.33) transforms to that in (5.18) and, additionally, a fourth term appears:

$$\tilde{G}_j^{(\nu)}(x, E, \mu) = G_j(x, E, \mu) + \sum_i \int_0^{2\pi} d\varphi \int_{-1}^{1} d\mu' \int_E^{E_0} dE' \Sigma_i(E')$$
$$\times F_{ij}(E', E, \mu_s)\Phi_i^{(\nu-1)}(x, E', \mu'), \tag{5.37}$$

where $\nu$ is the number of the outer iteration over the angles, and $\Phi_i^{(0)}(x, E, \mu) = \Phi_i(x, E)$ is the solution of Eqs. (5.35) and (5.36) in the zeroth iteration at $F_{ij} = 0$.

Substituting expression (5.37) into Eq. (5.35) and applying the Gaussian quadrature in the finite-sum method in the solution of Eq. (5.36), we find the recurrence relation

$$\Phi_j(x_n, E_l, \mu_k) = \sum_{m=1}^{n} \Bigg\{ w_m \exp\left[ -\Sigma_{j\,\mathrm{tot}}(E_l) \frac{x_n - x_m}{\mu_k} \right]$$
$$\times \frac{\tilde{G}_j(x_m, E_l, \mu_k)}{\mu_k} \tag{5.38}$$
$$+ \sum_{i=1}^{l} w_i \frac{\Sigma_j(E_i)}{\mu_k} K_{jj}(E_i, E_l)\Phi_j(x_m, E_i, \mu_k) \Bigg\},$$

which is used to calculate the function $\Phi_j$ at the nodal points $x_n$, $E_l$, and $\mu_k$, beginning with the end points $x = 0$, $E = E_0$, and $\mu = 1$. Here $w_i$ are the weights of the Gaussian quadrature.

The outer iterations in the angles and the inner iterations in the equations of system (5.31) and in the second term in Eq. (5.33) are completed if the following conditions are satisfied during the $s$th iteration:

$$\max_{\substack{x \in [0, H] \\ \mu \in [0, 1] \\ E \in [0, 4;\, E]}} \frac{|\Phi^{(s)}(x, E, \mu) - \Phi^{(s-1)}(x, E, \mu)|}{|\Phi^{s}(x, E, \mu)|} < \varepsilon, \tag{5.39}$$

where $\varepsilon$ is the relative calculation error. At $\varepsilon = 0.01$ the necessary number of iterations in each of the three cycles is no greater than five.

In the second energy interval, $15 \lesssim E \lesssim 400$ MeV, the system of equations (5.31) in the first stage reduces to the equations

$$\left.\begin{aligned}\hat{L}\Phi_n &= \hat{S}_{nn}\Phi_n + G_{0\,n};\\ \hat{L}\Phi_j - \frac{\partial}{\partial E}\beta_j\Phi_j &= G_{0\,j},\ j\neq n.\end{aligned}\right\} \tag{5.40}$$

The first equation in (5.40) is solved [after it is reduced to the form in (5.36)] by the discrete-ordinate method for the angular and energy variables. The equations such as (5.38) in this case are used for the entire angular interval: $-1\leqslant\mu\leqslant1$. The unscattered radiation and all hadrons with energies $E>0.4$ GeV contribute to the source $G_{0\,n}$. In the case of charged particles, the contribution to the source comes from the unscattered radiation, from all hadrons with energies $E>0.4$ GeV, and from neutrons with energies $15\leqslant E\leqslant400$ GeV. Applying the Laplace transform in the coordinate $x$ to the second equation in (5.40), we find a solution in the form

$$\Phi_j(x,E,\mu)=\frac{1}{\beta_j(E)}\int_E^{0.4}dE'G_{0\,j}\left[x-\mu\int_E^{E'}\frac{dE''}{\beta_j(E'')},E',\mu\right]$$
$$\times\exp\left[-\int_E^{E'}\frac{\Sigma_{j\,\mathrm{tot}}(E'')\,dE''}{\beta_j(E'')}\right]. \tag{5.41}$$

Using Eq. (5.41), we can find the contribution from the charged particles to the sources $G_{0\,n}$ and $G_{0\,j}$ by the iteration method. One or two iterations are necessary in this case.

The algorithm described here is the basis of the HAMLET computer program. Many comparisons of the results of calculations with the experimental data show a good agreement over a broad range of energies, materials, and layer thicknesses of the material.[8,112,113]

The method of successive collisions applied in the form suggested in Ref. 110 can be used effectively for certain types of problems. The distribution function can be written as a series in successive collisions

$$\Phi_j(x,E,\mu)=\lim_{N\to\infty}\sum_{n=0}^{N}\Phi_j^{(n)}(x,E,\mu),$$

where $\Phi_j^{(n)}$ is the distribution of $j$-type particles of the $n$ th generation. The recurrence relation $\Phi_j^{(n+1)}=\hat{B}_j\left[\sum_i\hat{S}_{ij}\times\Phi_j^{(n)}\right]$ holds in this case. In this relation the operator $\hat{S}$ representing the creation of particles is determined according to Eq. (5.32), while the operator $\hat{B}$ representing the transport of neutral particles is

$$\hat{B}_j\varphi_j=\frac{1}{\mu}\int_0^x dx'\exp\left[-\frac{\Sigma_j(E)(x-x')}{\mu}\right]\varphi_j(x',E,\mu),$$

and the operator for charged particles is

$$\hat{B}_j\varphi_j=\frac{1}{\beta_j(E)}\int_E^{E_0}dE'\exp\left[-\int_E^{E''}\frac{\Sigma_j(E')\,dE'}{\beta_j(E')}\right]\varphi_j(x'',E'',\mu),$$

where $x''=x-\mu\int_E^{E''}dE'/\beta_j(E')$.

The zeroth-generation functions are given by an expression of the type (5.7) or (5.8).

The decay of $\pi$ mesons, the multiple Coulomb scattering and the albedo part of the radiation, i.e., the angles at which the secondary hadrons are produced, $\theta > \pi/2$, are disregarded in the method of Ref. 110. In this method it is also assumed that the particle which interacts with the nucleus moves in the direction along the normal to the surface of the layer. A correction in this case is introduced into the scattering indicatrix such that, instead of Eq. (5.32), the following relation is used:

$$\hat{S}_{ij}\Phi_j^{(n)} = \int_0^1 d\mu' \int_E^{E_0} dE' \Sigma_i(E') \mathscr{P}_{ij}^{(n)}(E', E, \mu) \Phi_i^{(n)}(x, E'),$$

where $\mathscr{P}^{(n)}(E', E, \mu) = C_n \mathscr{P} [E', E, \mu_s = \cos(\theta \cdot 2^{-n/2})]$, and $C_n$ is an appropriate normalization factor.

Assuming additionally that $\Sigma_j(E) = \Sigma(E)$, we can construct a simple computer algorithm which will satisfactorily work at high energies in the case of reasonably thick layers.

## 5. Methods for the solution of transport equations for intermediate-energy hadrons

If the primary-hadron energy $E_0 < 1$ GeV or the hadron energy decreases to several hundred megaelectron-volts during the development of a high-energy hadron cascade in the medium, the qualitative and quantitative behavior of the interaction with matter will differ markedly from the behavior observed at high energies. Since at the indicated intermediate energies the total and differential cross sections depend essentially on the energy and atomic mass of the nucleus, and since the angular distributions are considerably less anisotropic and the ionization range is comparable to the range before the inelastic nuclear interaction, the neutrons in the total flux begin to dominate over the charged hadrons. The region of intermediate energies is of considerable practical interest in itself. It can be utilized in meson-factory accelerators, in heavy side shielding of proton accelerators with proton energies $E_0 > 1$ GeV and in space-vehicle shields.

Many attempts have been made to solve intermediate-energy nucleon transport equations: the method of spherical haromonics in $P_3$ approximation,[114] semi-analytic method,[115] the method of discrete ordinates,[116,117] the iteration scheme (5.40), 5.41) shown above, in which the method of discrete ordinates was used for the angular and energy variables. The particular features of the hadron–nucleus interactions in the intermediate energy region are used extensively in all these studies. Reliable results can be obtained to within $E \sim 10$ MeV by using theoretically substantiated numerical calculations.

A solution of the nucleon transport equations for nucleon energies in the range $0.1 \leqslant E_0 \leqslant 1$ GeV, which can be used to analyze the principal characteristic features, was proposed by Kazarnovskiĭ *et al.*[83] In the method used by Kazarnovskiĭ *et al.*[83] it was assumed that the proton distribution at a certain point situated some distance from the source which is greater than the ionization range is determined primarily by the neutron field near this point. Specifically, the following approximate relation holds:

$$\Phi_p(\mathbf{r}, E, \mathbf{\Omega}) = \frac{1}{\beta(E) + \Delta(E)} \left\{ \int_{4\pi} d\mathbf{\Omega}' \int_E^{E_0} dE' \tilde{\mathscr{P}}_{np}(E', E, \mu_s)\, \Phi_n(\mathbf{r}, E', \mathbf{\Omega}') \right.$$
$$\left. + \delta_{pj_0} \delta[\mathbf{r} - \mathbf{r}_0 - \mathbf{\Omega}_0 Q_0(E_0, E)] \delta(\mathbf{\Omega} - \mathbf{\Omega}_0) \exp[-Q_1(E_0, E)] \right\}, \tag{5.42}$$

where the unidirectional source (in the $\mathbf{\Omega}_0$ direction) and the monoenergetic source $E_0$ of the particles of species $j_0$ are located at the point $\mathbf{r}_0$;

$$Q_0(E', E) = \int_E^{E'} \frac{dE''}{\beta(E'') + \Delta(E'')};$$
$$Q_1(E', E) = \int_E^{E'} dE'' \frac{\Sigma(E'') + \overline{\mathscr{P}}_{pp}(E'') + \partial\Delta(E'')/\partial E''}{\beta(E'') + \Delta(E'')};$$
$$\overline{\mathscr{P}}_{pp}(E) = \int_E^{E_0} dE' \Sigma(E') \int_{4\pi} d\mathbf{\Omega}\, \mathscr{P}_{pp}(E', E, \mu_s);$$
$$\tilde{\mathscr{P}}_{np}(E', E, \mu_s) = \int_E^{E'} dE'\, \mathscr{P}_{np}(E', E'', \mu_s);$$
$$\Delta(E) = \int_E^{E_0} dE' \Sigma(E')(E' - E) \int_{4\pi} d\mathbf{\Omega}\, \mathscr{P}_{pp}(E', E, \mu_s).$$

Substituting Eq. (5.42) into Eq. (1.26) with $j = n$, we can obtain an approximate equation, after separating the system of kinetic equations, for the distribution of only the neutron component. To solve this equation, the distribution function and the scattering indicatrix must be written as a sum of the quasi-free and cascade components:

$$\left. \begin{aligned} \Phi_j(\mathbf{r}, E, \mathbf{\Omega}) &= \Phi_j^q(\mathbf{r}, E, \Omega) + \Phi_j^c(\mathbf{r}, E, \Omega); \\ \mathscr{P}_{ij}(E' E, \mu_s) &= \mathscr{P}_{ij}^q(E', E, \mu_s) + \mathscr{P}_{ij}^c(E', E, \mu_s), \end{aligned} \right\} \tag{5.43}$$

where expressions (3.33) and (3.34), respectively, are used to describe $\mathscr{P}_{ij}^q$ and $\mathscr{P}_{ij}^c$.

The one-dimensional equation for the transport of quasi-freely scattered neutrons is solved first. The small-angle approximation can be used if the strong anisotropy of the angular distribution of these neutrons in each $hA$-interaction event is taken into account. Using Fourier transform with respect to $x$, Mellin transform with respect to $E$, and Hankel transform with respect to $\theta$, we can then write the neutron transport equation in the form

$$(1 + ip)F - \hat{H}F = \Sigma^{-1}, \tag{5.44}$$

where

$$F(p, \nu, t) = \int_0^\infty \theta d\theta\, J_0(\theta\nu) \int_{-\infty}^\infty dx \exp(-ipx\Sigma) \int_0^\infty dE\,(E/E_0)^{t-1}$$

$$\times \Phi_n^q(x, E, \theta);\ \hat{H}$$

$$= -0.5\left(\frac{\partial^2}{\partial\nu^2} + \frac{1}{\nu}\frac{\partial}{\partial\nu}\right) - n_{nn}\frac{1+\xi}{\eta'}\exp(-\nu^2/4\eta');$$

$\eta' = t + \xi - 0.5$; and $J_0$ is zeroth-order Bessel function. It is assumed here that the vector of the primary-neutron beam, $\mathbf{\Omega}_0$, is normal to the plane of the source and that the energy dependence of the macroscopic cross section $\Sigma$ and of the indicatrix parameters $n_{nn}$ and $\xi$ is not included in Table 3.9.

Solving Eq. (5.44), we can find the energy distribution of quasielastically scattered neutrons

$$\Phi_n^q(x, E) = \int_0^\infty \Phi_n^q(x, E, \theta)\,\theta\, d\theta = \exp(-\Sigma x)\,(E/E_0)^{\xi - 1/2}$$

$$\times\left[\delta(E - E_0) + n_{nn}\frac{1+\xi}{E_0}\sqrt{\frac{\Sigma x}{\ln(E_0/E)}}\right. \tag{5.45}$$

$$\left.\times \sum_g{}' g^{-1/2} A(g)\,\varphi(g, 0)\, J_1(2\sqrt{\tau' g})\right],$$

where

$$\tau' = \Sigma x \ln(E_0/E),$$

$$A(g) = \int_0^\infty \varphi(g, x) \exp(-x^2/4)\, x\, dx,$$

$$\varphi(g, \nu\eta'^{-1/2})$$

and $g$ are the eigenfunctions of the operator $\hat{H}$ and the eigenvalues corresponding to them, and $\sum_g{}'$ is the summation over the discrete values of $g$ (for $g < 0$ there is at least one discrete value of $g$, $g = g_0$) and integration over the continuous values of $g$, with $0 \leqslant g \leqslant \infty$.

An approximate formula for the neutron spectrum, which is equal to the exact expression (5.45) in the limits $x \to \infty$ and $x \to 0$, has also been obtained[83]:

$$\Phi_n^q(x, E) = \exp(-\Sigma x)\,(E/E_0)^{\xi - 1/2}[\delta(E - E_0)$$

$$+ \frac{\tau'(\varkappa + 0.25)^2}{E_0 \ln(E_0/E)}\tilde{\varphi}\,[(\varkappa - 0.25)\sqrt{2\tau'}\,],$$

where

$$\tilde{\varphi}(y) = 1 + A(g_0)\,\varphi(g_0, 0)[I_1(y)/y - 1],$$

$$g_0 = 0.5(\varkappa - 0.25)^2,$$

$$\varkappa = \sqrt{2n_{nn}(\xi + 1)} - 0.25,$$

$$A(g_0)\,\varphi(g_0, 0) = 0.5\,(\varkappa + 0.125)\,(4\varkappa^2 + 7\varkappa + 4)/(\varkappa + 1)^3,$$

and $I_1(y)$ is a modified Bessel function of the first kind.

The cascade component $\Phi_n^c(x, E, \boldsymbol{\Omega})$ of the total distribution (5.43) can then be determined in the $P_1$ approximation of the spherical harmonic method, for example, if the distribution function of the quasi freely scattered neutrons is known.

In the general case of a point source in an infinite homogeneous medium, the function $\Phi_n(\mathbf{r}, E, \boldsymbol{\Omega})$ at long distances from the source can be estimated from the simple relation

$$\Phi_n(\mathbf{r}, E, \boldsymbol{\Omega}) \approx -\frac{1}{2\pi x}\frac{\partial}{\partial x}\Phi_n(x, E, \boldsymbol{\Omega})\bigg|_{x=|\mathbf{r}|},$$

where $\Phi_n(x, E, \boldsymbol{\Omega})$ is the neutron flux density distribution function found for the infinite planar source.

Substituting this expression into Eq. (5.42), we find the closed solution for the problem of intermediate-energy nucleon transfer.

The results of calculations based on the equations obtained are in fair agreement with the experiment and other calculated results. The approximate analytic solutions found in Ref. 83 are especially suitable for studying the asymptotic properties of the distribution functions and for use in combination with high-energy Monte Carlo programs.

## 6. Solution of the boundary value problem of low-energy neutron and secondary-photon transport

As a result of energy dissipation during the development of an electromagnetic cascade, the particles reach the low-energy region, $E < 15$ MeV, sooner or later. Since the probability for nuclear interaction of charged particles is negligible here ($R_{\text{ion}}/\lambda_{\text{in}} \leqslant 10^{-3}$), we consider, with the exception of the special cases, only the two-component radiation field: the neutrons and photons. The boundary value problem for the transfer of this radiation can be solved by using the thoroughly developed methods.[10,11,111] The methods[114,115] mentioned above and the HAMLET,[78] ANISN,[116] and POZ-400[117] programs include the algorithms for the solution of the low-energy radiation transfer problem. The approximate analytic solutions can, however, be used effectively because of the particular features of the situation under consideration.

### Neutrons

Since the sources of the slowing-down neutrons are distributed in the bulk of the shield and since they are nearly isotropic, the lower-order spherical harmonic method[78] and the age-diffusion approximation[118] can be used effectively, beginning at a certain distance from the central region of the hadron cascade. Let us consider in more detail the first of these methods which is used in the HAMLET program.[78,119]

Since in addition to the quasi-homogeneity and isotropy of the sources [$\overline{\Sigma}(E > 15$ MeV $\lesssim 0.5\,\overline{\Sigma}(E < 15$ MeV)] the inelastic scattering of low-energy neutrons is nearly isotropic, we can consider only the first two terms of the series in the expansion of the distribution function and scattering indicatrix (1.35) in Legendre polynomials and in a similar source-density expansion. Analyzing the medium with $A > 10$ and hence with a small elastic-scattering anisotropy, we reduce the equations for the zeroth and first harmonic to a quasi-diffusion multigroup equation where, in contrast with the classical case, the neutron transfer from the whole interval, 8 MeV$\leqslant E \leqslant E_0$, to all the groups up to the energy of 0.5 MeV is taken into account:

$$-D^q \frac{d^2\Phi^q(x)}{dx^2} + \Sigma_b^q \Phi^q(x) = B^q(x),\ q = 1, \ldots, M, \tag{5.46}$$

where

$$\Sigma_b^q = \Sigma^q - S^{q\to q};\ B^q(x) = \sum_{p=1}^{q-1} S^{p\to q}\Phi^p(x) + G^q(x);$$

$$D^q(x) = \frac{1}{\Phi^q(x)} \int_{E_q}^{E_{q-1}} dE D(E)\Phi_0(x, E);\ \Phi^q(x) = \int_{E_q}^{E_{q-1}} \Phi_0(x, E)\, dE;$$

$$S^{p\to q}(x) = \frac{1}{\Phi^q(x)} \int_{E_q}^{E_{q-1}} dE \int_{E_p}^{E_{p-1}} dE' \Sigma(E') S_0(E', E)\Phi_0(x, E);$$

$$\Sigma^q(x) = \frac{1}{\Phi^q(x)} \int_{E_q}^{E_{q-1}} dE \Sigma(E)\Phi_0(x, E);$$

$$G^q(x) = \frac{1}{\Phi^q(x)} \int_{E_q}^{E_{q-1}} dE G_0(x, E)\Phi_0(x, E);$$

$M$ is the number of energy groups; $\Phi_0$, $S_0$, and $G_0$ are the zeroth harmonics of the expansion in Legendre polynomials of the distribution function, the scattering indicatrices, and the source densities, respectively; $D(E) = 1/[3\Sigma_{tr}(E)]$ is the diffusion coefficient; $\Sigma_{tr}(E) = \Sigma(E) - \bar{\mu}(E)\Sigma_{el}(E)$; and $\bar{\mu}(E) \approx 2/(3A)$.

The density of the external sources is determined by the low-energy neutrons produced during the development of an electromagnetic cascade:

$$G_0(x, E) = 2\pi \sum_{k = p,n,\pi^\pm} \int_{E_{min}}^{E_0} dE' \Sigma_k(E') \frac{dN_{kn}}{dE}(E', E)\Phi_k(x, E'),$$

where $E_{min}$ is the lower energy threshold in the solution of the problem on the transfer of hadrons with $E > E_{min}$ ($E_{min} = 15$ MeV), $\Phi_k(x, E')$ are the energy distributions of these hadrons, and $dN_{kn}/dE$ is the sum of the spectra of the cascade and evaporative neutrons in the $hA$ collision.

The group constants $D^q$, $\Sigma^q$, and $S^{p\to q}$ usually depend only slightly on the neutron energy distribution $\Phi_0(x, E)$ and are nearly independent of $x$.[120] This circumstance makes it possible to use the previously compiled "standard" libraries for group cross sections (see, e.g., Refs. 120 and 121). The particular features of the neutron spectra, which are formed during the development of high-energy hadron cascades in the substance, can, however, be taken into account by averaging the group constants in the energy range $0.5 \leqslant E \leqslant 15$ MeV over the calculated spectra by an iteration method. For the first iteration it is easy to find[119]

$$\left.\begin{aligned} \Sigma_b^q &= \Sigma_{\text{in}}^q + \Sigma_{\text{el}}^q (4/A) E_q / (E_{q-1} - E_q); \\ S^{q \to q} &= \Sigma_{\text{el}}^q \left(1 - \frac{4}{A} \frac{E_q}{E_{q-1} - E_q}\right); \\ S^{p \to q} &= \Sigma_{\text{in}}^q \int_{E_q}^{E_{q-1}} dE' \frac{E'}{T_p^2} \exp(-E'/T_p) + \Sigma_{\text{el}}^p \frac{4}{A} \frac{E_p}{E_{p-1} - E_p} \delta_{q,p+1}, \end{aligned}\right\} \quad (5.47)$$

where $T_p$ is the temperature at which nucleus $A$ is excited by neutrons belonging to the $p$ th group.

Knowing the group constants[120,121] and (5.47), we can now solve Eqs. (5.46) by the Green's function method. Discarding the index $q$, we write

$$\left.\begin{aligned} -D\partial^2 f(x', x)/\partial x^2 + \Sigma_b f(x', x) &= \delta(x - x'); \\ f(x', x) \to 0 \ \text{as} \ x &\to \pm\infty. \end{aligned}\right\} \quad (5.48)$$

Applying to this equation the Fourier transform

$$\varphi(t) = \int_{-\infty}^{\infty} \exp(-itx) f(x', x)\, dx,$$

we find $\varphi(t) = \exp(-itx')/(\Sigma_b + Dt^2)$. Applying the inverse transformation to the function $\varphi(t)$, we find

$$f(x', x) = (2\sqrt{\Sigma_b D})^{-1} \exp(-\sqrt{\Sigma_b / D}|x - x'|)$$

or we finally find

$$\Phi^q(x) = \int_0^H \frac{B^q(x')}{2\sqrt{\Sigma_b^q D^q}} \exp\left(-\sqrt{\frac{\Sigma_b^q}{D^q}}|x - x'|\right) dx', \quad (5.49)$$

where $H$ is the thickness of the layer of the material under consideration.

A simple solution of (5.49) describes quite satisfactorily the low-energy neutron field outside the central region of the electromagnetic cascade.

## Photons

The principal processes for the production of photons in the development of an electromagnetic cascade in matter are[119,122] the decay $\pi^0 \to \gamma\gamma$, the radiative decay of thermal and slowing-down neutrons, the inelastic scattering of fast neutrons and the removal of residual nuclear excitation after the evaporative cascade interaction stage.

The density of photon sources of the first process is estimated in the calculation of the electron-photon shower (see Chap. 4). The angular distributions of photons in the remaining processes may be assumed isotropic and the source density can be written in the form

$$G(x, E, \mu) = \frac{1}{2}\left[N_\gamma(E)\sum_{q=1}^{M}\Sigma_\gamma^q\Phi^q(x) + \sum_{q=1}^{M_1}\Sigma_{\text{in}}^q(E)\Phi^q(x) + \sum_{k=p,n,\pi}\int_{E_{\min}}^{E_0} d'E'\Sigma_k(E')\mathscr{P}(E, E)\Phi_k(x, E')\right], \qquad (5.50)$$

where $\Phi^q(x)$ is given by Eq. (5.49), $\Phi_k(x, E)$ is the energy distribution of hadrons with $E > 15$ MeV, $M$ is the total number of groups of low-energy neutrons, $M_1$ is the number of groups of fast neutrons in the interval $0.5 \leqslant E \leqslant 15$ MeV, $N_\gamma(E)$ is the yield of the photons with an energy $E$ per neutron capture event, $\Sigma_\gamma^q$ is the cross section of the radiative capture in the $q$th group, $\Sigma_{\text{in}}^q(E)$ is the cross section of the inelastic scattering of neutrons of the $q$ th group with the yield of a photon with an energy $E$, $\mathscr{P}_\gamma(E', E)$ is the energy distribution of photons upon the removal of residual nuclear excitation, and $E_{\min} = 15$ MeV.

Upon neutron capture and the removal of residual excitation, the energy spectrum of photons can be written, according to the statistical theory, in the form $N(E_\gamma) \simeq BE_\gamma^3 \exp(-E_\gamma/T)$, where $T$ is the temperature of the emitting nucleus, and $B$ is the normalization constant.

In the case of inelastic scattering of fast neutrons, the photon spectrum can very accurately be assumed to be a $\delta$-function spectrum.

In the case of a known source (5.50), the transport equation for the photon energy flux density $I(x, E, \mu) = E\Phi(x, E, \mu)$ in a one-dimensional heterogeneous shield can be written, with the replacement of variables $E = m_e/\lambda$, in the form

$$\mu\frac{\partial I(x, \lambda, \mu)}{\partial x} + \Sigma(x, \lambda) I(x, \lambda, \mu) = n(x)\int_{-1}^{1} d\mu_s K(\lambda', \lambda) \times \int_0^\pi d\varphi' I(x, \lambda', \mu') + G(x, \lambda, \mu), \qquad (5.51)$$

where $\Sigma(x, \lambda)$ is the macroscopic cross section for the interaction of photons with atoms (electrons) of the medium at the point $x$; $\lambda$ is the photon wavelength; $n(x) = n_e(x) r_e^2/2$;

$$K(\lambda', \lambda) = \begin{cases} \left(\dfrac{\lambda'}{\lambda}\right)^2\left(\dfrac{\lambda}{\lambda'} + \dfrac{\lambda'}{\lambda} - 1 + \mu_s'\right), & \lambda_1 \leqslant \lambda \leqslant \lambda' + 2; \\ 0 \text{ otherwise;} \end{cases}$$

$\mu_s = 1 + \lambda' - \lambda$, is the cone of the scattering angle;

$$\mu' = \mu\mu_s + \sqrt{1-\mu_s^2}\sqrt{1-\mu^2}\cos\varphi';$$

and $n_e$ and $r_e$ are the electron density and the classical electron radius, respectively.

Equation (5.51) can be solved iteratively by a method of discrete ordinates.[122,123] In the collision integral we expand the function $I(x, \lambda', \mu')$ in a series in Legendre polynomials up to the $L$ th power and we expand the scattering indicatrix to the $M$ th power. The expansion coefficients of the indicatrix $K_m(\lambda)$ are calculated analytically and a quadrature formula is used for the differential flux harmonics:

$$\frac{2l+1}{4\pi} I_l(x, \lambda) = \sum_{q=1}^{q_m} a_{lq} I(x, \lambda, \mu_q); \; a_{lq} = \frac{2l+1}{2} P_l(\mu_q) W_q,$$

where $\mu_q$ and $W_q$ are the nodes and weights of the Gaussian quadrature.

Let us break up the interval of variation of $\lambda$ into $N_j$ subintervals of length $h = \Delta\lambda / N_j$ and apply the $N_k$-order Gaussian quadrature formula to the integral over $\mu_s$ in (5.51). After the $\nu$ th interaction the collision integral can then be written in the form

$$S^{(\nu)}(x, \lambda_j, \mu) = n(x) \sum_{q=1}^{2q_m} \sum_{k=0}^{K^*} B^j_{qk}(x, \mu) I^{(\nu-1)}(x, \lambda_j - kh, \mu_q),$$

where

$$K^* = \min\{N_k, N_j - j\}; N_k h = 2; h = \lambda_k - \lambda_{k-1} = \mu_{sk} - \mu_{sk-1}; B^j_{qk}(x, \mu)$$
$$= \sum_{l=0}^{L} a_{lq} P_l(\mu) \sum_{m=0}^{M} C^l_{mk} K_m(\lambda_j).$$

The matrices $C^l_{mk}$ are combinations of the expansion and weight indices of the Gaussian quadrature.

In the integration of the equation whose right side, $F = S + G$, is known, the interval of variation of the spatial coordinate is divided into several parts by the nodes $x_i$. The interfaces of various materials and the external boundaries in this case are included as nodal points. Assuming that $F$ is a linear function on the interval $[x_{i-1}, x_i]$ and that the macroscopic cross section $\Sigma(x, \lambda_j)$ on this interval is constant, we integrate Eq. (5.51) along the fixed ray $\mu_p$

$$I^{(\nu)}(x_i, \lambda_j, \mu_p) = \exp\left[ -\frac{\Sigma_i(\lambda_j)\Delta x_i}{\mu_p} \right] I^{(\nu)}(x_{i-1}, \lambda_j, \mu_p)$$
$$+ \frac{A^j_{ip}}{\mu_p} S^{(\nu-1)}(x_i, \lambda_j, \mu_p) + \frac{D^j_{ip} - A^j_{ip}}{\mu_p} S^{(\nu-1)}(x_{i-1}, \lambda_j, \mu_p),$$

where

$$A^l_{ip} = \frac{\Delta x_i - D^j_{ip}}{\Sigma_i(\lambda_j)\Delta x_i} \mu_p; \; D^j_{ip} = \frac{1 - \exp[-\Sigma_i(\lambda_j)\Delta x_i / \mu_p]}{\Sigma_i(\lambda_j)} \mu_p.$$

The integration is always started at the barrier boundary, at which the boundary conditions are specified. The iteration process is stopped if at the $\nu$ th iteration condition (5.39) is satisfied for all values of $x$, $\lambda$, and $\mu$.

The method described here leads to acceptable results even in strongly anisotropic cases with $L = 13$ and $M = 3$. The condition $\Sigma_i(\lambda_j)\Delta x_i \leqslant 0.5$ is a criterion for the choice of the reference grid.

The multicomponent radiation field in the parallel-plane layer of the material in the case of the development of an electromagnetic cascade in it can thus be calculated by using the methods discussed in this chapter.

# Chapter 6
# Study of electromagnetic cascades by the Monte Carlo method

## 1. Monte Carlo method in the transport of high-energy particles

An electromagnetic cascade is a process in which particles interact with matter in a random and independent manner. Knowing the laws governing the distribution of particles in an interaction event, the Monte Carlo method can be used to calculate a specific random occurrence of a process. This method is now essentially the only one that makes it possible to calculate the development of an electromagnetic cascade in any geometry in the presence of external electric and magnetic fields. It also makes it possible to fairly easily describe the interactions in detail and to solve the problems on the fluctuation, correlation, and small perturbations.

The Monte Carlo method is essentially a model-based experiment in which, in contrast with an actual experiment, the "experimental" conditions can be varied over a broad range. The feasibility of studying a particular physical effect and of identifying the unlikely processes by using the essential-selection method (see Sec. 1.5), along with the features mentioned above, make the Monte Carlo method the principal tool of high-energy physics at the planning stage of an experiment and at the stages of analysis and interpretation of its results in the solution of many radiation problems.

The general picture of the processes involving the passage of fast particles through matter is considered in Chap. 1, the simulation of electron–photon showers is examined in Chap. 4 and the Monte Carlo method is analyzed in Sec. 1.5. The systematic features and the specific schemes for stimulating high-energy hadron cascades are considered below. Certain general questions concerning the use of the Monte Carlo method for high-energy particles are discussed in Refs. 19, 45, 48, 124, and 125.

Two important peculiarities characterize the simulation of the propagation of high-energy hadrons in condensed matter:

(1) In contrast with the transport of low- and intermediate-energy hadrons through matter, the hadron cascade is a strongly branching process because of the multiple production of particles in the nuclear interactions at energies $E_0 \gtrsim 1$ GeV.

(2) In contrast with the other cascade process (the electron–photon shower; see Chap. 4), virtually all hadron-cascade calculations are based on the use of the phenomenological models and equations (see Sec. 3.4).

The first peculiarity leads to the fact that at the primary energies, $E_0 \gtrsim 50$ GeV, and at the kinetic *cutoff threshold energy in the calculation*, $E_{\text{thr}} \sim 10$ MeV, the "trajectory tree" branches out to such an extent that an electromagnetic cascade in an extended system, in which all the "branches" are scanned, cannot be calculated in a reasonable time even on present-day computers. This circumstance as well as the absence of complete information on the cross sections of all reaction channels at high energies and the difficulty of rigorously maintaining the energy-momentum conservation law in each simulated *hA* collision event provided an incentive for the development of the *inclusive schemes* for calculating electromagnetic cascades.

The second peculiarity is a consequence of the present state of the theory of multiple processes (see Chap. 3). In contrast with quantum electrodynamics, this theory does not give a unique description of the distributions occurring in high-energy nuclear reactions of hadrons in the entire range of kinematic variables, which is important for electromagnetic-cascade simulation. No single model can establish a quantitative agreement between the theory and the experimental data, even in a narrow phase interval, without the use of free parameters. Consequently, all the presently widely used methods and computer programs for calculating high-energy electromagnetic cascades, with the exception of individual efforts to accurately simulate the *hA* interactions from the phase space[45] or through the quasi-two-particle production of hadron resonances,[126,127] are based on phenomenological descriptions.* For these reasons and also because no single computer program for calculating electromagnetic cascades can apparently be singled out as the clearly dominant program, this chapter is structurally different from Chap. 4, where each physical process of the electron–photon shower can be accurately described. Unless otherwise stated, we shall consider in this chapter in the case of nonmultiplying media the proton, neutron, and $\pi^{\pm}$ meson cascades and the electron–proton showers which are initiated by the $\pi^0 \to \gamma\gamma$ decays and which accompany these cascades. The primary kinetic energy of hadrons and the kinetic energies of the cascade particles are denoted by $E_0$ and $E$, respectively.

## 2. Simulation schemes and computer programs

Among the presently widely used methods and computer programs for calculating high-energy electromagnetic cascades, three groups can be singled out, depending on the way in which the hadron–nucleus interactions are simulated:

*At $E_0 \gtrsim 3$ GeV the intranuclear-cascade models used in Refs. 126 and 128 should also be regarded as phenomenological models.

(1) *Exclusive approach*: HETC program, developed in ORNL (Ref. 126 and 127); SHIELD program, developed in JINR (Ref. 128); the programs for cosmic-radiation studies (see, e.g., Ref. 102).

(2) *Quasi-exclusive approach*: FLUKA program, developed in Leipzig and CERN (Refs. 85 and 129).

(3) *Inclusive approach*: MARS, a package of programs developed in MIFI and IHEP (Refs. 19 and 130–135); CASIM program developed in FNAL (Ref. 136); KASPRO program, developed in Leipzig and CERN (Refs. 80 and 137).

In the programs of the *first group* the production of any number of particles of different species which do not violate the conservation laws can be simulated directly and simultaneously in each hadron–nucleus interaction event. The evaporative cascade model is used in the HETC and SHIELD programs.[48,138] In calculating an electromagnetic cascade in each collision between a hadron and a nucleus it is necessary to simulate its interaction with intranuclear nucleons, to simulate possible reaction channels and subsequent evaporation of particles from the excited nucleus, and to determine the characteristics of the residual nucleus.

The HETC program traces the neutrons with an energy $E > 0.025$ eV, protons and $\pi^+$ mesons with $E \gtrsim 10$–$15$ MeV, and mesons until they are stopped. The multiple Coulomb scattering (the approximate) elastic interaction of nuclei, range straggling, meson decay, and the production of $\pi^0$ mesons, muons, photons, and heavy nuclear fragments in this case are taken into account. The photon transport and accordingly the development of electron–photon shower are not considered directly in the HETC program. There is a version of the program for the calculation of cascades with the participation of deuterons and $\alpha$ particles. The geometry and composition of the materials may be rather complicated. Because the evaporative cascade model has limited application and because of limitations of the computer capabilities, the maximum primary energy at which the HETC program can still be used effectively is $E_0 = 30$–$50$ GeV. The computation time and the required computer memory are very large (up to several hours of computation time; IBM 360/90, 575K memory).

The SHIELD program, which uses the evaporative cascade model[48] based on statistical approximation of the experimental data is basically similar to the HETC program. This program was written for the BESM-6 computer. The maximum hadron energy used in the calculations is $E_0 = 30$ GeV.

The HETC and SHIELD programs can be used to accurately calculate the nucleon–meson cascades in systems that are not too large ($\lesssim 5\lambda_{in}$) at energies $E_0 \lesssim 10$ GeV. These programs are used for studying cascades in multiplying media (in the electronuclear method, for example) and for calculating the probability distributions in detecting systems.

Programs such as those used in Ref. 102, written for very high energies, are based on many assumptions characteristic of cosmic-radiation studies (see Sec. 5.1). The distribution of the inelastic coefficients, which is identical for nucleons and $\pi$ mesons, is usually used as the principal characteristic of the elementary interaction event. In the cascade calculations the cutoff threshold energy from below is usually exceptionally high ($E_{\rm thr} \gtrsim 3$ GeV).

The programs of the FLUKA family belong to the *second group*.[89,129] In these programs the multiple production of hadrons is described by the phenomenological inclusive single-particle distributions (see Sec. 3.4). The exclusive process in this case is simulated approximately, making use of the algorithm which assures that the energy conservation law in each collision holds:

(1) In the simulation of an elementary $hA$ interaction event, it is necessary to initially choose a random order in which various types of secondary hadrons are simulated.

(2) Several counters $C$ are introduced: these are energy-balance counters which determine the energy balance in a given event which is initially assumed to be zero; the energy conservation law holds when $C=1$.

(3) The counters $D_{ijk}$ and $G_{ijk}$ are introduced: these are three-dimensional numerical data files used to make sure that the inelasticity coefficients in the laboratory frame are approximately equal to the mean values of $\langle K_{ij}\rangle$ during the simulation (see Chap. 3) when a secondary particle of species $j$ interacts with a hadron of species $i$ with a momentum $p_{ik}$ within the momentum interval $\Delta p_k$.

(4) At the beginning of the event the values of $\langle K_{ij}\rangle$ for all $p_{ik}$ prepared in advance are entered into the counters $D_{ijk}$.

(5) Each time a particle $j$ is chosen, its relative energy $E_j/E_i$ is added to the counters $C$ and $G_{ijk}$.

(6) The hadron production is simulated until the following condition no longer holds:

$$C \leqslant 1;\ G_{ijk} \leqslant D_{ijk}. \tag{6.1}$$

(7) If $C>1$ or $G_{ijk}>D_{ijk}$ for a regularly generated particle, then this particle cannot be produced in a given event and the value of $E_j/E_i$ corresponding to it must be subtracted from the counters $C$ and $G_{ijk}$.

(8) The momentum and angle of this particle are stored in special numerical data banks, $P_{ijk}$ and $A_{ijk}$, for use in the next event of this sort; the production of a new particle of type $ijk$ is not simulated until this particle, stored in the data banks $P_{ijk}$ and $A_{ijk}$, is used.

(9) After entering the parameters of the "unlucky" hadron into the $P_{ijk}$ and $A_{ijk}$ data files, the simulation of the analyzed event with the secondary hadron of the next type is resumed.

(10) The process terminates when one or both of the conditions in (6.1) are violated after choosing the last type of secondary hadrons; the particle in this case is assigned the exact value of the energy $E_j$ and momentum $p_j$, which upholds the conservation law $C=1$, and only the angle $\theta_j$ at which the hadron is emitted from the nucleus is simulated from the inclusive distributions.

(11) The following method is used to eliminate the displacement introduced in the preceding step by the nonrandom nature of the energy $E_j$ and momentum $p_j$: The element $I_{ijkl}$ of the supplementary numerical data bank increases by unity each time a specified momentum $p_j$ belonging to the sec-

ondary-momentum interval $\Delta p_l$ is assigned to the particle; at the beginning of this computation all elements of this data bank are assumed to be zero.

(12) If now the momentum $p_j$ drops out during the generation of the event, and the corresponding element of the data bank $I_{ijkl}$ is not zero, then the momentum $p_j$ drops out and the element $I_{ijkl}$ decreases by unity. This quite effective scheme can reproduce well the kinematic correlations but it cannot describe the dynamic correlations. In the FLUKA series program the proton, neutron, and $\pi^{\pm}$ meson energy range is 50 MeV$\leqslant E \leqslant$1000 GeV, where $E_{\text{thr}} = 50$ MeV. The spatial density distributions of hadronic interactions (stars) $S(\mathbf{r})$ and of the energy evolution $\epsilon(\mathbf{r})$ are calculated in complex geometries. The spatial distribution of the energy evolution from an electron–photon shower in the decay $\pi^0 \rightarrow \gamma\gamma$ is described by empirical formulas.

A common feature in the programs of the third group is the inclusive simulation of the hadron–nucleus interactions in which the statistical-weights method is used. A phenomenological description of the inclusive particle distributions is used in this case (see Sec. 3.4) and the energy-momentum conservation law is satisfied in many events on the average. The principal application of these programs is the solution of a broad range of physics problems involving proton accelerators with energies $E_0 \gtrsim 10$ GeV.

The programs of this group do not require very powerful computers and allow us to use the advantages of the essential-choice method to the fullest extent (in the problems with improbable reaction channels,[80,139] for example). In the CASIM and KASPRO programs the hadron energies are in the range from 50 MeV to 1000 GeV and in the MARS complex they range from 15 MeV to 5000 GeV and even up to $10^7$ GeV (Ref. 140). The computation time of a single history increases with the energy only logarithmically, in contrast with the first two groups of programs where this increase is linear.

An important feature of the third group of programs is that in them the electron-photon showers initiated by the decays $\pi^0 \rightarrow \gamma\gamma$ during the development of the hadron cascade can be described exactly. Modified versions of the MAXIM program,[141,142] MARS-8 program,[135] MARS-9 program,[143] and KASPRO program[144] have been developed. These programs were combined with the programs for the simulation of an electron–photon shower by the Monte Carlo method, the AEGIS program,[154] with its modifications[143] (see Sec. 6.3), and with the EGS program.[41] A direct or semidirect simulation of the accompanying electron–photon showers can be carried out directly during the hadron cascade computation. In particular, such a simulation makes it possible to correctly calculate the spatial distribution of the energy evolution in the substance for the high-energy electromagnetic cascades in the presence of external electromagnetic fields. Such calculations are especially important in the case of radiation-induced heating of superconducting magnets[134] and experimental studies with accelerators (the reader is referred to Chap. 8 for more details).

The physical processes considered in the proton, neutron, and $\pi^{\pm}$ meson transport (and also in some versions of the $K^{\pm}$, $\mu^{\pm}$, and $\bar{p}$ transport) are similar to the processes considered in HETC. The geometries, the compositions of the materials, and the external electromagnetic fields may be rather complex and the set of functionals to be calculated may be rather large.

The algorithms and the particular features of the MARS package of programs are described in Sec. 6.3.

In the CASIM program[136] and its modification, the MAXIM program,[141,142] the inclusive distributions of fast particles are described by the Hagedorn–Ranft thermodynamic model.[146] The phenomenological equations [Eqs. (3.32)] in Sec. 3.4 are used to describe the cascade nucleon distributions. Exactly one *transport hadron* ($p, n, \pi^{\pm}$) with a statistical weight, whose expectation value is equal to the total multiplicity, is generated at each nuclear interaction point. The sampling function for the sampling of these hadrons is proportional to the product of the energy in the laboratory frame of reference and the inclusive differential cross section. One or several *detectable hadrons*, whose sampling function is proportional to the differential cross section, can also be generated at this point. Most of the sampling functions in the program are given in the form of numerical tables. The macroscopic cross sections for $hA$ interaction are assumed to be independent of the energy. All particles are transported by a step-by-step method with fixed steps. The cutoff energy for hadrons is $E_{\mathrm{thr}} = 50$ MeV.

In the KASPRO program the inclusive hadron distributions are described by using a system of phenomenological equations[80] (see Sec. 4.3). Just as in the CASIM program, the production of only one fixed hadron is simulated in each elementary $hA$ interaction event. The program is intended primarily for the calculation of the emergence of hadrons and electrons from thick targets. The cutoff energy of hadrons is $E_{\mathrm{thr}} = 50$ MeV.

## 3. Simulation of the electromagnetic cascades in the MARS package of programs

All the common features of the inclusive programs discussed in the preceding section are characteristics of the MARS package of programs. The basic ideas and the early versions of this program package are described in Refs. 19, 199, and 130–133. Let us consider the *distinctive features* of the current system of MARS programs using primarily MARS-9 program as an example[143]:

(1) The range of hadron kinetic energies is 1 MeV$\leqslant E \leqslant$5000 GeV.

(2) At high energies the production of a rigorously fixed number of hadrons—one fast nucleon, one charged $\pi$ meson, and one slow cascade nucleon—is simulated in an inclusive manner in each $hA$ interaction event. The hadrons of different species ($p$ or $n$, $\pi^+$ or $\pi^-$) are simulated at random, and the statistical weights of the particles double in value, but the expectation values of the statistical weights are equal to the corresponding average hadron multiplicities.

(3) The characteristics of a produced $\pi^{\pm}$ meson with one-half the weight are used for the description of the production of a meson at the same vertex.

The $K^+$ mesons in problems where they are not studied separately or where the muon production is not considered are included in the $\pi^\pm$ meson component.

(4) Standard nuclear data libraries are used for the description of macroscopic cross sections of $hA$ interactions in the range of hadron kinetic energies form 15 MeV to 50 MeV (see Chap. 5). The data of Ref. 48 are used at energies 50 MeV$\leqslant E \leqslant$20 GeV and the results given in Table 3.2 are used at $E >$ 20 GeV. A power-law interpolation is used for nuclei with intermediate values of $A$.

(5) The elastic scattering of particles by hydrogen nuclei is described exactly (on the basis of the available data).

(6) The Coulomb scattering of charged particles and the elastic scattering of particles by nuclei with $A > 1$ are taken into account.

(7) The effect of the external electric and magnetic fields is taken into account.

(8) The electromagnetic cascade is calculated in heterogeneous three-dimensional systems which can contain up to three materials, each of which may be a mixture of up to six substances.

(9) Exact methods of analytic geometry are used (aside from the MARS-9 programs) for the description of the geometry of the problem and the particle transport. Up to 35 planes, 13 cylindrical surfaces, and 5 spherical surfaces can be examined simultaneously in the problem.[133]

(10) In MARS-9 programs, where the geometry is assumed to be essentially arbitrary, a step-by-step iteration method is used for the description of the particle transport. In principle, the number of surfaces is not limited.

(11) The expectation-value method is widely used.

(12) All four kinds of estimates described in Sec. 1.5 can be used simultaneously in each problem. In MARS-4 and MARS-9 programs the local estimate can be used to calculate the spatial-energy density of the particle flux and the density of the energy evolution in small detectors in the case of complex geometries.

(13) The quantities which are calculated are the spatial distributions of hadron–nucleus interactions (stars) $S_j(\mathbf{r})$, of the particle flux density $\Phi_j(\mathbf{r})$, energy release $\epsilon_i(\mathbf{r})$, equivalent dose $H(\mathbf{r})$, and of the heating $\Delta T(\mathbf{r})$, the statistical errors (1.41) and (1.47) of the differential distributions, hadron energy spectra $\Phi_{jk}(E)$ in certain regions $k$, and the loss spectra. The neutron spectra are estimated at energies between 0.025 eV and 15 MeV. Various convolutions with the participation of distributions mentioned above are calculated. Here the index $j = p, n, \pi^\pm$ and $i$ identifies the contributions to the energy evolution of hadrons in the development of the cascade, low-energy particles and an electron–photon shower from the decays $\pi^0 \rightarrow \gamma\gamma$.

The program package includes the following programs:

MARS-4 program[133] is used to calculate the hadron cascade with a description of the energy evolution from the electron–photon shower, with use of approximate semiempirical equations. The flux can be estimated locally.

This program is used principally to calculate the spatial distribution of the hadron flux in shields and accelerator structures and to calculate the induced activity.

MARS-5 program[139] is used to calculate the differential distribution of hadrons of the type $p$, $n$, $\pi^+$, $\pi^-$, $K^+$, and $K^-$ at the output of thick targets. It is also used to calculate the stopping densities of the negative hadrons $\pi^-$, $K^-$, $p^-$, and $\Sigma^-$.

MARSHI program[147] synthesizes the inclusive method used at high energies and the exclusive method (the intranuclear-cascade model) used at intermediate energies.

MARSU program[140] is the ultrarelativistic version of the MARS package of programs, which is used at energies $E > 5000$ GeV.

MARS-6 program[134] is used to calculate the hadron cascade in a substance in the presence of external electromagnetic fields. The particle trajectories are simulated numerically by the broken-spiral method. The azimuthal structure of the system is taken into account in a form that can easily be used.

MARS-7 program[135] is a modification of the MARS-6 program, in which the release of energy from the electron–photon showers formed as a result of the decays $\pi^0 \rightarrow \gamma\gamma$ is described by a semiempirical algorithm with allowance for the radial divergence of the shower. This program includes the transfer of evaporative protons and neutrons and also of nucleons produced as a result of capture of stopped $\pi^-$ mesons by the nuclei of the medium. $E_{\text{thr}} = 1$ MeV.

MARS-8 program[135] is similar to the MARS-7 program with the exception that it includes an indirect simulation of the electron–photon shower with use of the AEGIS algorithm.[145]

MARS-9 program[143] is a package of programs for calculating electromagnetic cascades in an essentially arbitrary geometry in the presence of large or small inhomogeneities, and external electric field (**E**) and magnetic field (**B**). The description of the physical processes and most of the computational procedures used in MARS-8 program are modified considerably in this program. The algorithm for particle transfer and the method of obtaining the estimates are completely changed both in the hadron part and the electron–photon part of the programs. This program package includes the following programs: MARS-9M program, which is primarily an exact calculation of the spatial distribution of the energy release $\epsilon(\mathbf{r})$ (including the distribution in the limit $r \rightarrow 0$) in any field $\mathbf{B}(\mathbf{r})$ and the calculation of the escape of electrons and photons from thick targets bombarded by hadron beams.

MARS-9S is a program for calculating electromagnetic cascades in arbitrary large bulks of material, for calculating a large number of radiation functionals, and for calculating the escape of hadrons from thick targets. It can give a local estimate of the flux of various types of hadrons, make a spatially dependent choice of the particle-transport method, and estimate the functionals.

MARS-9T program is similar to the MARS-9M program, which is intended for calculation of electromagnetic cascades in complex magnetic structures of large proton accelerators. It is included in the MARTUR package of programs,[153] where it can be used with the familiar TURTLE program.[148]

In calculating the hadron cascade in the material the *trajectory tree* is formed by the vertices—the points of the inelastic $hA$ interaction with three "weighted" long-range final-state hadrons (see the discussion above), which are connected by straight or curved paths of these particles. The $\pi^0$ mesons which are produced and which give rise to many electron–photon showers that develop in parallel are assumed to decay because of their short lifetime ($\tau_{\pi^0} \sim 10^{-16}$ s) directly into two photons at the point of their production.

To reduce the computation time, a roulette-type procedure can be used in certain cases.[9] A maximum statistical weight in the $hA$ collision of the $k$th generation, $W_{max}(k)$, is accumulated in the current history from all preceding histories for each type of particle. A number $P \ll 1$ is chosen, and the trajectory, whose statistical weight is $W^{(j)}(k) < W^{j}_{max}(k)\alpha$, ends with the probability $1-P$. Here $\alpha \ll 1$. With the probability $P$ the trajectory is scanned further, but now with the weight $W^{(j)}(k)/P$. A similar procedure is followed if the statistical weight of the particle becomes very small, e.g., $W < 10^{-12}$. For $P = 0.01$, $\alpha = 10^{-4}$, and a primary energy $E_0 \gtrsim 50$ GeV the counting time of electromagnetic cascades in large systems may decrease by a factor of 3–10 if the statistical errors are the same.

The *electromagnetic interactions* of hadrons are described by the equations in Chap. 2. The cross section for the decay of $\pi^\pm$ mesons $\pi \to \mu + \nu$ is included in the macroscopic cross section: $\Sigma^\pi(E) = \Sigma^\pi_{abs}(E) + m_\pi / c\tau p$, where $E$, $p$, and $m_\pi$ are cross sections, momentum, and rest mass of the $\pi$ meson, respectively; $c\tau = 781$ cm.

## Inelastic nuclear interactions

If the energy of the hadron which interacts with the nucleus is $E_0 > 5$ GeV, then a fast nucleon ($p$ or $n$), a $\pi$ meson ($\pi^+$ or $\pi^-$), and a slow cascade nucleon ($p$ or $n$) with the corresponding doubled statistical weights are, as was indicated above, generated at each vertex.

Such a scheme as compared, for example, with the algorithm for the production of only a single particle at the vertex (CASIM, KASPRO) considerably reduces the fluctuations of the product of the statistical weight of the sampling and the particle energy $W_j E_j$. This circumstance is particularly important in the calculations of the spatial distribution of the energy release $\varepsilon(\mathbf{r})$. The sum $\frac{1}{N} \sum_{n=1}^{N} \sum_{j} W_j^{(n)} E_j^{(n)}$ now converges close enough to the primary energy $E_0$ at $N \sim 50$–100. The characteristics of fast particles are simulated in the cm frame by using a pair of kinematic variables $x_F$ and $p_\perp$, which are then scaled to the laboratory values of the variables $E$ and $\theta$.

Let us assume that the incident particle is a nucleon: $i = p,n$. The nucleon spectrum in this case is normalized to two nucleons. If the following conditions are satisfied:

$$\gamma \leqslant 0.135/2(A+2)^{0.355};\ j=i;\ E_0>15\ \text{GeV}, \tag{6.2}$$

where $\gamma$ is a random number from the interval (0,1), a double diffraction nucleon of a random species $j$ will then be generated by the inverse-function method [(1.43)] from the PPP and PPR terms of the three-Reggeon description (see Chap. 3). If conditions (6.2) do not hold, the nuclear production is simulated by the method of inverse functions from the phenomenological equations [Eqs. (3.28)]. Relations (1.3)–(1.6) must in this case, as everywhere below, be taken into account, and which variable, $x_F$ or $p_\perp$, must be simulated first, is determined randomly.

Equation (3.29) for $d^2N/dxdp_\perp^2$ is used in describing the production of secondary pions and, effectively, the $K$ mesons. The transverse momentum is simulated by the inverse-function method, and the sampling function $f_1(x_F; p_\perp^2) = C_0(p_\perp^2)\exp(-C_1|x_F|)$, where $C_0$ is a normalization factor, and the parameter $C_1$ depends on the type of problem encountered, is used for the variable $x_F$. The additional statistical weight is

$$W = \frac{2d^2N(x_F,p_\perp)}{dx_F dp_\perp^2}\quad f_1(x_F;\ p_\perp^2). \tag{6.3}$$

Let us now assume that the incident particle is a $\pi$ meson: $i=\pi, \pi^-$. The fast nucleon spectrum in this case is normalized to unity. The secondary-nucleon production is described by Eqs. (3.28) and (3.30), and the variables $x_F$ and $p_\perp$ are directly simulated by the inverse-function method. The production of secondary $\pi$ mesons is simulated on the basis of Eq. (3.31) for $d^2N/dxdp_\perp^2$ and the sampling function

$$f_2(x, p_\perp^2) = C_0 \exp(-C_1|x|)[\exp(-C_2 p_\perp^2) + C_3 \exp(-C_4 p_\perp^2)],$$

where $C_0, ..., C_4$ are certain parameters. The corresponding statistical weight is similar to (6.3).

The production of slow nucleons ($p$ or $n$) with an energy $E \leqslant 1$ GeV is simulated in the laboratory frame. The energy and the angles of emission of the particles are determined by using the numerical tables and the method of inverse functions from the phenomenological equations [Eqs. (3.32)] and the energy and the angles of emission of the evaporative nucleons are determined from Eqs. (3.40). The statistical weight of the cascade nucleons is approximately equal to their total multiplicity: $\langle n(E_0 > 5\ \text{GeV})\rangle \simeq 0.514\sqrt{A}$.

If the kinetic energy of the primary hadron in $hA$ interaction is confined to the interval 15 MeV$\leqslant E_0 \leqslant$5 GeV, then the simulation can be carried out in the laboratory frame. At $E_0 > 0.4$ GeV two particles are generated: A nucleon ($p,n$) and a $\pi$ meson ($\pi^+, \pi^-$); at lower energies only one nucleon is generated: $p$ or $n$. At fixed maximum $E_0$ and minimum $E_{\text{th}}$ energies a normalized sampling function is used:

$$f_3(E,\theta) = C_0[1 + C_1(E_0, E_{\text{tanh}})/E]C_2(E,\tau)\exp(-E\theta/\tau), \tag{6.4}$$

where $C_0$, $C_1$, and $C_2$ are parameters, and $\tau = 0.2E_0(0.5+E_0)^{-1}$. At $i \neq j$ only the second term remains in the square brackets in Eq. (6.4). Using Eq. (3.36) for the description of the inclusive spectra of nucleons ($p$ or $n$) and $\pi$ mesons ($\pi^+$ or $\pi^-$), we can write the statistical weight of this sample in the form

$$W = \frac{2\pi \sin\theta}{f_3(E,\theta)} 2 \frac{d^2N}{dEd\Omega}(E,\theta).$$

We also evaluate Eq. (3.40).

The azimuthal angle $\varphi$, more exactly, the functions $\cos\varphi$ and $\sin\varphi$ are simulated according to the Neumann algorithm. Two random numbers, $\gamma_1$ and $\gamma_2$, are generated and the quantities $a = 2\gamma_1 - 1$ and $b^2 = a^2 + \gamma_2^2$ are calculated. If $b^2 \geqslant 1$, this pair of random numbers is discarded and a new pair is chosen. Otherwise, it is assumed that

$$\cos\varphi = (a^2 - \gamma_2^2/b^2; \sin\varphi = 2a\gamma_2/b^2. \tag{6.5}$$

## Elastic nuclear scattering

The high-energy elastic and quasielastic scattering of hadrons by nuclei with $A > 9$ is simulated directly from Eqs. (3.41)–(3.43).

The data on the total, elastic, and differential cross sections obtained in Refs. 149–152 are used for the simulation of the elastic scattering of hadrons by the nuclei of the hydrogen atoms.

At $E < 0.8$ GeV ($\sigma$ denotes cross section in mb, and $E$ denotes the energy in GeV) the following equations are used:

$$\sigma_{\text{tot}}^{(pp)} = \begin{cases} 1.1748\times10^{-3}/E^2 + 3.0885/E + 5.3107, & E < 0.04; \\ 9.3074\times10^{-2}/E^2 + 1.1148\times10^{-2}/E + 22.429, & 0.04 \leqslant E \leqslant 0.31; \\ 8.8737\times10^{-4}/E^2 + 53.31E + 3.5475, & 0.31 < E \leqslant 0.8; \end{cases}$$

$$\sigma_{\text{tot}}^{(np)} = \begin{cases} 5.0574\times10^{-3}/E^2 + 9.0692/E + 6.9466, & E < 0.04; \\ 0.23938/E^2 + 1.802/E + 27.147, & 0.04 \leqslant E \leqslant 0.4; \\ 34.5, & 0.4 < E \leqslant 0.8; \end{cases}$$

$$\sigma_{\text{el}}^{(pp)} = \begin{cases} \sigma_{\text{tot}}^{pp}, & E < 0.31; \\ 24, & 0.31 \leqslant E \leqslant 0.4; \\ 23.5, & 0.4 \leqslant E \leqslant 0.8; \end{cases}$$

$$\sigma_{\text{el}}^{(np)} = (1/2)[2\sigma_{\text{tot}}^{(np)} - \sigma_{\text{tot}}^{(pp)}]\sigma_{\text{el}}^{(pp)}/\sigma_{\text{tot}}^{(pp)} + (1/2)\sigma_{\text{el}}^{(pp)}.$$

At $E < 2.5$ GeV the cross sections of the $\pi p$ interaction are taken in tabulated form from Ref. 151. At higher energies the following equations are used:

$$\sigma_{\text{tot}}^{(pp)} = 37.5 + 7E^{-1/2}, \quad E > 3.5 \text{ GeV};$$

$$\sigma_{\text{el}}^{(pp)} = 7 + 21.03E^{-0.837}, \quad E > 3.5 \text{ GeV};$$

$$\sigma_{\text{tot}}^{(np)} = \begin{cases} 42.0, & 3.5 \leqslant E \leqslant 8 \text{ GeV}; \\ 37.5 + 26.45E^{-0.852}, & E > 8 \text{ GeV}; \end{cases}$$

$$\sigma_{\text{el}}^{(np)} = \begin{cases} -0.222E + 12.48, & 3.5 \leqslant E \leqslant 8 \text{ GeV}; \\ 6.0 + 73.144E^{-1.32}, & E > 8 \text{ GeV}; \end{cases}$$

$$\sigma_{\text{tot}}^{(\pi+p)} = \begin{cases} -4.66E + 42.95, & 2.5 \leqslant E \leqslant 3 \text{ GeV}; \\ 0.75E + 31.24, & 3 \leqslant E \leqslant 6 \text{ GeV}; \\ 21.5 + 18.91E^{-0.716}, & E > 6 \text{ GeV}; \end{cases}$$

$$\sigma_{\text{tot}}^{(\pi-p)} = \begin{cases} -1.66E + 37.34, & 2.5 \leqslant E \leqslant 5 \text{ GeV}; \\ 23.6 + 25.51E^{-0.959}, & E > 5 \text{ GeV}; \end{cases}$$

$$\sigma_{\text{el}}^{(\pi+p)} = \begin{cases} -1.4E + 10, & 2.5 \leqslant E \leqslant 3 \text{ GeV}; \\ -0.166E + 6.298, & 3 \leqslant E \leqslant 6 \text{ GeV}; \\ 3.5 + 9.93E^{-0.953}, & E > 6 \text{ GeV}; \end{cases}$$

$$\sigma_{\text{el}}^{(\pi-p)} = 3.78 + 7.27E^{-0.896}, \quad E \geqslant 2.5 \text{ GeV}.$$

At energies in the range $0.8 < E < 3.5$ GeV the hadron–nucleon interaction cross sections are found by means of a power-law interpolation. At energies $E < 3$ GeV the differential scattering cross section can be approximated by

$$d\sigma/d\Omega \sim A \cos^4 \theta^* + B \cos^3 \theta^* + 1,$$

where $\theta^*$ is the scattering angle in the cm frame. For these energies the cosine of the scattering angle is simulated uniformly and the weight of the scattered hadron is $W = (A \cos^4 \theta^* + B \cos^3 \theta^* + 1)/(1 + 0.2A)$; the weight of the recoil proton is $W = (A \cos^4 \theta^* - B \cos^3 \theta^* + 1)/(1 + 0.2A)$.

Depending on the energy, the coefficients $A$ and $B$ can be calculated in accordance with the procedure of Ref. 151. At $E > 3$ GeV the differential cross section of the elastic nucleon–nucleon scattering is approximated by the expression

$$d\sigma/dt = \exp(a + bt).$$

Here the variable $t$ is related to the cosine of the scattering angle in the cm frame by $\cos \theta^* = 1 - |t|/2K^2$, where $K^2 = S/4 - [(m_1^2 + m_2^2)/2 + (m_1^2 - m_2^2)^2]/(4S)$, $S$ is the square of the total collision energy in the cm frame, and $m_1$ and $m_2$ are the masses of the incident particle and the particle at rest, respectively. The parameter was determined in Ref. 152:

$$b = \begin{cases} 7.26 + 0.0313 p_0 \text{ for } pp \text{ and } np, \\ 7.3 \text{ for } \pi p; \end{cases}$$

here $p_0$ is the primary momentum.

The cosine of the scattering angle in the laboratory frame is determined from the kinematic relations for the particles of equal masses.

For the scattering of a nucleon we have

$$\cos \theta = \sqrt{1 + E_0/2m}(1 + \cos \theta^*)/[2 + (E_0/2m)(1 + \cos \theta^*)]$$

and for the scattering of a recoil proton we have

$$\cos \theta = \sqrt{1 + E_0/2m}(1 - \cos \theta^*)/[2 + (E_0/2m)(1 - \cos \theta^*)].$$

After the scattering the nucleon energy is $E = 0.5E_0(1 + \cos \theta^*)$ and the recoil proton energy is $E' = E_0 - E$. After the elastic scattering the angle and energy of $\pi^{\pm}$ mesons are determined in a similar manner.

In the case of the development of an electromagnetic cascade the release of energy in the substance is determined by three factors: (1) the electromagnetic interaction of hadrons with $E > E_{thr}$, (2) the charged particles with $E < E_{thr}$ and particles which are produced as a result of removal of the residual excitation of the nucleus, and (3) the particles of the electron–photon shower which is initiated by the decay $\pi^0 \to \gamma\gamma$.

The energy $\Delta E = E_1 - E_2$, which is lost along the path length $l$ by a charged hadron $j$ with initial energy $E_1$, can be determined from the relation

$$R_j(E_2) = R_j(E_1) - l \tag{6.6}$$

by means of direct and inverse power-law interpolation from the previously calculated range tables $R_j(E)$ (see Chap. 2).

In all the studies carried out in the USSR and abroad which were published before the appearance of the study by Mokhov and Van Ginneken[142] and which dealt with the calculations of high-energy electromagnetic cascades, the charged hadrons that reach the kinetic energy $E < E_{thr}$ were assumed to be hadrons that have stopped and that release their energy locally. Mokhov and Van Ginneken[142] showed that such an approach in some problems may lead to large errors and that the proton transport and $\pi^{\pm}$ meson transport should be traced to energies on the order of 1 MeV. MARS-7, MARS-8, and MARS-9 programs include the algorithm for the calculation of the transport and the release of energy from charged hadrons with energies in the range 1 MeV$\leqslant E \leqslant E_{thr} = 10$–20 MeV.

The average excitation energy (GeV) of nuclei is calculated from the equations of Ref. 85, which were modified for light nuclei[119]

$$E^* = \begin{cases} CE_0, & E_0 \leqslant 0.05 \text{ GeV;} \\ 0.05C + \left[\dfrac{E_0 - 0.05}{990}\right](A - 10C), & 0.05 < E_0 < 5 \text{ GeV;} \\ 0.005A, & E_0 \geqslant 5 \text{ GeV,} \end{cases}$$

where $E_0$ is the kinetic energy (GeV) of the incident hadron; for a $\pi^{\pm}$ meson $E_0$ is its total energy;

$$C = \begin{Bmatrix} 1, & A \geqslant 26; \\ (A-2)/24, & 3 < A < 26. \end{Bmatrix}$$

In the approximate calculations of the energy release it is assumed that one-half of the excitation energy $E^*$ is used to produce neutrons, one-fourth of it is used to produce protons and one-fourth of it is used to produce secondary nuclei heavier than hydrogen. The neutron transport and the neutron energy release are then modulated isotropically in a sphere of radius $\lambda_{in}$ ($E \sim 20$ MeV). The isotropically emitted protons are traced until they are stopped in a manner similar to the particles with $E < E_{thr}$. The heavy nuclei, by definition, release their energy locally at the production point.

Instead of two photons one photon with a doubled statistical weight is considered in the simulation of an electron–photon shower from the decay of neutral $\pi$ mesons. The sampling function is used in the form $f(E_\gamma) = 2E_\gamma / p_\pi$

$E_\pi$, where $p_\pi$ and $E_\pi$ are the momentum and energy of the $\pi^0$ meson, respectively. The energy, the angle, and the statistical weight of the photon are then determined by the relations

$$\left.\begin{aligned} E_\gamma &= [p_\pi E_\pi \xi + (1/4)(E_\pi - p_\pi)^2]^{1/2}; \\ \cos\theta_\gamma &= (E_\pi/p_\pi)(1 - m_\pi^2/2E_\gamma E_\pi); W = E_\pi/2E_\gamma; \\ \xi &\in (0,1). \end{aligned}\right\} \tag{6.7}$$

The azimuthal angle $\varphi$ is simulated, as in all the processes below, as an isotropic angle according to Eqs. (6.5).

All the programs for the simulation of the transmission of photons through a medium (with the exception MARS-8, MARS-9M, and MARS-9T) use the algorithm[136] which was modified by Baĭshev *et al.*[133] and Mokhov.[143] The photon range before the production of the $e^+e^-$ pair is simulated from the distribution

$$p(x) = (7/9t_r)\exp(-7x/9t_r), \tag{6.8}$$

where $t_r$ is the radiation length (see Chap. 2).

The sampling function for the sampling of the electron (positron) energy and the corresponding doubled statistical weight, with allowance for the distributions from Chap. 2, can be written in the form

$$\left.\begin{aligned} f(E_e) &= 2E_e/E_\gamma^2; \\ W &= (9/7)[(4/3)(E_e/E_\gamma - 1) + E_\gamma/E_e]. \end{aligned}\right\} \tag{6.9}$$

The angle of emission $\theta$ of an electron is simulated in the following way:

$$\left.\begin{aligned} \theta &= (Y - \alpha^2)^{1/2}; \\ Y &= \{\alpha^{-2} - \xi[\alpha^{-2} - (\alpha^2 + \pi^2)^{-1}]\}^{-1}; \\ \alpha &= (m_e/\pi E_e)\ln(E_\gamma/m_e); \xi \in (0,1). \end{aligned}\right\} \tag{6.10}$$

Following the procedure used by Van Ginneken,[136] we can use the variable $\lambda_0$ (g/cm$^2$)

$$\lambda_0 = 325\ln(1000E_e)/(\ln Z)^{1.73}, \tag{6.11}$$

where $Z$ is the atomic number of the medium.

In the case of such a choice of the unit length, the approximate longitudinal distribution of the energy release $F(x/\lambda_0)$ does not depend on the depth $x$ and the electron energy at energies $E_e > 6$ GeV. At lower energies the distribution deformation can be taken into account by using the function $C_0F(x'/\lambda_0)$, instead of $F(x/\lambda 0)$, where $x' = x + 1.2(1000E_e)^{-1/2}$, and $C_0$ is the normalization constant. The function $C_0F(x'/\lambda_0)$ was tabulated in Van Ginneken's program.[136] The radial divergence of the electron–photon shower (see Chap. 2) can be taken into account by introducing the effective angle $\theta = r/x$. The spatial distribution of the energy release of the electron–photon shower, which starts at the point found from Eq. (6.8) and which is initiated by an electron with an energy and angle chosen from distributions (6.9) and (6.10), can then be written in the form

$$\left.\begin{aligned}\varepsilon(x,\theta = r/x) &= C_0F(x'/\lambda_0)g(\theta);\\ g(\theta) &= (0.46\lambda_0/x_M)\exp(-0.46\lambda_0\theta/x_M),\end{aligned}\right\} \quad (6.12)$$

where $x_M = 0.0212t_r/E_c$ is the Molière length, and $E_c$ is the critical energy (GeV) of the material of the medium; here $F(x' = /\lambda_0) = 0$ for $x < 0$. The vector $\Omega$, which is determined by the superposition of the angles from Eqs. (6.7), (6.10), and (6.12) and from the corresponding isotropic azimuthal distributions, is simulated in this manner and the shower energy is distributed along this ray in accordance with the function $C_0F(x'/\lambda_0)$.

In a heterogeneous medium consisting, with respect to the ray, of $N$ segments of various materials in terms of their properties the depth of the substance in units of (6.11) is calculated from the equation $x/\lambda_0 = \sum_{i=1}^{N} x_i/\lambda_{0i}$, and the angles are scaled accordingly at each boundary on the basis of Eq. (6.12).

In the programs MARS-8, MARS-9M, and MARS-9T the electron–photon shower is simulated indirectly, after the determination of the photon parameters from Eqs. (6.7), by the Monte Carlo method on the basis of the AEGIS algorithm.[145] In this program the principal difference between this simulation and direct simulation is that the electron–photon shower is represented as a nonbranching process. At each vertex of the trajectory tree, instead of the production of two particles ($e^+e^-$ or $\gamma e^-$), the production of only a single random particle with a doubled statistical weight is simulated. In such a scheme the energy is conserved on the average in the collision series and additional fluctuations are included in the results. The computing time and the required computer memory are, however, reduced radically in this case. At $E_0 \gtrsim 100$ GeV only on the basis of this scheme can an electron–photon shower be simulated nearly directly during the development of a hadron cascade. In the MARS-9M and MARS-9T programs a new algorithm for the electron, positron, and photon transport, which is given below, is used in describing the electron–photon shower, the accuracy of plotting trajectories in a magnetic field has been improved, and several computational procedures have been changed.

In all the programs of the MARS program package, except MARS-9, trajectories are simulated concurrently with the estimate of the functional in a geometry which is arbitrarily defined by the oriented planes, by the cylindrical surfaces with the axes running parallel to the coordinate axes, and by the spherical surfaces. The intersection of the ray (1.46) with all the surfaces can be calculated exactly by using an analytical geometry procedure.

The block of material under consideration is assumed to be comprised of a finite number $N$ of homogeneous zones. Each zone is replaced by a corresponding control point: this can be any point of the zone not part of its boundaries. The specification of this point determines all the characteristics of the zone: the index of the material of the zone, the list of surfaces that bound the zone, etc.

The index that identifies the segment of the ray $\mathbf{r}$ with zone $i$ can be determined as follows. When the ray crosses the internal boundary, the point which is displaced along the ray in the forward direction is connected by straight-line segments with the control points that are systematically sorted out. If the surfaces bounding zone $i$ are not crossed in segment $i$ or if a surface is crossed an even number of times, then this segment assumes the index that identifies it with zone $i$.

Let us examine the simulation of the range for neutrons and charged hadrons separately in this geometry. We choose a random number $\gamma$ which is uniformly distributed in the interval (0,1).

The probability that a neutron with an energy $E$ will not interact inelastically with the nuclei when it passes $n$ zones of the substance is

$$P_n = \exp\left[-\sum_{i=1}^{n} \Sigma_i(E) l_i\right], \tag{6.13}$$

where $\Sigma_i(E)$ is the macroscopic cross section for the absorption of a neutron in zone $i$ and $l_i$ is the length of the segment of the ray in zone $i$.

The number of the zone in which the neutron experiences an interaction is determined from the condition $P_n < \gamma P_{n-1}$ or from the condition

$$\sum_{i=1}^{n-1} \Sigma_i l_i < -\ln\gamma < \sum_{i=1}^{n} \Sigma_i l_i,\ n = 2,\ldots,N. \tag{6.14}$$

The neutron range will then be

$$R = \sum_{i=1}^{n-1} l_i + \frac{1}{\Sigma_n}\left(-\ln\gamma - \sum_{i=1}^{n-1} \Sigma_i l_i\right).$$

If condition (6.14) does not hold for any of the values of $n$, the neutron is assumed to have escaped from the substance.

*An exponential transformation* must be used to obtain data for distances far from the source: In the case of simulation the path length $\lambda_i = \Sigma_i^{-1}$ is deliberately increased by a factor of $C_i$ in zone $i$ and an additional statistical weight is introduced,

$$W = \begin{cases} C\gamma^{C-1} \ \text{for}\ C_i = C = \text{const}, \\ C_n \exp\left[-\dfrac{R - \sum_{i=1}^{n-1} l_i}{\lambda_n C_n}(C_n - 1) - \sum_{i=1}^{n-1} \dfrac{l_i}{C_i\lambda_i}(C_i - 1)\right] \\ \text{for}\ C_i = f(i). \end{cases} \tag{6.15}$$

We introduce the following notation for charged hadrons:

$$Q_i = \int_0^E \frac{\Sigma_i(E')}{\beta_i(E')}\,dE'; \Delta Q_i = Q_i(E_{i-1}) - Q_i(E_i), \tag{6.16}$$

where $\beta_i(E)$ is the stopping power of the substance: the average energy loss of a particle due to electromagnetic processes (see Chap. 2) in zone $i$ for protons or $\pi^\pm$ mesons. Condition (6.14) can then be written in the form

$$\sum_{i=1}^{n-1} \Delta Q_i < -\ln\gamma < \sum_{i=1}^{n} \Delta Q_i. \tag{6.17}$$

Let us evaluate the following quantity: $Q_n(E') = \ln \gamma + Q_n(E_{n-1}) + \sum_{i=1}^{n} \Delta Q_i$.

The energy $E'$ at which the nuclear interaction occurs can be found from the pretabulated values of $Q_i(E)$ on the basis of $Q_n(E')$ by means of a quadratic interpolation. A part of the range $\Delta R$ before the nuclear interaction in zone $n$ and the mean free path $R = \sum_{i=1}^{n-1} l_i + \Delta R$ are found from the difference $\Delta E = E_{n-1} - E'$ and from the curve for the range versus energy.

Let us assume that condition (6.17) does not hold for any of the values of $n$. If the ionization range is greater than the distance from the external boundary along the ray, the particle can then be assumed to have escaped from the substance. If the ionization range is shorter than the distance from the external boundary, the particle can be assumed to have been absorbed in the relevant zone.

In MARS-9 programs the trajectories begin to be simulated with the simulation of the hadron range $R$ before the nuclear interaction. The neutron range is calculated from the relation $R = -\Sigma_i^{-1}(E)\ln \gamma$, where $\Sigma_i$ is the macroscopic cross section for the inelastic interaction of a neutron with the nucleus of the $i$th material. For charged hadrons with energies $E > 5$ GeV the range is modulated in a similar way, at energies in the range $0.05 \leqslant E \leqslant 5$ GeV algorithm (6.17) is used, and at energies $E < 0.05$ GeV it is assumed that $R$ is equal to the ionization range $R_i^{\text{ion}}(E)$.

The *step method* is used for the construction of a three-dimensional trajectory of a particle. The multiple Coulomb scattering of particles, the effect of external electromagnetic fields, and other effects can easily and naturally be simulated by using this method. This particular system may have a geometry of any complexity and materials of any composition. Any number of coordinate systems, including curvilinear coordinate systems, can be specified simultaneously. One of the principal advantages of this method is that only the segment of the trajectory need be identified with a particular zone and that the coordinates of the points at which the trajectory intersects the surfaces need not be calculated. The linking of a trajectory segment with a particular zone is a much simpler procedure that requires much less computation time.

In a standard step method any boundary localizes within a step $l$, which introduces a certain error to the result. Since the use of large values of $l$ everywhere greatly distorts the picture of the process, and since the use of small values of $l$ leads to an unacceptably long computing time, an *iterative step method*[143,153] was developed for MARS-9 package of programs, which can be described as follows.

The step $l$ is assumed to be equal to the simulated range $R$ or a special range due to the division of $R$ by an integer $L$. We assume that the starting point $O_n$ of the segment $l$ belongs to the zone with the number $n$ and that the end point belongs to the zone with the number $k$. To localize the point at which the trajectory segment $l$ crosses the boundary with the prescribed accuracy $\delta$, it would then be necessary:

•To take a step of length $l/2^m$ from $O_k$ in the direction of $O_k$–$O_n$ and to determine the number of the zone where this step ends, $O_i$.

•To take a step of length $l/4$ from $O_i$ in the direction of $O_n$–$O_k$ if $O_i$ belongs to zone $n$ or in the direction of $O_k$–$O_n$ if $O_i$ belongs to zone $k$.

This procedure is continued $m$ times until $l/2^m$ is less than $\delta$, after which the iteration is continued with a step equal to $\delta$. The process is terminated when the point $O_i$ reaches zone $k$ after a step of length $\delta$ in the direction of $O_n$–$O_k$. The lengths of the segments $O_nO_i$ and $O_iO_k$ in this case are known within $\delta$ or better. If it turns out during the iteration that the point at which the segment $l$ crosses the boundary is separated from the point $O_n$ a distance smaller than $\delta$, then the entire segment is assumed to belong to zone $k$, and vice versa.

In this scheme the changes in the energy of charged particles in the step and the estimate of the energy-release functionals are accomplished according to algorithm (6.6). The vacuum spaces are crossed with increased step lengths. If the boundary between material $i$ and material $k$ is crossed during the iteration, then at the boundary the step can be transformed in accordance with $l_k = l_i\Sigma_i(E)/\Sigma_k(E)$, where $\Sigma_i$ and $\Sigma_k$ are the macroscopic cross sections of the $hA$ interaction for substances $i$ and $k$, respectively; and $E$ is the energy measured at the time of the crossing. For charged hadrons with $E < 0.05$ GeV we have $l_k = l_iR_k^{\text{ion}}(E)/R_i^{\text{ion}}(E)$.

Calculations showed that the computing time based on this method is nearly independent of $\delta$ in the interval $\delta \sim 0.005$–$0.05$ cm and that this scheme is highly effective in the calculations of hadron and electron–photon cascades, especially in the case of a complex geometry.

If the problem includes a magnetic field $\mathbf{B}(\mathbf{r}) = \{B_x(\mathbf{r}), B_y(\mathbf{r}), B_z(\mathbf{r})\}$, the charged particle will change its direction upon passing a segment $l$ of the trajectory. Suppose that a particle with a momentum and charge $q$ moves in the direction $\mathbf{\Omega} = \{\omega_x,\omega_y,\omega_z\}$. The new direction of motion $\mathbf{\Omega}' = \{\omega_x',\omega_y',\omega_z',\}$ can then be determined, after the passage of step $l$, in the following way:

$$\left.\begin{aligned} \omega_x' &= D^{-1}[\omega_x(1+\alpha^2-\beta^2-\gamma^2)+2\omega_y(\gamma+\alpha\beta)+2\omega_z(\alpha\gamma-\beta)];\\ \omega_y' &= D^{-1}[2\omega_x(\alpha\beta-\gamma)+\omega_y(1-\alpha^2+\beta^2-\gamma^2)+2\omega_z(\alpha+\beta\gamma)];\\ \omega_z' &= D^{-1}[2\omega_x(\beta+\alpha\gamma)+2\omega_y(\beta\gamma-\alpha)+\omega_z(1-\alpha^2-\beta^2+\gamma^2)], \end{aligned}\right\} \quad (6.18)$$

where $\alpha = \xi B_x$, $\beta = \xi B_y$, $\gamma = \xi B_z$, $\xi = 0.0015lq/p$, $D = 1+\alpha^2+\beta^2+\gamma^2$, $l$ is the step (cm), $B$ is the magnetic field introduction ($T$), and $p$ is the momentum of the particle (GeV/$c$).

Similar equations can be obtained for the motion in an electric field. The step $l$ is assumed to be so small that throughout its length the change in the field components and the decrease in the particle energy in the substance due to the ionization loss can be disregarded.

The algorithm for the simulation of the Coulomb scattering in the step $l$ is described in Chap. 2. In changing from the old direction $\mathbf{\Omega}$ to the new one $\mathbf{\Omega}'$, just as in all the discrete nuclear interactions described above, the angles are scaled by means of the following equations (for the *right-handed coordinate system!*):

$$\left.\begin{aligned} \omega'_x &= \omega_x\mu + (\beta/\alpha)(\omega_x\omega_z \cos\varphi - \omega_y \sin\varphi); \\ \omega'_y &= \omega_y\mu + (\beta/\alpha)(\omega_y\omega_z \cos\varphi + \omega_x \sin\varphi); \\ \omega'_z &= \omega_z\mu - \alpha\beta\cos\varphi, \end{aligned}\right\} \tag{6.19}$$

where $\mu = \cos\theta$ is the cosine of the scattering angle in the laboratory frame, $\varphi$ is the azimuthal scattering angle, $\alpha = (1-\omega_z)^{1/2}$, and $\beta = (1-\mu^2)^{1/2}$.

Let us consider, in conclusion, the algorithm for the local estimate of the particle flux, which is used in MARS-4 and MARS-9 programs. The energy distribution of the particles of species $j$ in detector $m$ is described as the expectation value by the equation

$$\Phi_j(\mathbf{r}^*_m, E) = \mathbf{M}\left\{\sum_{k=0}^{K} W_k \frac{P_j(E_k, r)}{r^2} \sum_{i=p,n,\pi} \frac{d^2N_{ij}}{dEd\Omega}\right.$$

$$\left. \times(\mathbf{r}_k, E_{k-1}, E_k, \boldsymbol{\mu}_k)\eta_j\eta_\Omega(\Delta\boldsymbol{\Omega}_{km} - \boldsymbol{\omega}_k)\eta_E(\Delta E_{km} - E_k)\right\}, \tag{6.20}$$

where $W_k$ is the statistical weight of the $k$th $hA$ interaction: $W_k = \prod_{n=1}^{k-1}\prod_{s=1}^{S_n} W_{ns}$, which is defined as the product of the statistical weights of all $S_n$ samplings in each of the preceding $(k-1)$ interactions; $P_j(E,r)$ is probability (6.13) for neutrons or probability (6.16) for charged hadrons after replacing $\Sigma_i l_i$ by $\Delta Q_i$; $r = |\mathbf{r}^*_m - \mathbf{r}_k|$ is the distance between the interaction point $k$ and the detector $m$ at the point $\mathbf{r}^*_m$; $\mu_k = (\boldsymbol{\Omega}_m, \boldsymbol{\omega}_k)$; $\omega_k = (\mathbf{r}^*_m - \mathbf{r}_k)/r$; $d^2N_{ij}/dEd\boldsymbol{\Omega}$ is the inclusive distribution of hadrons of species $j$ in the interaction with the nuclei of the hadrons of species $i$ at the point $\mathbf{r}_k$; and

$$\eta_j = \begin{cases} 1, & j = n; \\ \dfrac{\beta_j(E_k)}{\beta_j(E)}, & j = p\pi; \end{cases} \qquad \eta_\Omega = \begin{cases} 1 & \text{if } \boldsymbol{\omega}_k \in \Delta\boldsymbol{\Omega}^j_{km}; \\ 0 & \text{otherwise;} \end{cases} \qquad \eta_E = \begin{cases} 1 & \text{if } E_k \in \Delta E^i_{km}; \\ 0 & \text{otherwise.} \end{cases}$$

Here $\beta_j$ is the stopping power of the material, $\Delta\boldsymbol{\Omega}^j_{km}$ is the kinematically resolved region of the directions of the motion of particles of species $j$, and $\Delta E^j_{km}$ is the interval of resolved energies. For neutrons this interval is determined by the kinematic limits; for charged hadrons the energy loss along the ray segment $r$ in the material is also taken into account. The distribution of the energy release and of the other functionals in the local detectors is found from Eq. (6.20) by means of appropriate convolutions like (1.40).

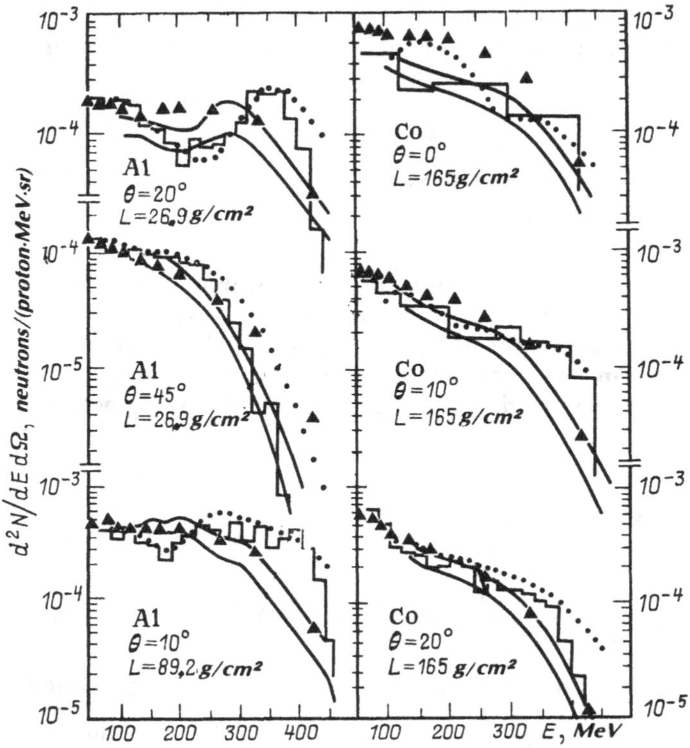


**Figure 6.1. Differential neutron distribution behind an aluminum barrier and cobalt barrier of thickness *L* produced as a result of normal incidence of a broad 400-MeV proton beam. Curves, experimental corridor (Ref. 154); calculations, histograms (Ref. 128); ●, Ref. 155; ▲, MARS-4.**

# 4. Certain characteristics of the electromagnetic cascades

Let us consider the systematic features of the development of electromagnetic cascades in condensed matter, which were determined with the MARS package of programs. To demonstrate the quality of the programs, we will give, where possible, the experimental results of other authors.

In those cases where the longitudinal dimensions of the layer of a substance are small ($L \ll \lambda_{abs}$), the hadronic cascade changes the spectra of the emitted particles only slightly. Their differential distributions are approximately equal to the corresponding distributions in the $hA$ interactions with isolated nuclei. This conclusion follows from the analysis of Figs. 6.1–6.3, which show the results for various targets at proton energies between 0.4 GeV and 50 GeV.

At a high primary energy $E_0$ the longitudinal distributions of the secondary-particle flux density in thick layers of the substance take on the charac-

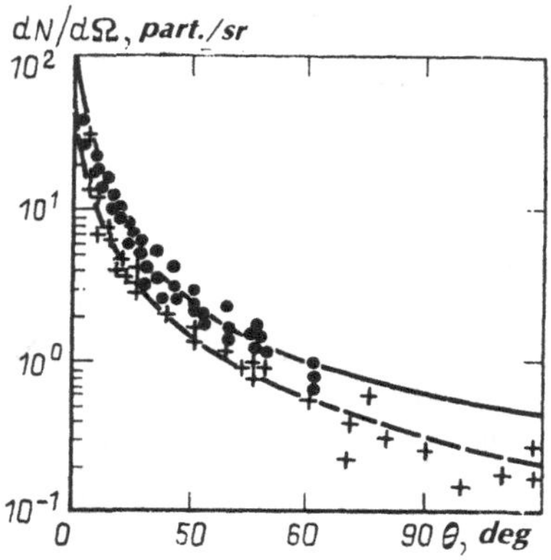


**Figure 6.2.** Angular distribution of hadrons with $E \geqslant 35$ MeV from a 5-cm-long copper target bombarded by protons with momenta $p_0 = 8$ GeV/$c$ (--) and 24 GeV/$c$ (—). Experimental point, from Ref. 156; curves, calculation based on MARS program.

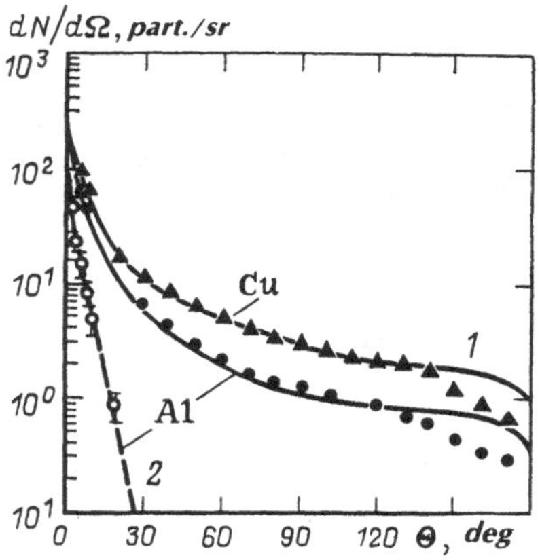


**Figure 6.3.** Angular distributions of hadrons with energies above 20 MeV (1) and above 2 GeV (2) from an aluminum target and copper target 57 cm and 45 cm long, respectively. $p_0 = 50$ GeV/$c$; points, experimental data taken from Ref. 157; curves, calculation based on MARS program.

teristic shape of a cascade curve. Figures 6.4 and 6.5 show the calculated distributions, along with the experimental data in the energy range from 9 GeV to 150 GeV. In Fig. 6.4 the *track density* is taken to mean the total flux density of protons with a kinetic energy $E > 0.5$ GeV and of $\pi^{\pm}$ mesons with a kinetic energy $E > 0.08$ GeV.

The maximum of the longitudinal distributions of the particle flux density, which are integrated in the transverse plane (*the transverse integral*),

$$\Phi(z) = 2\pi \int_0^{\infty} \Phi(z,r) r \, dr \tag{6.21}$$

increases with the energy in a nearly linear manner. After going through a maximum and the transition region, the attenuation length $\Lambda$ [the exponential function of the quasiexponential decay of the distribution (6.21)] increases monotonically with energy, becoming nearly constant at energies $E_0 \gtrsim 30$–50 GeV. In the distributions of secondary hadrons with energies $E > 20$ MeV $\Lambda \simeq (1.5$–$1.7)\lambda_{abs}$ for such energies.

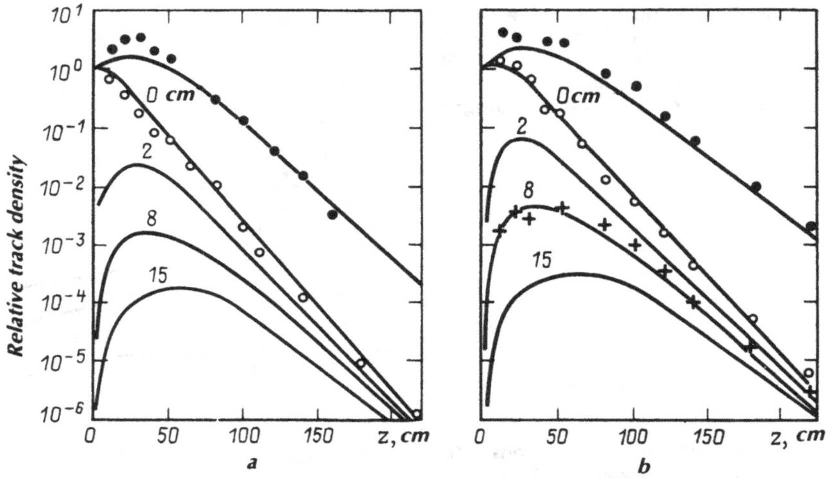


**Figure 6.4.** Spatial distribution of tracks in nuclear emulsions in an iron barrier bombarded by a narrow proton beam. Upper curve and dark points, transverse integral (6.21); the remaining curves and points, longitudinal distribution at various distances $r$ (cm) from the beam axis; curves, MARS calculation: (a) $E_0 = 9.1$ GeV, experiment (Ref. 158); (b) $E_0 = 18.3$ GeV, experiment (Ref. 159).

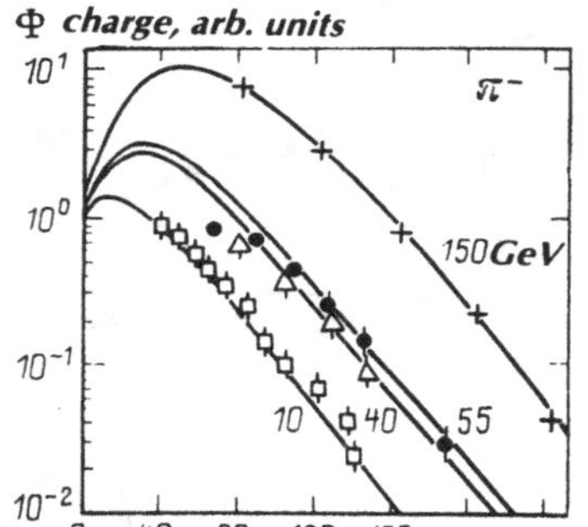


**Figure 6.5.** Distribution (6.21) of the flux density of charged hadrons in an iron barrier bombarded by $\pi^-$ mesons of various energies ●, △, □, experimental data taken from Ref. 112, +, Ref. 160; curves, MARS calcuations.

At high energies (the $\langle p_\perp \rangle$ is restricted) the consequence of a strong scattering anisotropy is the appearance of a strong radial flux density and radial star density gradient (Fig. 6.6). [*The star density* $S(z)$ is the number of inelastic nuclear interactions initiated by hadrons with $E > 50$ MeV in a unit volume of the substance.] At high energies the following behavior of the distribution in the transverse plane is valid in all cases: $\Phi(r) \sim (1/r)\exp(-r/\lambda_r)$, where $\lambda_r = \lambda_r(E_0,z)$ is a function of the absorber thickness and of the primary energy.

The energy distribution of the electromagnetic-cascade particles depends strongly on the primary energy $E_0$, the structure of the absorber and the materials it contains, the initial conditions, and the spatial domain. However, outside the central region of the cascade and especially in the direction

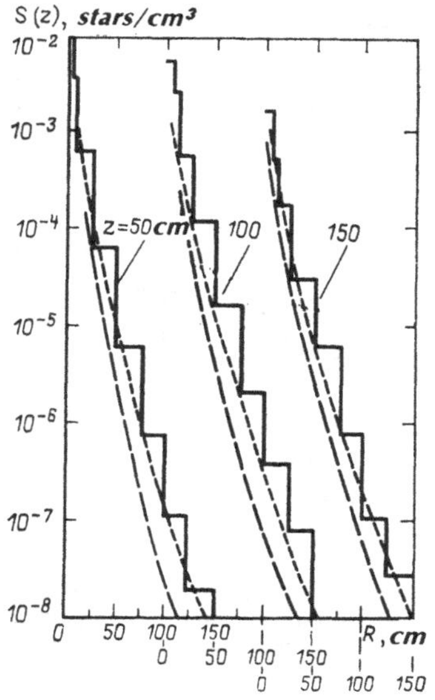


**Figure 6.6. Radial distributions of the star densities at various depths of a steel absorber bombarded with 200-GeV protons. Dashed curves, results of calculations based on Monte Carlo programs: – – –, FLUKA (Ref. 129); -- CASIM (Ref. 136); histograms, MARS-4.**

perpendicular to the primary-beam axis, these functional dependences weaken considerably. The dependence on the material remains largely the same in the energy range $E < 20$ MeV. At energies $E < 1$ GeV the neutrons take precedence over all other particles. In the interval $1 < E < 10$ MeV the flux of secondary photons is strong.

Figure 6.7 shows the results of calculations of the neutron spectra based on the MARS and LENTMC programs.[161] For convenience, the energy distributions are multiplied by the neutron energy. Also shown in this figure are the results of Armstrong and Alsmiller's[162] calculations. In the distributions two maxima are clearly discernible in the neutron energy ranges $E \approx 100$ MeV and $E \sim 0.2$ MeV. An increase in the shielding thickness in the radial direction from 20 cm to 100 cm does not change the spectrum shape appreciably.

An increase in the energy of primary protons has also only a slight effect on the neutron spectrum shape in a similar situation (Fig. 6.8). Here the distributions are given for the part behind the lateral surface of the cylindrical shield with a 50-cm iron and 100-cm concrete radial cladding. A proton beam with a linear density of 1 proton/$m$ strikes the inner wall uniformly in the longitudinal direction at an angle $\theta \simeq 3$ mrad. The shielding of iron with concrete changed the shape of the distribution dramatically: at $E \sim 150$ MeV only one high-energy maximum remained and at low energies the distribution began to behave in a manner similar to the $\sim E^{-1}$ law.

Figure 6.9 shows the results of calculations of the passage of neutrons with an initial spectrum of the type $CE^{-1}$ at $E \leqslant 200$ MeV and with a spectrum of the type $0.2CE^{-2}$ at $E > 200$ MeV through a concrete layer. At kinetic energies $E < 1$ GeV the shape of the energy distributions of protons and $\pi^{\pm}$ mesons

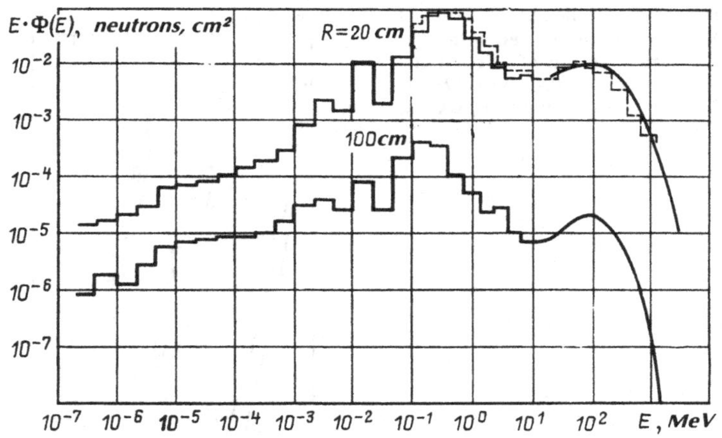


**Figure 6.7. Energy spectra of neutrons behind the lateral surface of an iron cylinder of radius *R*, along the axis of which 3-GeV protons with a linear density of 1 proton/cm interact longitudinally in a uniform manner (dashed curves, calculation carried out in Ref. 162).**

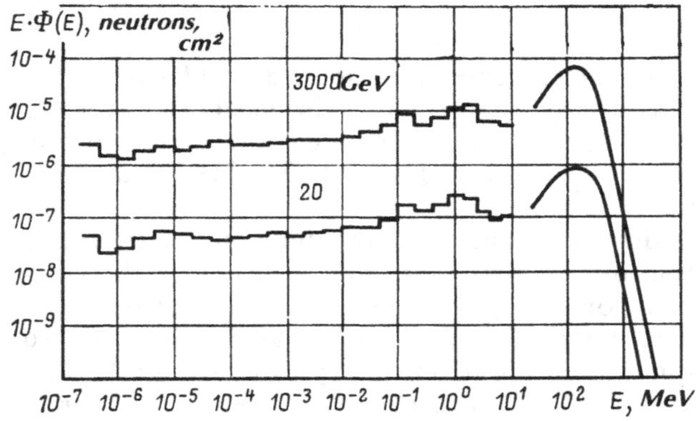


**Figure 6.8. Energy spectra of the neutron loss from heterogeneous cylinder for two values of the primary-proton energy.**

differs sharply from that of the neutron spectrum. The relative contribution to the total flux of hadrons with energies $E > 10$ MeV behind the barrier is nearly independent of the angle of incidence of the source neutrons and the barrier thickness at least in the range 50–200 cm. The total particle flux is comprised of approximately 91.5% neutrons, 7.5% protons, and 2% $\pi^{\pm}$ mesons.

The distribution of the intensity of photons produced in the development of an electromagnetic cascade has a characteristic shape (Fig. 6.10). The distribution anisotropy decreases with decreasing energy. At energies $E \sim 7$ MeV the maximum is caused by the capture photons (see Sec. 5.6). For these

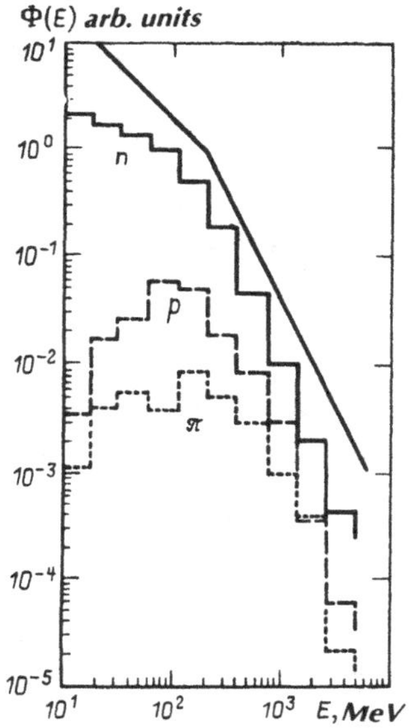


**Figure 6.9. Energy spectra of hadrons behind a 100-cm-thick concrete barrier (curve, model-based spectrum of the source neutrons).**

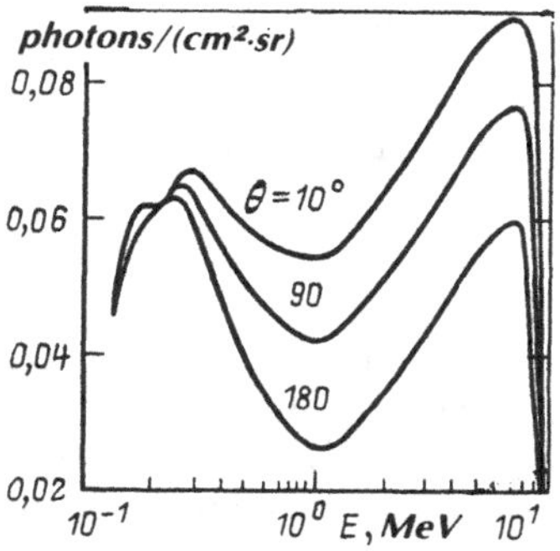


**Figure 6.10. Energy distribution of the intensity of the photon flux at various angles at a depth of 180 cm of an iron barrier bombarded by 10-GeV protons.**

thicknesses ($\sim$180 cm Fe) the high-energy part of the spectrum, which is determined by the electron–photon shower from the decays $\pi^0 \rightarrow \gamma\gamma$, is several orders of magnitude lower than this maximum at moderately high proton energies.

Let us consider some *integral characteristics* of electromagnetic cascades. The energy dependence of the total number of stars, $N_s$ in iron is shown in Fig. 6.11. Over a broad energy range these results are approximated by the function $N_s$ ($E >$ 50 MeV) = 5.9 $E_0^{0.771}$, where $10^{-1} \leqslant E_0 \leqslant 10^4$ GeV for neutrons and $5 \leqslant E_0 \leqslant 10^4$ GeV for protons. Also shown in this figure are the results of other theoretical studies: $N_s$ ($E >$ 100 MeV) = $2E_0$ (Ref. 163) and $N_s$ ($E >$ 15 MeV) = $4.36E_0 + 75$ at $E_0 >$ 40 GeV (Ref. 164). The results of Ref. 164, which are based on an evaporative cascade model, are too high at high energies.

*The quasi-albedo* $\xi$ of the hadrons with energies $E >$ 20 MeV is shown in Fig. 6.12. Let us summarize the principal properties of the quasi-albedo.

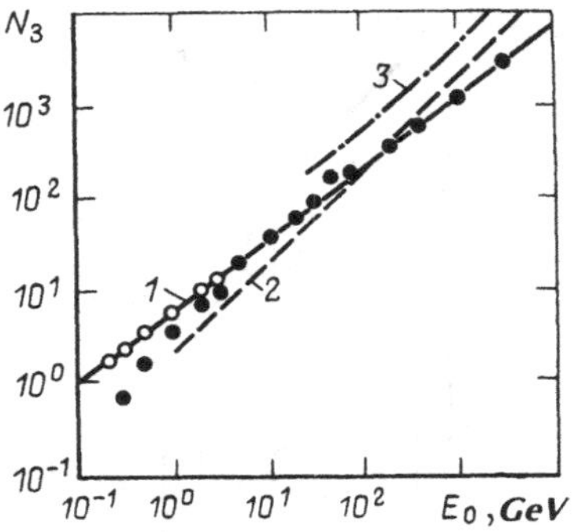


**Figure 6.11. Total number of stars in a semi-infinite layer of iron when a single proton (●) or a single neutron (○) with an energy $E_0$ is incident on it. 1, MARS-6 calculation for a minimum energy $E_{th} = 50$ MeV; 2, predictions (Ref. 163) for protons and $E_{th} = 100$ MeV; 3, HETC calculation (Ref. 164) for protons and $E_{th} = 15$ MeV.**

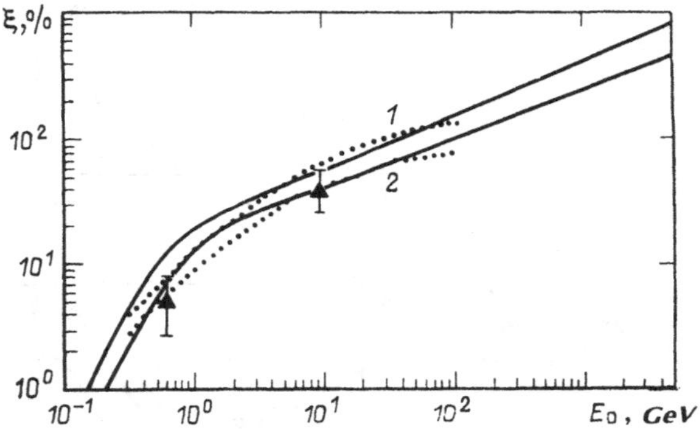


**Figure 6.12. Total inverse hadron yield (quasi-albedo) from a semi-infinite iron barrier (1) and a semi-infinite concrete barrier (2) vs the primary proton energy. —, MARS-4 calculation; ...., HAMLET calculation (Ref. 77); ▲, experimental data taken from Refs. 165 and 166 for concrete.**

(1) The value of $\xi$ increases rapidly with increasing energy up to several gigaelectron-volts, and then this functional dependence becomes weaker.

(2) At $E_0 \gtrsim 5$ GeV the results of calculation of $\xi$ using the MARS-6 program can be described as follows:

$$\xi,\% = \begin{cases} 19E_0^{0.447} & \text{for iron,} \\ 16.6E_0^{0.39} & \text{for concrete.} \end{cases}$$

(3) At $E_0 > 40$ GeV (iron) and $E_0 > 100$ GeV (concrete) the value of $\xi$ is greater than 100%, i.e., each incident hadron causes more than one hadron with an energy $E > 200$ MeV to be released from the block.

(4) The quasi-albedo spectrum has virtually no particles with energies $E > 0.7$–1 GeV, regardless of the primary energy.

(5) The charged hadron contribution to $\xi$ is $\lesssim 10\%$.

(6) At $E_0 \gtrsim 10$ GeV the quasi-albedo is nearly independent of the type of primary hadron.

Energy release due to the development of an electromagnetic cascade. The component composition of the total energy release calculated from the MARS-6 program is shown in Fig. 6.13. The following groups of particles, which lose their energy in the electromagnetic processes, are singled out:

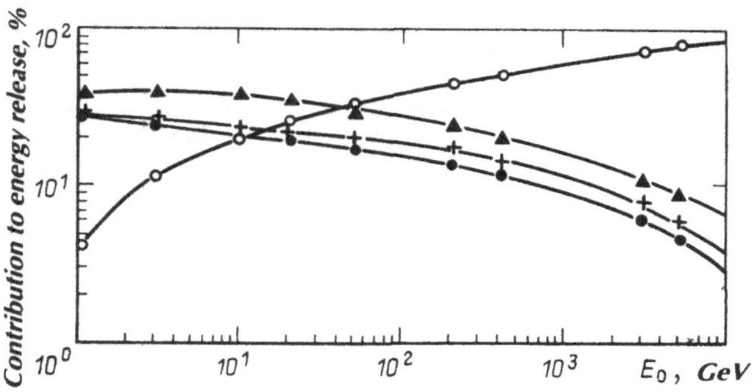


**Figure 6.13. Contribution of various radiation components to the total energy release in an iron block vs the energy of the incident protons. ▲, $p$, $\pi^{\pm}$, and $\mu^{\pm}$; (+), charged particles from nuetrons with $E < 20$ MeV; ●, evaporative protons and fragments ($d$, $t$, $\alpha$...); ○, electron–photon shower from the decay $\pi^0 \to \gamma\gamma$.**

(1) Charged hadrons and muons which participate in the development of an electromagnetic cascade;

(2) charged particles produced by neutrons with kinetic energies $E < 20$ MeV;

(3) evaporative protons and fragments ($d,t,\alpha$...);

(4) electron–photon shower particles which initiate the decays $\pi^0 \to \gamma\gamma$.

The electron–photon shower component amounts to about 4% at the lower boundary of the kinetic energy range under consideration, $E_0 = 1$ GeV, while the contributions from other groups are approximately equivalent. With an increase in the energy the electron–photon-shower fraction increases approximately to 90% at $E_0 = 10^4$ GeV, while the contributions of the other groups decrease monotonically.

The role of the electron-photon shower in the distribution of the energy release $\varepsilon(\mathbf{r})$ is even more important in the central region of the electromagnetic cascade. In intermediate and heavy substances the electron–photon shower determines almost completely the maximum values of $\varepsilon(\mathbf{r})$ at energies $E_0 \gtrsim 100$ GeV and short distances from the beam axis.

A sharp increase in the maximum of the longitudinal distribution of the energy release and a shift of this maximum toward the beginning of the target as its atomic number is increased are attributable to the increasing role of the electron–photon shower (Fig. 6.14).

The radial distributions of the energy release are shown in Fig. 6.15. In the transverse plane the particle distribution in the beam is a Gaussian distribution with the standard deviation $\sigma = 1.25$ mm. The radial gradient of $\varepsilon$, which is much larger than that mentioned previously for the flux density, is related here to the particular features of the energy release from the electron–photon shower initiated by the decay of neutral $\pi$ mesons. At very high energies this component completely determines the maximum energy re-

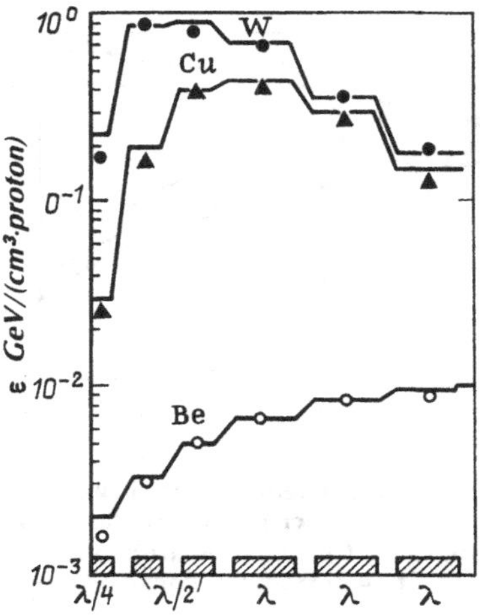


**Figure 6.14. Density profile of the energy release along the length of various targets which are equally partitioned, in units of $\lambda_{abs} = \lambda$. Target diameter, 2.54 cm; primary photon energy, $E_0 = 300$ GeV; histograms, MARS-4 calculation; points, experimental data taken from Ref. 167.**

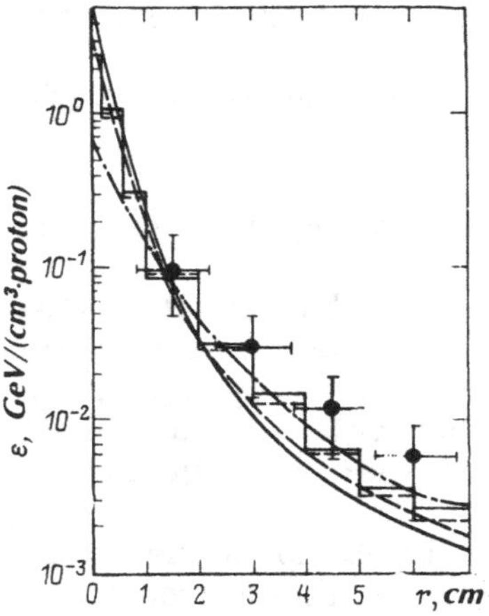


**Figure 6.15. Radial energy-release density profiles for a depth of 45 cm in a copper target bombarded by a proton beam with $E_0 = 400$ GeV. Points, experimental data taken from Ref. 168; Monte Carlo calculations: —·—·—, FLUKA (Ref. 129); ---, CASIM (Ref. 136); —, MAXIM (Refs. 141 and 143); solid histogram, MARS-9M; dashed histogram, MARS-9S.**

lease $\varepsilon_{max}$ even in light materials. At energies above 100 GeV an exact description of the electron-photon shower is therefore particularly important for reliable calculations at the maximum of the functions $\varepsilon(\mathbf{r})$. Aside from the first one or two $\pi^0$-meson production events, the hadron component of the electromagnetic cascade does not have a controlling effect on $\varepsilon_{max}$.

The results of calculations of the energy release obtained on the basis of various programs are compared in Fig. 6.16. The proton distribution in the beam was assumed to be a Gaussian distribution in the vertical and horizontal directions with deviations (dispersions) $\sigma_v = 0.07$ cm and $\sigma_h = 0.14$ cm, respectively. The results of calculations based on various MARS and CASIM program series are largely in fair agreement with each other.

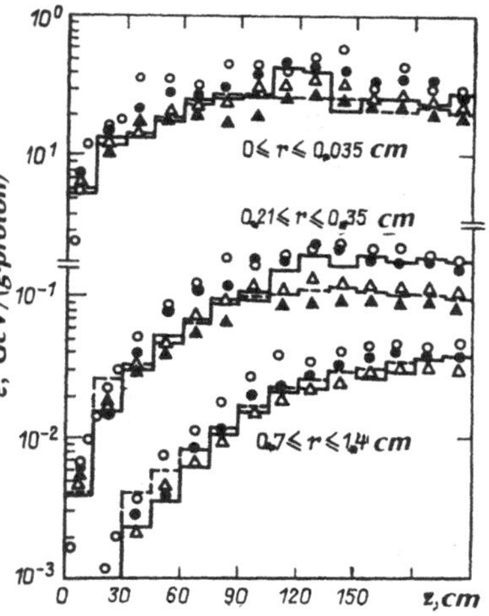


**Figure 6.16.** Longitudinal energy-release density profile in three radial intervals of a graphite target ($\rho = 1.71$ g/cm³) bombarded by protons with $E_0 = 1000$ GeV; calculations (Ref. 135) were carried out according to the following programs: △, MARS-7; ○, MARS-8; ▲, CASIM; ●, MAXIM; histograms: —, calculations based on MARS-9M program; ---, MARS-9S.

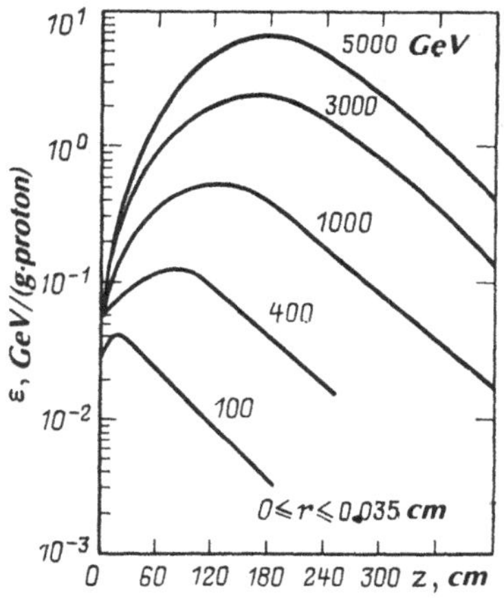


**Figure 6.17.** Longitudinal density profile of the maximum energy release in a graphite target at various primary proton energies ($\sigma_v = 0.07$ cm, $\sigma_h = 0.14$ cm).

The results of MARS-8-program based calculations[135] of the spatial distributions of the energy release of 100- to 5000-GeV protons in graphite are shown in Figs. 6.17 and 6.18. The vertical dispersion ($\sigma_v$) and horizontal dispersion ($\sigma_h$) of the beam are also shown in these figures. Here we see a deformation in the distributions with an increase in the energy and an absence of radial dependence when $r < 0.5\ \sigma_{\min}$. This result, first obtained in Ref. 142, implies that this range of radii corresponds to the true maximum, $\varepsilon_{\max}$. At energies $E_0 \gtrsim 1000$ GeV and radius $r > 1$ cm, the energy-release density, divided by $E_0$, does not depend on the primary energy (scaling). The following relation holds in this region: $\varepsilon/E_0 = 4.1\times10^{-5} r^{-2.32}, g^{-1}$, where $1 < r < 10$ cm for graphite, $\sigma < 1$ cm, and $E_0 \gtrsim 1000$ GeV.

The maximum energy-release density satisfies the simple law $\varepsilon_{\max} = CE_0^n$ over a broad energy range, where the parameters $C$ and $n$ depend at

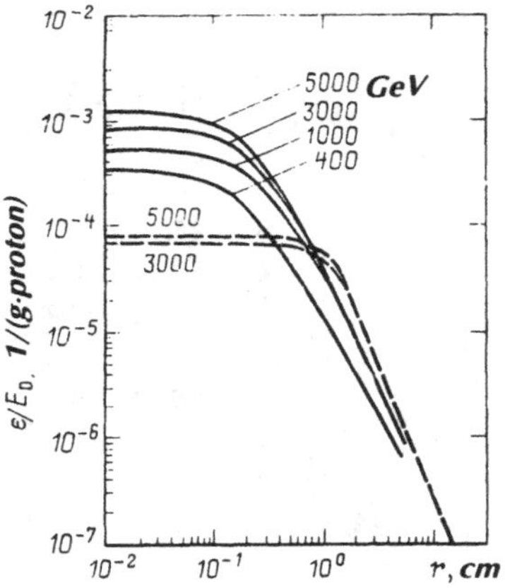


**Figure 6.18.** Energy-release density divided by the corresponding primary proton energy vs the radius of the graphite target at the cascade maximum. —, $\sigma_v = 0.07$ cm, $\sigma_h = 0.14$ cm; ---, $\sigma_v = 0.7$ cm, $\sigma_h = 1.4$ cm.

$E_0 > 200$–$300$ GeV only on the target's material and the beam size. For graphite, for example, $n = 1.44$ when $\sigma_v = 0.07$ cm and $\sigma_h = 0.14$ cm and $n = 1.2$ when $\sigma_v = 0.7$ cm and $\sigma_h = 1.4$ cm.

Interestingly, the maximum energy release density, $\varepsilon_{max}$, depends exclusively on the beam area, but not its shape, i.e., on the relationship between $\sigma_v$ and $\sigma_h$ (Refs. 135 and 142).

# Chapter 7
# High-energy muons

## 1. Muon production and systematic features of muon interactions

The study of high-energy muons and neutrinos deals with the solution of the key problems in elementary particle physics.[1,3,30,169] In all latest-generation accelerators, such as the 3000-GeV proton accelerator-storage ring,[170] the neutrino experiments hold a prominent place in the experimental programs. In large accelerators the muons, the most penetrating component, determine a considerable part of the volume of the radiation shield, whose cost represents an appreciable fraction of the cost of the entire facility. Muons play a prominent role in cosmic-ray physics.

It was shown in Sec. 1.2 that the main muon source is the decay of long-lived mesons:

$$\pi^{\pm} \rightarrow \mu^{\pm} + \nu(\tilde{\nu}); \quad K^{\pm} \rightarrow \mu^{\pm} + \nu(\tilde{\nu}) . \tag{7.1}$$

It was also noted in Sec. 1.2 that in many applications the role of the decay of short-lived mesons $\rho, \omega, \varphi, D$, and $J/\psi$, increases with increasing energy by one or two so-called direct muons. Finally, the inclusive reaction

$$\nu A \rightarrow \mu X , \tag{7.2}$$

whose cross section increases linearly with the energy, is important in high-energy neutrino experiments.

The kinematics of the decays mentioned above is described by the relations in Chap. 2. The production of direct muons has recently been studied extensively (see, e.g., Ref. 5). The results are usually given as a ratio of the muon yield to the $\pi^{\pm}$ -meson yield in a single $hA$-interaction event. The experimental data on the yield of direct-muon pairs, for example, are approximated well by the expression[6]

$$N(\mu^{+} + \mu^{-})/N(\pi^{+} + \pi^{-}) = C(S)\, B(x_F) \times 10^{-4} , \tag{7.3}$$

where

$$C(S) = \begin{cases} -1.91 + 0.88 \ln\sqrt{S}, & \sqrt{S} > 15.44 \text{ GeV} \\ 0.5, & \sqrt{S} < 15.44 \text{ GeV}; \end{cases}$$

$$B(x_F) = \begin{cases} 1 - 2x_F, & x_F < 0.4; \\ (1 - x_F)/3, & 0.4 \leqslant x_F \leqslant 1; \end{cases}$$

$S$ and $x_F$ are the square of the total energy of the $pp$ collision and the Feynman variable for the muon, respectively.

During the passage through matter the produced muon loses its energy due to ionization and due to the excitation of atoms in the direct production of $e^+e^-$ pairs, in bremsstrahlung, and in the nuclear interaction through a virtual photon. These processes are denoted by $i$, $p$, $r$, and $n$, respectively. The average specific energy loss of a muon can be represented as follows.

$$-(dE/dx)_{\text{tot}} = a_i(E) + E[a_p(E) + a_r(E) + a_n(E)] ,$$

where the functions $a_{i,p,r,n}$ are given in Chap. 2.

The average specific energy losses of muons in various materials[3] are given in Table 7.1. Also given in this table are the mean ranges of muons:

$$R = \int_0^{E_0} \frac{dE}{-(dE/dx)_{\text{tot}}} .$$

If at moderately high energies the principal loss mechanism is the ionization and excitation of atoms, the dominant role of the direct production of $e^+e^-$ pairs and bremsstrahlung can clearly be seen in the ground at $E_0 \gtrsim 800$ GeV, in iron at $E_0 \gtrsim 300$ GeV, and in uranium at $E_0 \gtrsim 100$ GeV.

Actually, the energy loss is the result of the order in which the discrete events occur. If a few events of a given type occur in the segment of the muon trajectory under study or if the effective cross section of large bulks of energy transfer, $\Delta E$, is small, the fluctuation of the energy loss may be substantial. The expressions describing the differential cross sections of the processes $i$, $p$, $r$, and $n$ for muons are given in Chap. 2.

Figures 7.1–7.4 show the principal characteristics of the interaction of muons with matter at various energies.[3]

Analysis of the results presented here clearly shows that the continuous-retardation approximation cannot be used in the precision calculations of the passage of high-energy muons through matter.

## 2. Calculation of the passage of ultrarelativistic muons through matter. Energy-loss fluctuations

The main systematic features of the propagation of muons of moderately high energy in matter can be identified by solving a one-dimensional muon-transport equation in the continuous-energy-loss approximation.[101,169,171–173] The solution methods used in cosmic-ray physics were discussed in Sec. 5.1.

The equation for the transport of muons in a plane one-dimensional barrier was solved for the first time by Eyges,[171] without regard for the decay of muons and their angular deviation due to bremsstrahlung and the production of $e^+e^-$ pairs and without regard for the inelastic nuclear interaction. The expression for the radial distribution of the muon flux at a fixed depth $x$ is (see also Sec. 2.6)

**Table 7.1. Average, total, and partial energy losses of muons due to ionization and excitation of atoms (*i*), direct production of $e^+e^-$ pairs (*p*), bremsstrahlung (*r*), nuclear interaction (*n*), and total mean free paths of muons in the ground ($\rho = 1.8$ g/cm³), in concrete ($\rho = 2.35$ g/cm³), in iron ($\rho = 7.87$ g/cm³), and in uranium ($\rho = 18.72$ g/cm³).**

| | *dE*/*dx*, MeV·cm²/g | | | | | |
|---|---|---|---|---|---|---|
| *E*, GeV | tot | *i* | *p* | *r* | *n* | *R*, g/cm² |
| | | | Ground | | | |
| 10 | 2.187 | 2.179 | — | 7.806 − 3 | — | 4.931 + 3 |
| 15 | 2.243 | 2.230 | — | 1.276 − 2 | — | 7.187 + 3 |
| 20 | 2.281 | 2.263 | — | 1.801 − 2 | — | 9.395 + 3 |
| 30 | 2.369 | 2.306 | 3.352 − 2 | 2.915 − 2 | — | 1.368 + 4 |
| 40 | 2.424 | 2.335 | 4.856 − 2 | 4.090 − 2 | — | 1.783 + 4 |
| 60 | 2.535 | 2.372 | 8.091 − 2 | 6.566 − 2 | 1.625 − 2 | 2.585 + 4 |
| 80 | 2.626 | 2.398 | 1.154 − 1 | 9.164 − 2 | 2.182 − 2 | 3.360 + 4 |
| 100 | 2.714 | 2.417 | 1.513 − 1 | 1.185 − 1 | 2.743 − 2 | 4.110 + 4 |
| 150 | 2.932 | 2.456 | 2.456 − 1 | 1.886 − 1 | 4.156 − 2 | 5.870 + 4 |
| 200 | 3.140 | 2.478 | 3.444 − 1 | 2.618 − 1 | 5.582 − 2 | 7.518 + 4 |
| 300 | 3.558 | 2.509 | 5.500 − 1 | 4.145 − 1 | 8.459 − 2 | 1.051 + 5 |
| 400 | 3.980 | 2.531 | 7.624 − 1 | 5.732 − 1 | 1.136 − 1 | 1.315 + 5 |
| 600 | 4.836 | 2.562 | 1.199 | 9.035 − 1 | 1.721 − 1 | 1.770 + 5 |
| 800 | 5.705 | 2.584 | 1.644 | 1.246 | 2.311 − 1 | 2.149 + 5 |
| 1.0 + 3 | 6.583 | 2.601 | 2.094 | 1.597 | 2.905 − 1 | 2.475 + 5 |
| 1.5 + 3 | 8.808 | 2.633 | 3.230 | 2.505 | 4.401 − 1 | 3.127 + 5 |
| 2.0 + 3 | 1.104 + 1 | 2.655 | 4.376 | 3.417 | 5.909 − 1 | 3.631 + 5 |
| 3.0 + 3 | 1.538 + 1 | 2.686 | 6.679 | 5.125 | 8.950 − 1 | 4.392 + 5 |
| 4.0 + 3 | 1.973 + 1 | 2.708 | 8.988 | 6.833 | 1.202 | 4.961 + 5 |
| 6.0 + 3 | 2.842 + 1 | 2.739 | 1.361 + 1 | 1.025 + 1 | 1.320 | 5.795 + 5 |
| 8.0 + 3 | 3.711 + 1 | 2.761 | 1.825 + 1 | 1.367 + 1 | 2.443 | 6.415 + 5 |
| 1.0 + 4 | 4.581 + 1 | 2.778 | 2.228 + 1 | 1.708 + 1 | 3.069 | 6.897 + 5 |
| 2.0 + 4 | 8.928 + 1 | 2.831 | 4.605 + 1 | 3.416 + 1 | 6.238 | 8.432 + 5 |
| 3.0 + 4 | 1.328 + 2 | 2.863 | 6.923 + 1 | 5.125 + 1 | 9.443 | 9.334 + 5 |
| 4.0 + 4 | 1.763 + 2 | 2.885 | 9.240 + 1 | 6.833 + 1 | 1.267 + 1 | 9.996 + 5 |
| 6.0 + 4 | 2.634 + 2 | 2.916 | 1.388 + 2 | 1.025 + 2 | 1.918 + 1 | 1.091 + 6 |
| 8.0 + 4 | 3.505 + 2 | 2.938 | 1.851 + 2 | 1.367 + 2 | 2.574 + 1 | 1.157 + 6 |
| 1.0 + 5 | 4.376 + 2 | 2.955 | 2.315 + 2 | 1.708 + 2 | 3.234 + 1 | 1.208 + 6 |
| | | | Concrete | | | |
| 10 | 2.211 | 2.205 | — | 5.618 − 3 | — | 4.819 + 3 |
| 15 | 2.260 | 2.251 | — | 9.159 − 3 | — | 7.056 + 3 |
| 20 | 2.295 | 2.282 | — | 1.291 − 2 | — | 9.249 + 3 |
| 30 | 2.368 | 2.323 | 2.435 − 2 | 2.086 − 2 | — | 1.353 + 4 |
| 40 | 2.415 | 2.351 | 3.530 − 2 | 2.923 − 2 | — | 1.769 + 4 |
| 60 | 2.510 | 2.388 | 5.884 − 2 | 4.687 − 2 | 1.651 − 2 | 2.578 + 4 |
| 80 | 3.585 | 2.414 | 8.392 − 2 | 6.535 − 2 | 2.217 − 2 | 3.362 + 4 |
| 100 | 2.655 | 2.433 | 1.101 − 1 | 8.447 − 2 | 2.787 − 2 | 4.126 + 4 |
| 150 | 2.828 | 2.473 | 1.789 − 1 | 1.343 − 1 | 4.223 − 2 | 5.940 + 4 |
| 200 | 2.989 | 2.495 | 2.511 − 1 | 1.862 − 1 | 5.671 − 2 | 7.760 + 4 |
| 300 | 3.309 | 2.527 | 4.015 − 1 | 2.946 − 1 | 8.593 − 2 | 1.084 + 5 |
| 400 | 3.629 | 2.549 | 5.571 − 1 | 4.072 − 1 | 1.154 − 1 | 1.371 + 5 |
| 600 | 4.274 | 2.581 | 8.772 − 1 | 6.413 − 1 | 1.749 − 1 | 1.878 + 5 |
| 800 | 4.926 | 2.603 | 1.204 | 8.839 − 1 | 2.348 − 1 | 2.312 + 5 |

$3.352 - 2 \equiv 3.352 \times 10^{-2}$
$1.361 + 1 \equiv 1.361 \times 10^{1}$

**Table 7.1. (continued)**

| $E$, GeV | $dE/dx$, MeV·cm²/g tot | $i$ | $p$ | $r$ | $n$ | $R$, g/cm² |
|---|---|---|---|---|---|---|
| 1.0 + 3 | 5.584 | 2.621 | 1.536 | 1.133 | 2.951 − 1 | 2.693 + 5 |
| 1.5 + 3 | 7.249 | 2.652 | 2.374 | 1.776 | 4.471 − 1 | 3.475 + 5 |
| 2.0 + 3 | 8.987 | 2.675 | 3.219 | 2.493 | 6.003 − 1 | 4.094 + 5 |
| 3.0 + 3 | 1.227 + 1 | 2.706 | 4.919 | 3.739 | 9.092 − 1 | 5.042 + 5 |
| 4.0 + 3 | 1.556 + 1 | 2.729 | 6.625 | 4.986 | 1.221 | 5.758 + 5 |
| 6.0 + 3 | 2.213 + 1 | 2.761 | 1.004 + 1 | 7.478 | 1.849 | 6.821 + 5 |
| 8.0 + 3 | 2.870 + 1 | 2.783 | 1.347 + 1 | 9.971 | 2.482 | 7.612 + 5 |
| 1.0 + 4 | 3.528 + 1 | 2.800 | 1.680 + 1 | 1.246 + 1 | 3.118 | 8.247 + 5 |
| 2.0 + 4 | 6.815 + 1 | 2.855 | 3.403 + 1 | 2.493 + 1 | 6.337 | 1.025 + 6 |
| 3.0 + 4 | 1.010 + 2 | 2.886 | 5.116 + 1 | 3.739 + 1 | 9.594 | 1.144 + 6 |
| 4.0 + 4 | 1.339 + 2 | 2.909 | 6.830 + 1 | 4.985 + 1 | 1.288 + 1 | 1.229 + 6 |
| 6.0 + 4 | 1.998 + 2 | 2.940 | 1.026 + 2 | 7.478 + 1 | 1.949 + 1 | 1.351 + 6 |
| 8.0 + 4 | 2.657 + 2 | 2.963 | 1.369 + 2 | 9.971 + 1 | 2.615 + 1 | 1.437 + 6 |
| 1.0 + 5 | 3.316 + 2 | 2.980 | 1.711 + 2 | 1.246 + 2 | 3.285 + 1 | 1.505 + 6 |
| | | | Iron | | | |
| 10 | 1.941 | 1.925 | — | 1.599 − 2 | — | 5.595 + 3 |
| 15 | 1.998 | 1.972 | — | 2.630 − 2 | — | 8.132 + 3 |
| 20 | 2.040 | 2.002 | — | 3.726 − 2 | — | 1.061 + 4 |
| 30 | 2.173 | 2.042 | 6.979 − 2 | 6.059 − 2 | — | 1.533 + 4 |
| 40 | 2.255 | 2.069 | 1.010 − 1 | 8.524 − 2 | — | 1.981 + 4 |
| 60 | 2.424 | 2.104 | 1.681 − 1 | 1.373 − 1 | 1.509 − 2 | 2.828 + 4 |
| 80 | 2.579 | 2.127 | 2.394 − 1 | 1.921 − 1 | 2.026 − 2 | 3.630 + 4 |
| 100 | 2.733 | 2.145 | 3.137 − 1 | 2.488 − 1 | 2.547 − 2 | 4.387 + 4 |
| 150 | 3.125 | 2.181 | 5.081 − 1 | 3.971 − 1 | 3.860 − 2 | 6.078 + 4 |
| 200 | 3.516 | 2.201 | 7.110 − 1 | 5.520 − 1 | 5.183 − 2 | 7.586 + 4 |
| 300 | 4.316 | 2.230 | 1.132 | 8.752 − 1 | 7.854 − 2 | 1.016 + 5 |
| 400 | 5.134 | 2.251 | 1.564 | 1.213 | 1.055 − 1 | 1.227 + 5 |
| 600 | 6.804 | 2.280 | 2.449 | 1.915 | 1.598 − 1 | 1.562 + 5 |
| 800 | 8.508 | 2.300 | 3.349 | 2.644 | 2.146 − 1 | 1.823 + 5 |
| 1.0 + 3 | 1.024 + 1 | 2.316 | 4.256 | 3.393 | 2.697 − 1 | 2.037 + 5 |
| 1.5 + 3 | 1.463 + 1 | 2.345 | 6.543 | 5.320 | 4.086 − 1 | 2.441 + 5 |
| 2.0 + 3 | 1.863 + 1 | 2.366 | 8.841 | 6.871 | 5.487 − 1 | 2.738 + 5 |
| 3.0 + 3 | 2.699 + 1 | 2.395 | 1.345 + 1 | 1.031 + 1 | 8.311 − 1 | 3.174 + 5 |
| 4.0 + 3 | 3.535 + 1 | 2.415 | 1.808 + 1 | 1.374 + 1 | 1.116 | 3.504 + 5 |
| 6.0 + 3 | 5.208 + 1 | 2.444 | 2.733 + 1 | 2.061 + 1 | 1.690 | 3.960 + 5 |
| 8.0 + 3 | 6.881 + 1 | 2.465 | 3.659 + 1 | 2.748 + 1 | 2.268 | 4.301 + 5 |
| 1.0 + 4 | 8.554 + 1 | 2.480 | 4.585 + 1 | 3.435 + 1 | 2.850 | 4.561 + 5 |
| 2.0 + 4 | 1.692 + 2 | 2.530 | 9.218 + 1 | 6.870 + 1 | 5.792 | 5.376 + 5 |
| 3.0 + 4 | 2.529 + 2 | 2.559 | 1.385 + 2 | 1.031 + 2 | 8.769 | 5.847 + 5 |
| 4.0 + 4 | 3.366 + 2 | 2.579 | 1.848 + 2 | 1.374 + 2 | 1.177 + 1 | 6.198 + 5 |
| 6.0 + 4 | 5.040 + 2 | 2.608 | 2.775 + 2 | 2.061 + 2 | 1.781 + 1 | 6.671 + 5 |
| 8.0 + 4 | 6.715 + 2 | 2.629 | 3.702 + 2 | 2.748 + 2 | 2.390 + 1 | 7.023 + 5 |
| 1.0 + 5 | 8.390 + 2 | 2.645 | 4.628 + 2 | 3.435 + 2 | 3.003 + 1 | 7.286 + 5 |

**Table 7.1. (continued)**

| $E$, GeV | $dE/dx$, MeV·cm²/g tot | $i$ | $p$ | $r$ | $n$ | $R$, g/cm² |
|---|---|---|---|---|---|---|
| | | | Uranium | | | |
| 10 | 1.599 | 1.566 | — | 4.217 − 2 | — | 6.965 + 3 |
| 15 | 1.668 | 1.598 | — | 7.008 − 2 | — | 1.002 + 4 |
| 20 | 1.724 | 1.624 | — | 9.995 − 2 | — | 1.296 + 4 |
| 30 | 2.021 | 1.659 | 1.988 − 1 | 1.638 − 1 | — | 1.810 + 4 |
| 40 | 2.200 | 1.681 | 2.874 − 1 | 2.315 − 1 | — | 2.286 + 4 |
| 60 | 2.575 | 1.710 | 4.769 − 1 | 3.752 − 1 | 1.256 − 2 | 3.124 + 4 |
| 80 | 2.951 | 1.730 | 6.776 − 1 | 5.268 − 1 | 1.687 − 2 | 3.841 + 4 |
| 100 | 3.336 | 1.745 | 8.859 − 1 | 6.842 − 1 | 2.121 − 2 | 4.489 + 4 |
| 150 | 4.332 | 1.775 | 1.420 | 1.096 | 3.213 − 2 | 5.755 + 4 |
| 200 | 5.355 | 1.792 | 1.991 | 1.528 | 4.316 − 2 | 6.797 + 4 |
| 300 | 7.466 | 1.816 | 3.151 | 2.433 | 6.539 − 2 | 8.396 + 4 |
| 400 | 9.636 | 1.833 | 4.337 | 3.378 | 8.782 − 2 | 9.550 + 4 |
| 600 | 1.409 + 1 | 1.857 | 6.747 | 5.349 | 1.331 − 1 | 1.121 + 5 |
| 800 | 1.864 + 1 | 1.874 | 9.185 | 7.398 | 1.787 − 1 | 1.243 + 5 |
| 1.0 + 3 | 2.325 + 1 | 1.887 | 1.164 + 1 | 9.507 | 2.246 − 1 | 1.346 + 5 |
| 1.5 + 3 | 3.501 + 1 | 1.911 | 1.779 + 1 | 1.497 + 1 | 3.402 − 1 | 1.513 + 5 |
| 2.0 + 3 | 4.426 + 1 | 1.929 | 2.397 + 1 | 1.791 + 1 | 4.568 − 1 | 1.635 + 5 |
| 3.0 + 3 | 6.585 + 1 | 1.953 | 3.635 + 1 | 2.686 + 1 | 6.919 − 1 | 1.819 + 5 |
| 4.0 + 3 | 8.745 + 1 | 1.970 | 4.874 + 1 | 3.582 + 1 | 9.289 − 1 | 1.950 + 5 |
| 6.0 + 3 | 1.307 + 2 | 1.994 | 7.353 + 1 | 5.372 + 1 | 1.407 | 2.136 + 5 |
| 8.0 + 3 | 1.739 + 2 | 2.011 | 9.833 + 1 | 7.163 + 1 | 1.888 | 2.268 + 5 |
| 1.0 + 4 | 2.171 + 2 | 2.024 | 1.231 + 2 | 8.954 + 1 | 2.373 | 2.373 + 5 |
| 2.0 + 4 | 4.331 + 2 | 2.065 | 2.472 + 2 | 1.791 + 2 | 4.822 | 2.690 + 5 |
| 3.0 + 4 | 6.492 + 2 | 2.089 | 3.712 + 2 | 2.686 + 2 | 7.301 | 2.879 + 5 |
| 4.0 + 4 | 8.653 + 2 | 2.106 | 4.952 + 2 | 3.581 + 2 | 9.798 | 3.010 + 5 |
| 6.0 + 4 | 1.298 + 3 | 2.131 | 7.433 + 2 | 5.372 + 2 | 1.483 + 1 | 3.199 + 5 |
| 8.0 + 4 | 1.730 + 3 | 2.148 | 9.914 + 2 | 7.163 + 2 | 1.990 + 1 | 3.330 + 5 |
| 1.0 + 5 | 2.162 + 3 | 2.161 | 1.239 + 3 | 8.954 + 2 | 2.500 + 1 | 3.435 + 5 |

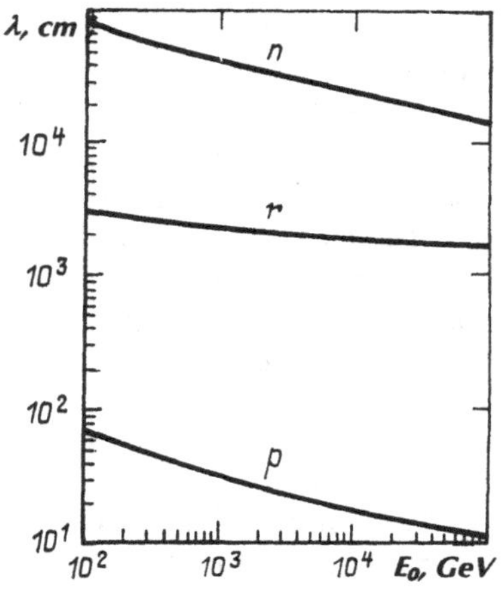


**Figure 7.1. Energy dependence of the muon range before the production of $e^+e^-$ pairs ($p$), before bremsstrahlung ($r$), and before nuclear interaction ($n$) in iron.**

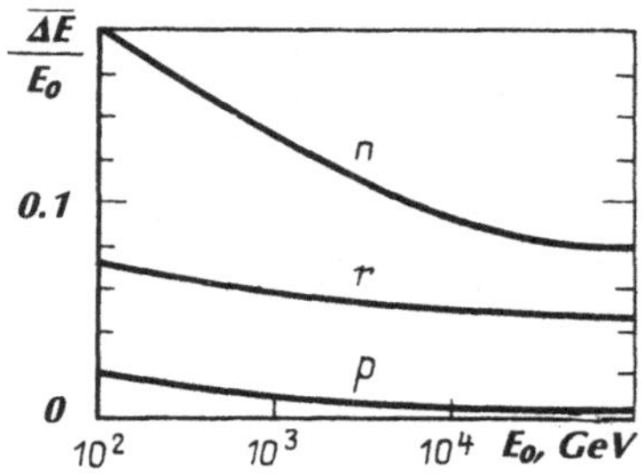


**Figure 7.2. Relative energy loss of a muon due to the production of $e^+e^-$ pair (*p*), due to bremsstrahlung (*r*), and due to nuclear interaction (*n*).**

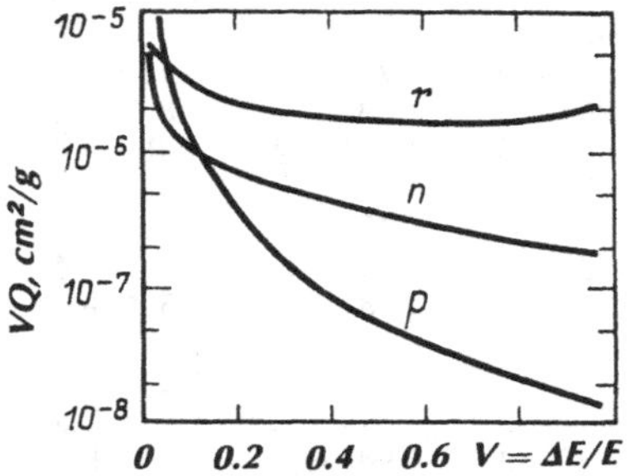


**Figure 7.3. Differential probabilities for the energy loss of a 2000-GeV muon due to the production of $e^+e^-$ pairs (*p*), due to bremsstrahlung (*r*), and due to nuclear interaction (*n*) in the ground.**

$$\Phi\,(E_0, x, r) = \frac{1}{4\pi A_2\,(E_0, x)}\exp\left[-\frac{r^2}{4A_2\,(E_0, x)}\right], \tag{7.4}$$

where $A_2\,(E_0, x) = \int_0^x \chi^2\,[E_\mu(E_0, x')]\,(x - x')^2 dx'$, $E_0$ is the energy of a muon incident along the normal to the barrier, $E_\mu$ is the energy of a muon at a depth $x$ (this energy can be found by solving the equation $x = \int_{E_\mu}^{E_0} \frac{dE'}{-\,[dE/dx(E')]}$), and $\chi^2(E_\mu)$ is the mean-square angular deviation of the muon per unit length of the path; this deviation is related to the mean angle of the multiple Coulomb scattering (see Chap. 2) by $\chi^2 = (1/4)\langle\theta_s^2\rangle$.

Analysis has shown[3,172] that at energies $E_0 \lesssim 100$–200 GeV the assumptions on which the solution of Eq. (7.4) for a thick layer of the material were based are sound. The particular features of the interaction of ultrarelativistic ($E_0 \gtrsim 100$ GeV) muons with matter, which were mentioned at the end of Sec. 7.1, require, however, that the individual emission of bremsstrahlung photons by a muon, the direction production of $e^+e^-$ pairs and the inelastic nuclear interaction be described in detail. The fluctuations in the energy loss and in the angular deviations of muons, whose role increases with increasing energy, markedly change the spatial and energy distributions of the muon flux in matter.[2,3]

The following algorithm for simulation of the passage of high-energy muons through matter was developed by Mokhov *et al.*[3]

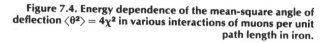

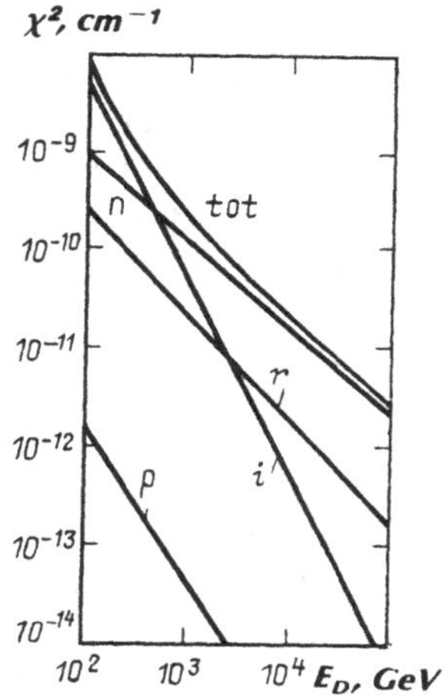


**Figure 7.4.** Energy dependence of the mean-square angle of deflection $\langle\theta^2\rangle = 4\chi^2$ in various interactions of muons per unit path length in iron.

The muon range before the emission of a bremsstrahlung photon or before a photonuclear interaction is simulated according to the known mean ranges $\lambda$ (see Fig. 7.1). The type of interaction and the energy and angle of the muon after the interaction, found from the distributions given in Chap. 2, are simulated at this point. The angle of deviation of the muon at the interaction point can be simulated approximately from the normal distribution with the appropriate dispersion (see Fig. 7.4). The azimuthal angle $\varphi$ can be simulated by using algorithm (6.15).

The muon transport between the indicated discrete events is brought about by means of the step method (see Sec. 6.3). At each step $l \leqslant \lambda_{r,n}$ the following processes are simulated:

(1) The multiple Coulomb scattering (see Chap. 2).

(2) The effect of an external electric field or an external magnetic field [relations (6.18)].

(3) The loss of muon energy due to ionization and excitation of atoms from the Landau distribution[35] (see Chap. 2).

(4) The loss of muon energy due to the production of $e^+e^-$ pairs in the continuous-retardation approximation or from the algorithm described below.

In the presence of external electromagnetic fields the muon step $l$ in complex systems is chosen under the assumption that the following three requirements are satisfied[174]:

(1) The energy the particle loses, $\Delta E$, must be low in comparison with the instantaneous energy, allowing, in particular, the use of the small-angle approximation.

(2) The induction and direction of the field $\mathbf{B}$ must change only slightly at each step.

(3) The transverse displacement $\Delta r$ of the muon due to the multiple Coulomb scattering and the magnetic field must be small in comparison with the minimum size $t$ of the system's elements in the direction of the displacement $\Delta r = l\Delta\varphi \ll t$, where $\Delta\varphi = l/R$ is the magnetic-field-induced change in the direction of motion of the muon at the step $l$, and $R$ is the radius of curvature of the trajectory in a magnetic field. We impose the requirement that the transverse displacement of the muon $\Delta r$ be constant and independent of the step $l$. Since $R \propto E$, we can then write the following expression from which the muon step with the energy $E$ can be determined[174]: $l = C\sqrt{E}$, where the value of the parameter $C$ is chosen under the assumption that conditions (1)–(3) are satisfied. In the special case where these conditions are satisfied, the step $l$ may be equal to the simulated range $\lambda_r(\lambda_n)$ or the distance from the boundary between the media along the ray.

The change in the coordinates and angles at the chosen step is represented as a sum of two components, one of which is stipulated by the multiple Coulomb scattering and the other arises as a result of integration of the equation of motion of the muon in a magnetic field.[175] In the transformation from the initial direction $\mathbf{\Omega}$ to the new one $\mathbf{\Omega}'$ the conversion of the angles is accomplished, just as in the simulated discrete events, on the basis of Eqs. (6.19).

The fate of the muon in each event is traced until it escapes the system under study or until the kinetic energy decreases below the threshold energy, $E_{\text{thr}}$. Usually, $E_{\text{thr}} = 20$ MeV in the MUTRAN program which is responsible for the method described here.

The range distributions of the muons before stopping, calculated in accordance with this program, are shown in Fig. 7.5. With an increase in the energy and atomic mass of the substance, the distribution width increases. Calculations show that the fluctuations of the energy loss due to the bremsstrahlung and the photonuclear interactions are the key factors in the cases under consideration.

The fluctuations in the energy loss lead to an increase in the average muon range in the substance at energies higher than several hundred gigaelectron-volts. The energy dependences of the average muon ranges (Table 7.1) and the ranges which allow for the fluctuations—thicknesses which 1% of muons incident on the absorber might reach—are shown in Fig. 7.6 for three materials. It can be seen from the figure that the difference may be as high as 50%.

The energy distribution of ultrarelativistic muons that have passed a layer of substance was obtained by Mokhov *et al.*[2] They have also considered the fluctuations simultaneously in all four muon interaction processes ($i$, $p$, $r$, and $n$) and studied the effect of each one of them on the formation of the energy spectra behind substances of various thicknesses.

Let us assume that the layer $x$ of the substance is such that the energy loss $\Delta$ in it is small in comparison with the initial energy $E$. Solving, by analogy with Ref. 35, the corresponding transport equation with allowance for all

**Figure 7.5. Range distribution of muons of various energies in iron and ground.**

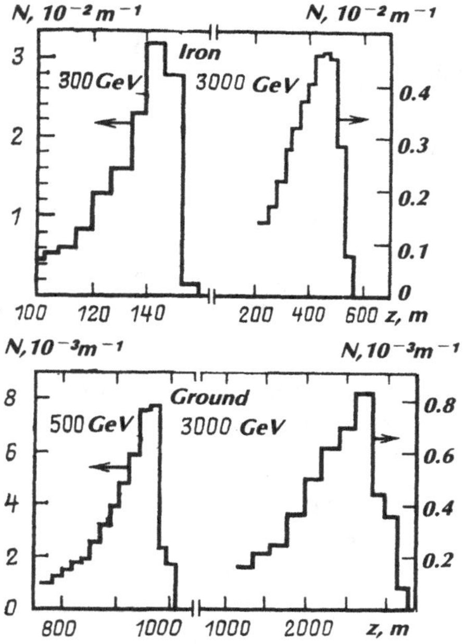

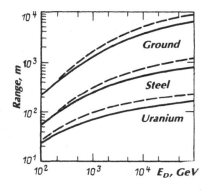

**Figure 7.6. Energy dependence of the mean free path of muons (solid line) and the mean free path in which the fluctuations are taken into account (dashed line).**

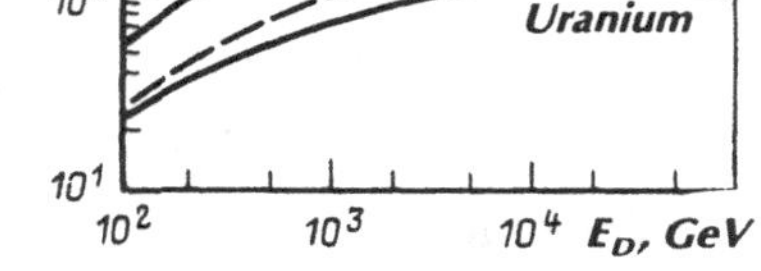


the processes mentioned above, we find the muon energy-loss distribution in a thin layer of the substance

$$f(\Delta, E, x) = (1/2\pi i)\int_{\delta - i\infty}^{\delta + i\infty} dp \exp\left\{p\Delta - x \sum_{k = i, p, r, n} I_k(E, p)\right\}; \tag{7.5}$$

$$I_k(E, p) = \int_0^E W_k(\varepsilon, E)[1 - \exp(-p\varepsilon)]d\varepsilon\,, \tag{7.6}$$

where $W_k(\varepsilon, E)$ is the probability density for the loss of energy $\varepsilon$ per unit path length due to a process of type $k$. Since the principal functional dependences of the cross sections for bremsstrahlung ($r$) and for inelastic nuclear interaction ($n$) of muons are identical, we will henceforth consider these processes jointly:

$$W_{r+n}(\varepsilon, E) = W_r(\varepsilon, E) + W_n(\varepsilon, E) .$$

It is obvious that

$$I_k(E, p) \leqslant E_k(E) = 1/\lambda_k , \tag{7.7}$$

where $W_k(E)$ is the macroscopic interaction cross section of type $k$, and $\lambda_k$ is the range before the interaction. Since $\Sigma_i \gg \Sigma_p \gg \Sigma_{r+n}$, we find reasonable estimates from relations (7.5)–(7.7): If the layer $x$ of the substance is such that

$$x/\lambda_k \ll 1 , \tag{7.8}$$

a process of type $k$ can be disregarded in determining the muon energy spectrum behind this layer. Let us assume that the $p$ in Eq. (7.5) are such that

$$2m_e p \ll 1, \quad z = pE \gg 1 . \tag{7.9}$$

The integration in Eq. (7.6) can then be extended from 0 to $\infty$. For the ionization losses we have[35]

$$xI_i = p\bar{h}_i + p\xi\,[1 + \beta^2 - C - \ln(p\varepsilon_{\max})] ,$$

where $\bar{h}_k$ is the average energy loss along the path $x$ due to a process of type $k$; $\xi = Bx$; $B = 0.3(m_e/\beta^2)Z/A$; $A$ and $Z$ are the atomic mass and the atomic number of the substance, respectively; $m_e$ is the electron mass; $\varepsilon_{\max}$ is the maximum energy that can be transferred to the atomic electron as a result of ionization; $\beta$ is the muon velocity; and $C = 0.5777$ is Euler's constant. For the ultrarelativistic muons we can write, on the basis of the results of Chap. 2, the differential cross sections of the remaining processes in the form

$$W_{r+n}(\varepsilon, E) = b_{r+n}/\varepsilon; \quad W_p(\varepsilon, E) = b_p a(1+a)/[\varepsilon(\varepsilon/E + a)^2] ,$$

where

$$b_k \simeq \left[-\frac{dE}{dx}(E)\right]_k E; \left(-\frac{dE}{dx}\right)_k = \frac{\bar{h}_k}{x}; \quad a = 0.0071 .$$

The quantities $(-dE/dx)_{r+n,p}$ can be calculated from the equations given in Chap. 2. Integrating (7.6), we obtain

$$\sum_{k=p,r,n} I_k = b_{r+n}(C + \ln z) + \frac{b_p(1+a_p)}{ai}\,[C + \ln(az) - (1-az)\exp(az)\,Ei(-az)] ,$$

so that the energy loss distribution function takes the form ($u = apE$)

$$f(\Delta, E, x) = \frac{1}{2\pi i a E}\int_{\delta - i\infty}^{\delta + i\infty} du\, \exp\left\{\frac{1}{b}\,[\lambda u + u\ln u - D_p\Phi_1(u) - aD_{r+n}\Phi_2(u)]\right\} , \tag{7.10}$$

where $\Phi_1(u) = C + \ln u - (1-u)\exp(u)Ei(-u)$, $\Phi_2(u) = C + \ln(u/a)$, $b = aE/\xi$, $D_k = \bar{h}_k/\xi$, and $\lambda = (\Delta - \bar{h}_i)/\xi - 1 - \beta^2 + C + \ln(\varepsilon_{\max}/aE)$ or in actual variables

$$f(\Delta, E, x) = \frac{1}{\pi a E}\int_0^\infty dy \exp\left\{-\frac{1}{b}[\lambda y + y\ln y + D_p\Phi_3(y) + D_{r+n}a\Phi_2(y)]\right\}\sin\left\{\frac{\pi}{b}[y + D_p\Phi_4(y) + aD_{r+n}\right\}, \tag{7.11}$$

where $\Phi_3(y) = C + \ln y - (1+y)\exp(-y)\,Ei(y)$, and $\Phi_4(y) = 1 - (1+y)\times\exp(-y)$.

Since the largest contribution to the integral in (7.11) comes from the first period of the sine under the integral sign, the integral is determined by such $y$ that $y + D_p\ \Phi_4(y) + aD_{r+n} \sim b$. Therefore, if $D_p \gg b$ (the layers of the substance are thick), then $y$ in the integral in (7.11) is important, $y \sim \sqrt{2(b - aD_{r+n})/D_p}$. If the layer of the substance is thin ($D_p \ll b$), the integral in (7.11) is determined by the integration range, $y \sim b$. Since $y = apE$, assumption (7.9) reduces to the conditions

$$\xi \gg 2m_e\,; \tag{7.12}$$

$$\sqrt{(E - 2\bar{h}_{r+n})/a\bar{h}_p} \gg 1\,. \tag{7.13}$$

Condition (7.13) does not impose any constraints on the range of applicability of expression (7.11), where it was assumed from the outset that the energy loss is small in comparison with the initial energy. If the layer thickness is such that condition (7.12) does not hold, $2m_e$ must be used as the lower limit of integration in the calculation of $I_p$ in (7.6). The energy loss of muons in this case is described by the expression

$$f(\Delta, E, x) = \frac{1}{2\pi i a E}\int_{\delta - i\infty}^{\delta + i\infty} du \exp\left\{\frac{1}{b}[\lambda u + u\ln u - D_p\Phi_5(u,q) - aD_{r+n}\Phi_2(u)]\right\}, \tag{7.14}$$

where $\Phi_5(u,q) = Ei(-qu) - \ln q - 1 + \exp(-qu) - (1-u)\exp(-u)Ei(-u)$, and $q = 2m_e/(aE)$. This expression transforms to (7.10) if condition (7.12) holds. If $x \sum_{k=r,p,n} I_k \ll 1$, expression (7.14) becomes the Landau distribution. In the integral in (7.14) the important quantities in this case are $u \sim b$ (Ref. 35). For a substance of such a thickness ($x \ll \lambda_p$) we have

$$b = aE/\xi \gg aE/\lambda_p B = (-dE/dx)_p/B \sim 10^{-3}Z^2E \sim 1\,,$$

and from (7.14) we obtain an estimate for the thickness $x$, which is more accurate than (7.8) and in which it is sufficient to take only the ionization loss into account in the solution:

$$\frac{b_p x}{a}\left[Ei\left(-\frac{2m_e}{\xi}\right)-\ln\left(\frac{2m_e}{\xi}\right)+\exp\left(-\frac{2m_e}{\xi}\right)+\ln\left(\frac{aE}{\xi}\right)\right]\ll 1\,. \quad (7.15)$$

If the layer $x$ of the substance is such that

$$x\sum_{k=r,n} I_k \ll 1\,, \quad (7.16)$$

the bremsstrahlung and the nuclear interaction of muons can be ignored in Eqs. (7.11) and (7.14) in determining the muon energy-loss distribution function. Condition (7.16) is satisfied if $D_p \sim b$. In such a case, $u \sim 1$ are the important quantities in (7.14), and for the layer thickness $x$, beyond which the energy spectrum is determined by the energy loss exclusively due to the ionization and direct production of the $e^+e^-$ pairs, we have

$$x \ll 1/[b_{r+n}(C-\ln a)]\,. \quad (7.17)$$

The solution can be simplified in this case. If $\xi \gg 2m_e$, for example, we have

$$f(\Delta, E, x) = \frac{1}{2\pi i a E}\int_{\delta - i\infty}^{\delta + i\infty} du \exp\left\{\frac{1}{b}\left[\lambda u + u\ln u - D_p \Phi_1(u)\right]\right\}\,. \quad (7.18)$$

The function (7.18) can easily be used to simulate muon transport by the Monte Carlo method in a scheme which was described at the beginning of this section (Sec. 7.2) if the step $l$ satisfies condition (7.17).

Figure 7.7 shows the energy loss of muons obtained from the equations given above and calculated from the modified MUTRAN program.[3] In this version of the program, the energy loss in the direct production of $e^+e^-$ pairs is viewed as a discrete process with a corresponding differential cross section. The energy-loss distributions are in satisfactory agreement in the case of a thin layer of the substance (Fig. 7.7). With thicker layers, where the agreement, as expected, worsens (the lower plot in Fig. 7.7), the results of a direct simulation of the transfer process should be used. As the thickness of the layer is increased, the distributions acquire a Gaussian shape and the value of the most probable energy loss approaches the mean energy loss $\bar{\Delta}$ (Fig. 7.8).

We can therefore draw the following conclusions:

(1) If the layer thickness of the substance is such that condition (7.15) is satisfied, the muon energy spectrum beyond the layer is determined exclusively by the ionization loss. The Landau distribution can be used to obtain this spectrum.

(2) The muon energy spectra behind the layers of the substance, whose thickness satisfies condition (7.17), can be calculated from expression (7.18), ignoring the bremsstrahlung and the photonuclear interaction of muons.

(3) In the last two processes the fluctuations play a dominant role in the formation of the muon energy spectra behind the thick ($x \gtrsim 3$–$4\lambda_n$) layers of the substance.

Let us consider the effect of the energy-loss fluctuations on the distribution of the integral muon flux in the substance. It can be seen in Fig. 7.9 (on the left side) that at moderately high energies the effect of the energy-loss fluctuations on this functional is determined largely by the ionization strag-

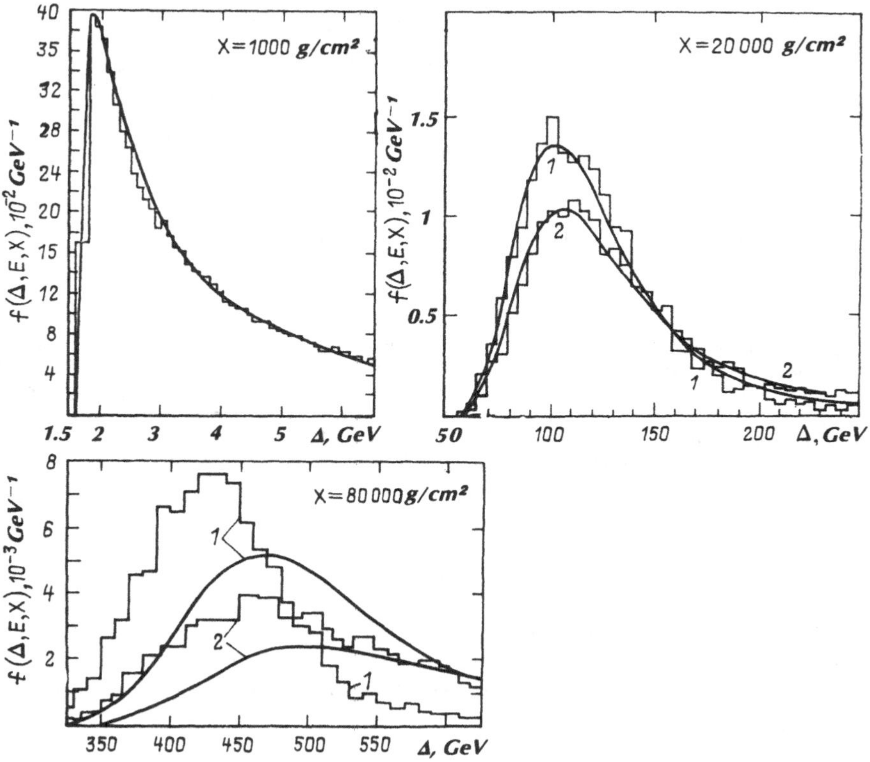


**Figure 7.7.** The probability that a 1000-GeV muon will lose energy $\Delta$ at various depths in iron $x$. 1, calculation based on Eq. (7.10) (curve) and on the Monte Carlo method (histogram) ignoring bremsstrahlung and nuclear interaction; 2, calculation based on Eq. (7.18) and on the Monte Carlo method with allowance for all processes.

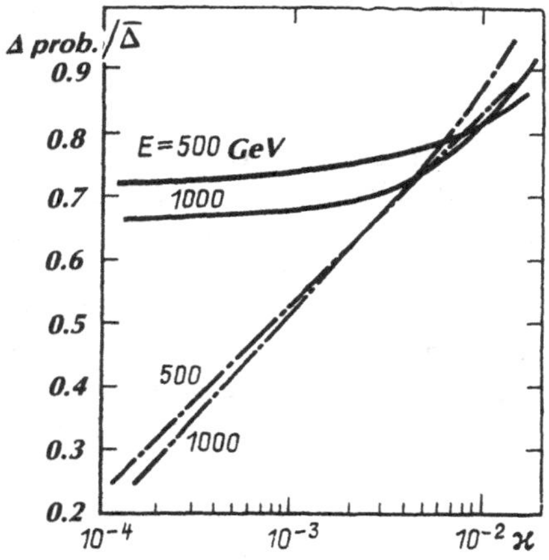


**Figure 7.8.** Ratio of the most probable energy loss of a muon in the material to the mean loss in that material at a particular depth, calculated in the continuous-slowing-down approximation as a function of $\kappa = \xi/\varepsilon_{max}$. Solid curves, calculation with allowance for the fluctuations due to bremsstrahlung and nuclear interactions; collision with the atomic electrons and direct pair production were analyzed in the continuous-slowing-down approximation. Dot-dashed curves, calculation with allowance for fluctuations in all four processes.

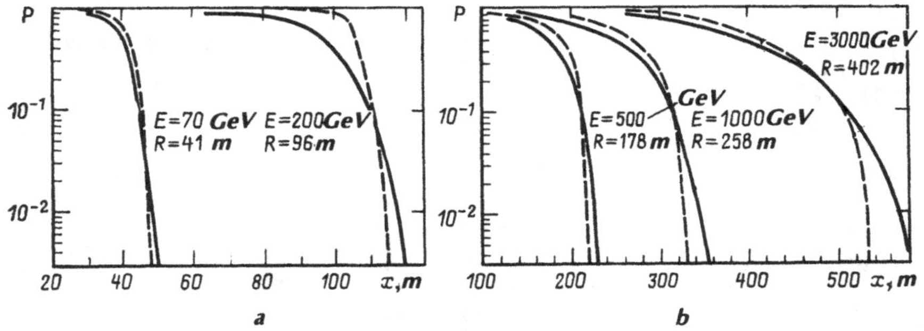


**Figure 7.9. The probability *P* that a muon with initial energy *E* will reach a depth *x* in an iron absorber. *R* is the range of the muon before stopping, with allowance for all processes in the continuous-slowing-down approximation. Solid curves take into account the fluctuations in all processes; dashed curves correspond to case (a) in which in the continuous-slowing-down approximation the bremsstrahlung, pair production, and nuclear interactions are considered and case (b) in which the pair production and collisions with atomic electrons are considered.**

gling. At $E > 300$–500 GeV the dominant fluctuations are those involving the energy loss due to the bremsstrahlung and nuclear interaction of muons. The difference in the curves in Fig. 7.9 (on the right side) stems from the fact that the fluctuations in the energy loss due to the direct production of $e^+e^-$ pairs were ignored in some of the calculations. Allowance for the ionization straggling, on the other hand, does not change the function $P(x,E)$ in iron[3] at $E > 300$ GeV.

## 3. Muons and the development of hadronic cascades

As was noted in Secs. 1.2 and 7.1, muons are produced in the decay of long-lived and short-lived mesons in matter, in the so-called *decay intervals*, and in the reactions of the type (7.2). Since before the decay of a muon "parent" the range $\lambda_D$ increases linearly with the energy, the probability for high-energy decay, in which the most penetrating component is produced, is very small. Since direct simulation of a decay, especially in matter, is highly ineffective, the following algorithm for the simulation of muon production was developed in Refs. 3, 173, and 174.

The hadronic cascade in a target and in absorbers can be calculated by the inclusive method (Secs. 6.2 and 6.3). Each inelastic hadron–nucleus-interaction event generates six hadrons ($p, n, \pi^{\pm}, K^{\pm}$) with the statistical weights whose expectation values are the same as the corresponding average multiplicities. The production of direct muons can be simulated in the same interaction by using relation (7.3).

Before the new nuclear interactions occur, the ranges $\lambda$ of the produced hadrons are simulated on the basis of the scheme in Sec. 6.3. The production

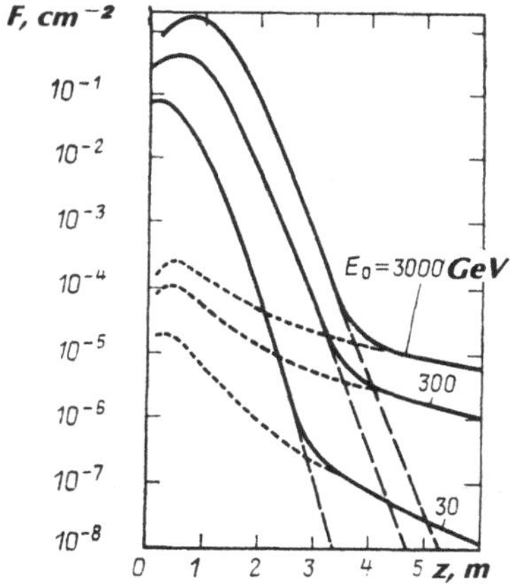


**Figure 7.10. Longitudinal distribution of the differential density of particles with $E > 20$ MeV along the axis of the iron absorber on which protons of various energies are incident: Solid curves, total distribution; dashed curves, hadrons; long dashes, muons.**

**Table 7.2. The absorber thicknesses (cm) for which hadron and muon flux densities are the same along the axis of the proton beam with energy $E_0$.**

| Material | $E_0$, GeV | | |
|---|---|---|---|
| | 30 | 300 | 3000 |
| Concrete | 617 | 830 | 916 |
| Iron | 300 | 355 | 380 |
| Uranium | 240 | 250 | 275 |

of muons with a statistical weight $W_\mu = W_0\Sigma_D'(E)/\Sigma(E)$ and the inelastic nuclear interaction with a weight $W_0' = W_0\Sigma_{abs}(E)/\Sigma(E)$ are simulated, with allowance of the kinematics, for each $\pi$ or $K$ meson at these points. Here $E$ and $W_0$ are the energy and statistical weight of a hadron at the end of the path $\lambda$, $\Sigma = \Sigma_{abs} + \Sigma_D$, and $\Sigma_{abs}$ and $\Sigma_D$ are the macroscopic cross sections for the absorption and decay of $\pi$ and $K$ mesons, respectively.

The hadrons that are produced participate in the further development of the cascade by producing new generations of hadrons and leptons. But the fate of the muons can be traced on the basis of the algorithms described in Sec. 7.1. The tree for the trajectories in the substance should be considered before the third generation, since the higher-generation hadrons contribute essentially nothing to the formation of the muon field.[3,173]

If the hadron trajectory crosses the vacuum (or the air) space, then a forced decay of $\pi$ and $K$ mesons will occur in this space, with the corresponding statistical weight. The decay point in the interval $d$ can be found from the distribution density

$$p(x) = \exp(-x/\lambda_D)\lambda_D^{-1}[1-\exp(-d/\lambda_D)]^{-1}.$$

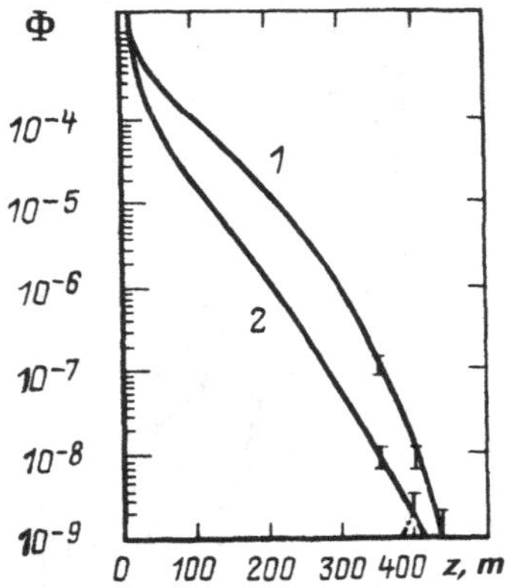


**Figure 7.11. Longitudinal distribution of the muon flux in a circle of radius $R = 250$ cm upon incidence of 3000-GeV protons on an iron absorber. 1, calculation (Ref. 174); 2, calculation in which the "direct" muons are ignored; the statistical calculation error is included.**

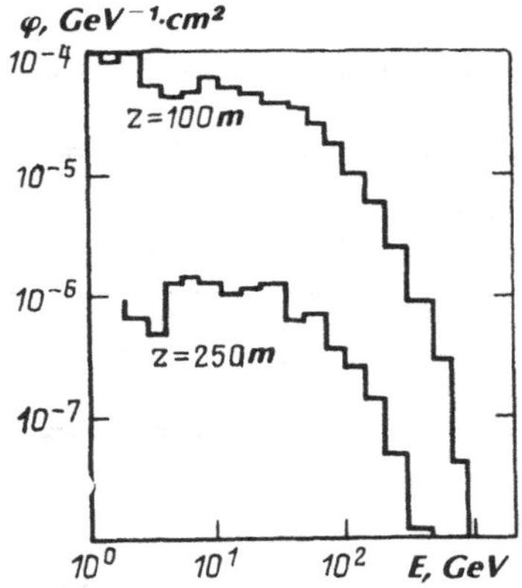


**Figure 7.12. Energy spectra of muons at two depths of the absorber. 3000-GeV protons interact with the target at a distance $d = 100$ m from the absorber.**

The statistical weight of the produced muon is $W_\mu = \eta W_0[1 - \exp(-d/\lambda_D)]$, where $\eta = 1$ for the decays $\pi^\pm \to \mu^\pm + \nu(\tilde{\nu})$ and $\eta = 0.65$ for the decays $K^\pm \to \mu^\pm + \nu(\tilde{\nu})$.

The hadron, with the remaining weight $W_0' = W_0 - W_\mu$, can be further traced before reaching the substance in which it interacts with the atomic nuclei. The energy and emission angle of the muon at the decay point can be found from the kinematic relations in Chap. 2. At high energies the muon production in reaction (7.2) can also be simulated in thick absorbers.

We present below some results of calculations, based on the described algorithm, of the muon fields in matter produced in the development of high-energy hadron cascades.[3,174] The functionals studied are the differential flux density of muons $F(z,r)$ and the longitudinal distribution of the muon flux in a circle of radius $R$ $\Phi(z) = \int_0^R F(z,r)\, 2\pi r dr$.

Figure 7.10 shows the contributions from the hadron and muon components to the function $F(z,r)$ in the limit $r \to 0$. It can be seen that muons begin

**Figure 7.13. Radial distribution of the muon flux density at a depth $z = 100$ m of an iron absorber. 3000-GeV protons interact with the target at a distance $d$ from the absorber.**

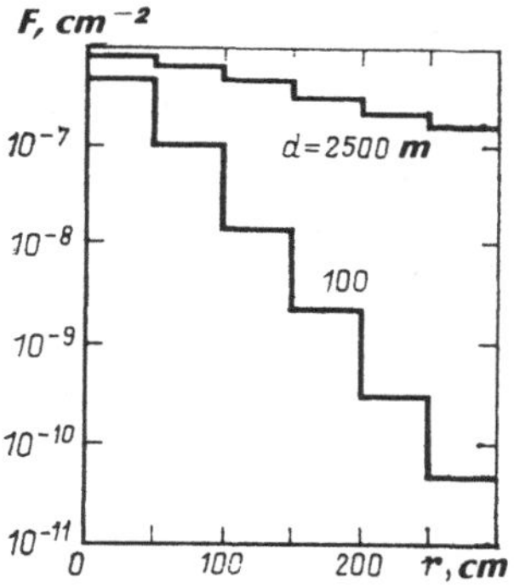


**Figure 7.14. Integrated muon flux at various depths $z$ of an iron absorber vs the decay interval. The energy of the protons that interact with the target is $E_0 = 3000$ GeV; solid curves, $R = 250$ cm; dashed curve, $R = 50$ cm.**

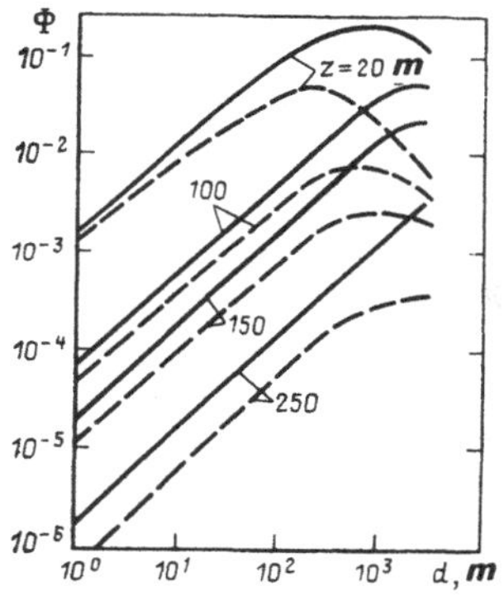


to dominate completely at a certain absorber thickness. These thicknesses are given in Table 7.2 for hadrons and muons with energies[3] $E > 20$ MeV.

Figure 7.11 shows the results of calculations of the function $\Phi(z)$, in the case in which "direct" muon production is ignored[3] and in which it is taken into account[174] in expression (7.3). Analysis showed that at energies $E_0 \gtrsim 1000$ GeV, when a hadron beam is directly incident on a solid absorber, disregarding the direct muon production or describing it incorrectly may result in appreciable errors in the case of a thick absorber.

The energy spectra of muons along the axis of a steel absorber are shown in Fig. 7.12. In the case of thin absorbers, the spectra always change appreciably with an increase in $z$. The quasi-equilibrium muon energy spectra become stabilized at certain values of $z$ and then again change at a maximum absorber thickness permissible for the given energy. The dependence of the function $F(z,r)$ on the radius decreases as the decay interval $d$ and the thickness $z$ are increased (Fig. 7.13).

In the case of a moderately large distance $d$ between the target and the absorber, the function $\Phi(z)$ increases almost linearly with increasing $d$ (Fig. 7.14). At $d > 50$–$100$ m, this function changes appreciably. The nature of the given dependences is determined primarily by two competing factors: the

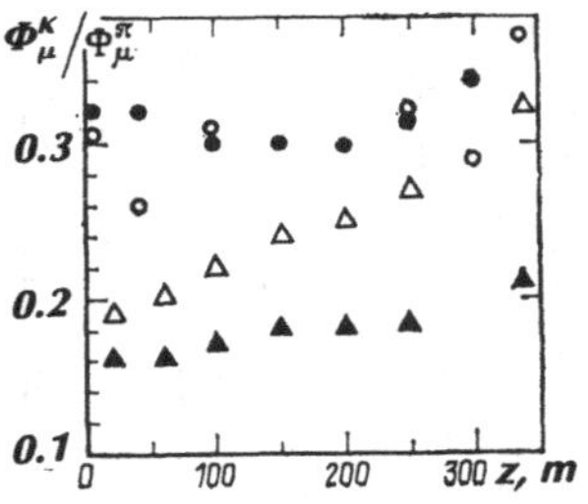


**Figure 7.15.** Longitudinal distribution of the ratio of the flux of muons produced in the decay of $K^\pm$ mesons to the flux of muons produced in the decay of $\pi^\pm$ mesons in a steel absorber. At a distance $d$ from this absorber a 3000-GeV proton interacts with the target. ●, ○—$d = 100$ m; ▲, △—$d = 2500$ m; dark symbols—$R = 50$ cm; light symbols—$R = 250$ cm.

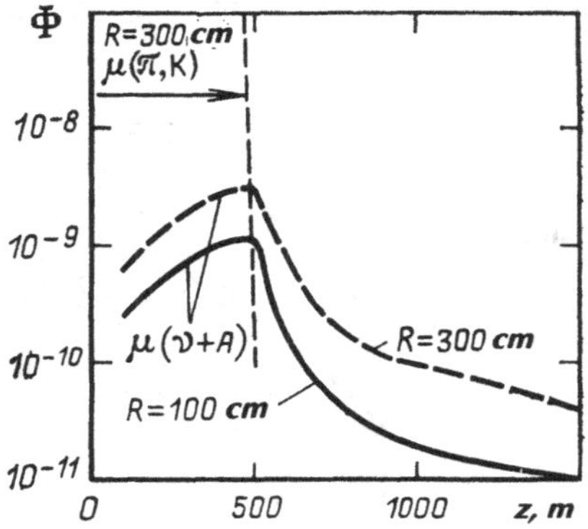


**Figure 7.16.** Longitudinal distribution of the muon flux in a circle with two radii in a steel absorber ($z \leqslant 500$ m) and behind the steel absorber ($z > 500$ m) in the case of the interaction of a 3000-GeV proton with the target at a distance $d = 2500$ m from the absorber.

geometric factor and the probability for the decay of $\pi$ and $K$ mesons. Note that if the angles at which the muons deviate from the original trajectory at the points at which the $\pi$ and $K$ mesons decay are ignored in the case of large decay intervals, the error in determining the differential muon flux density may be very large.

The muon contribution from the kaon decay to the total flux depends on the ratio of the $\pi$- and $K$-meson yields and on the decay interval. It generally increases with increasing thickness of the absorber. The muon-flux ratios in Fig. 7.15 vary over the range 0.15–0.45 and are nearly independent of the energy of the primary protons and the absorber material. In the analysis of the data shown in Fig. 7.15 the following factors must be taken into account: the probability for the kaon decay is approximately 7 times greater than that for the pion decay; in the relevant region the $\pi$-meson spectrum is steeper than the $K$-meson spectrum; the kaons produced in the hadron–nucleus interactions have the greater production angle; the muon deviation angle in the $K$-meson decay is approximately 30 times greater than in the $\pi$-meson decay.

At $d \gtrsim 2$–$3$ m the greatest contribution to the formation of the muon field in matter comes from decays (7.1) in this interval. In the case of a large primary energy $E_0$, large decay interval $d$, and large absorber thickness $z$, however, the muon flux generated in the bulk of a neutrino absorber in reactions (7.2) may be greater than the flux of decay muons, beginning at a certain value of $z_0$ (Fig. 7.16). At $z > z_0$ the flux of decay muons decreases sharply as the absorber thickness is increased. The energetic neutrinos which are produced in decays (7.1) over a decay interval $d$ and which travel enormous distances in matter because of the small $\nu A$ interaction cross section continue to generate muons even when the absorber thickness is much greater than $z_0$. This effect is important in experiments involving large proton accelerators and in cosmic-ray experiments.

# Chapter 8
# Applications of the methods and results of experimental studies of electromagnetic cascades

## 1. Calorimeters (total-absorption spectrometers)

### Types of calorimeters and the principle of their operation

Upon entering condensed matter a high-energy particle, depending on its nature, initiates an electromagnetic or hadronic cascade (see Chap. 1). The cascade characteristics such as the release of energy in the region of the development of a shower and the total length of the tracks of charged secondary particles are proportional to the primary-particle energy and can be used to determine the energy of the primary particle that has initiated the cascade. Devices allowing the energy of a primary particle to be determined by measuring the characteristics of the cascade produced by this particle are called *calorimeters* or *total-absorption spectrometers*, keeping in mind that the particle's energy can be determined precisely if its total energy released in the form of a cascade can be absorbed in the sensitive volume of the calorimeter.

Calorimeters are divided into two categories: electromagnetic and hadronic, depending on the type of cascades they detect. They are also distinguished by the principle of operation upon which they are based, i.e., the cascade characteristic they detect and the method used to detect it. They can, for example, measure the ionization produced by secondary charged shower particles or the intensity of the scintillations caused by a developing cascade in a scintillation material, or the intensity of Cerenkov radiation in a transparent homogeneous medium, which is proportional to the total length of the secondary charged-particle tracks. The energy of a primary particle can in principle be determined by measuring an exotic characteristic of the cascade generated by this particle. Such a characteristic may be the emission of a sound wave by a shower (see, e.g., Ref. 140).

Various types of electromagnetic and hadronic calorimeters[176] are used to solve the particular problems that arise in high-energy physics and cosmic-ray physics experiments. Calorimeters are used typically to (1) determine the primary-particle energy, (2) determine the energy carried off by a particle bunch (a jet), (3) determine the direction of motion of the primary particle and the coordinates of the point at which it enters the calorimeter, and (4) identify the particles.

All calorimeters, however, have a common feature, regardless of the nature of the problem to be solved, the calorimeter type, or the principle of its operation: Calorimeters essentially make use of the basic laws that govern the development of cascades. Here we briefly summarize these systematic features which were examined in detail in the preceding chapters.

(1) During the development of a cascade, the energy of the primary particle is distributed among many secondary particles and finally is absorbed in a certain effective region in which the cascade develops. The thickness of the layer in which the energy released in the cascade is absorbed is nearly independent of the primary-particle energy ($\sim \ln E_0$). The absorbed energy is proportional to the primary energy.

(2) The linear dimension of a cascade is much greater than its transverse dimension. The secondary particles of the cascade are grouped in the direction of their motion near the shower axis, which runs in the direction of the primary particle momentum. This occurs because transverse momentum imparted to the secondary particles in the multiple production is small. In the case of hadronic interactions $\langle p_\perp \rangle \sim 0.3$ GeV/$c$ and in the case of the interaction of electrons and photons $\langle p_\perp \rangle \sim m_e c$.

(3) The spatial distribution of secondary particles of a cascade and its release of energy fluctuate considerably. As the energy is increased, the relative fluctuations decrease as $E_0^{-1/2}$, since the number of cascade secondary particles increases with increasing energy as $N_e \sim E_0$.

(4) The electromagnetic cascades (electron–photon showers) and hadron cascades (nuclear cascades) have several common features in their behavior. However, because these processes are of an entirely different nature and because the secondary particles in them are different, their behavior also differs markedly in many ways (Table 8.1).

Calorimeters can be divided into two basic classes. The first class of calorimeters use one active material—homogeneous calorimeters (e.g., a scintillator or a Cerenkov-radiation source). This class of detectors can measure the energy of primary particles very accurately. Scintillator-based calorimeters (e.g., NaI calorimeters) are sensitive to the total ionization energy loss of secondary charged shower particles, while Cerenkov-radiation calorimeters (e.g., lead-glass-based source) are sensitive to the total length of the tracks of secondary charged shower particles (the track length depends linearly, as was shown in Chap. 4, on the energy of the primary particle). The second class of calorimeters use multilayer (multiple-plate) detectors (the so-called sandwich-type calorimeters). The multilayer calorimeters are widely used to detect hadronic and electromagnetic cascades, whereas homogeneous calorimeters are used to detect primarily electron–photon showers. A sandwich-type calorimeter consists of a set of active detecting layers alternating with layers of passive solid material, in which the cascade develops inten-

**Table 8.1. Comparison of the properties of electromagnetic and hadron cascades.**

| Characteristic | Electromagnetic cascade | Hadron cascade |
|---|---|---|
| Multiplication processes | Production of $e^-e^+$ pairs and electron bremsstrahlung | Inelastic hadron–nucleus collisions |
| Mean free path before the interaction | $(9/7)t_r$ for photons, $t_r/\ln(E_e/\omega_\gamma)$ for electrons | Mean free path of a hadron before the inelastic interactions: $\lambda_{in} \approx A/N_A\sigma_{in}$ |
| Average inelastic interaction coefficient | $\langle K\rangle \sim 1$ for photons, $\langle K\rangle \sim 0.07$ for electrons | $\langle K\rangle \sim 0.5$ |
| Secondary particles | Electrons and photons | Principally $\pi$ mesons and nucleons |
| Longitudinal development of the cascade | Depends weakly on matter if the depth $t$ is expressed in units of radiation length, $x = t/t_r$. | Depends weakly on matter if the depth $t$ is expressed in units of the mean free path of a hadron before inelastic interaction, $x = t/\lambda_{in}$. |
| Position of the maximum of the cascade | $\sim \ln E_0$ | $\sim \ln E_0$ |
| Characteristic length of the cascade decay | $\Lambda = (3\text{–}4)\, t_r$, does not depend on $E_0$ | $\Lambda = (1\text{–}2)\, \lambda_{in}$, increases with increasing $E_0$ |
| Length of the effective region of development of the cascade | $t_{eff} = (10\text{–}30)\, t_r$<br>$t_{eff} \sim \ln E_0$ | $t_{eff} = (50\text{–}10)\, \lambda_{in}$<br>$t_{eff} = \ln E_0$ |
| Transverse development of the cascade | Depends weakly on matter if the distance from the shower axis is expressed in Moliere units, $r_M = E_s t_r/\varepsilon_c$ | Depends weakly on matter if the distance from the cascade axis is expressed in $g/cm^2$ |
| Transverse dimension of the effective region of development of the cascade | $r_{eff} \simeq 2r_M$ | $r_{eff} \sim \lambda_{in}$ |
| Undetectable energy | None | Energy expended on the excitation and spallation of nuclei; energy carried away by the neutrinos |
| Ineffectively detectable energy | Low-energy photons | Low-energy neutrons and photons, nuclear fragments, muons |
| Principal sources of fluctuations | Depth of the first interaction | Energy transferred to $\pi^0$ mesons in the first interaction and the depth of the interaction |

sively. This arrangement assures a reasonable size and cost of the calorimeter even for hadronic cascade detection. Such cascade characteristics as the number of times the charged-particle tracks cross the layers of active material (see, e.g., Chap. 4) or the energy loss by ionization in these layers can be measured with multilayer calorimeters with a high degree of sensitivity.

Calorimeters are generally used primarily to determine particle energy. To perform this task effectively, the size of a calorimeter (regardless of its type) must satisfy the following requirements[97]:

(1) For a particular species of primary particles at a given energy the signal received from the calorimeter must be independent of the coordinates of the point of initial interaction.

(2) For a given size of the calorimeter its signal must depend linearly on the primary-particle energy.

(3) For a given size of the calorimeter its *energy resolution* $\sigma_E/E$ must be determined primarily by the statistical fluctuations of the cascade characteristics that are measured ($\sigma_E/E \sim E^{-1/2}$).

(4) If the energy $E$ of the primary particle is fixed, the energy resolution of a calorimeter, beginning with a certain size of the calorimeter, will not improve with further increase of its size.

(5) The distribution of the signal amplitudes must be symmetrical. If the size of the calorimeter is inadequate, the fluctuation of that part of the cascade energy which is released outside the sensitive region of the calorimeter will cause the distribution of the signal amplitudes to be asymmetric, since these fluctuations do not satisfy the Gaussian distribution law.

### Electromagnetic calorimeters

Both sandwich-type calorimeters and homogeneous calorimeters with one active material are used to detect electron–photon showers. The key characteristic of a calorimeter is its energy resolution. From the practical standpoint, it is also important that the signal depend linearly on the primary-particle energy. This goal is always achieved if the amount of shower energy released from the sensitive volume of the calorimeter is small.

In an infinite-dimensional *homogeneous calorimeter* (the case of total energy absorption) the accuracy in determining the energy is restricted by the statistical fluctuation in the number of elementary processes that occur in the development of an electron–photon shower. The total ionization loss and the total length of the electron (positron) tracks in an electron–photon shower, as well as the fluctuation in their number, depend on the relation $\xi = E_{\text{thr}}/\varepsilon_c$, where $E_{\text{thr}}$ is the energy threshold for the detection of secondary electrons, and $\varepsilon_c$ is the critical energy of the active material of the calorimeter (see Chap. 4). For a NaI calorimeter ($\xi = 0.5/12.5 = 0.4$) the relative accuracy in determining the energy is given by[177]

$$\sigma_E/E \simeq 0.007/\sqrt{E}\,. \tag{8.1}$$

It is assumed, unless otherwise stated, that the energy $E$ is given in GeV.

**Figure 8.1.** 1, experimentally measured dependence of the energy resolution of a NaI calorimeter of length 24$t$, on the energy of the primary particle (Ref. 97). 2, the dependence calculated from Eq. (8.1).

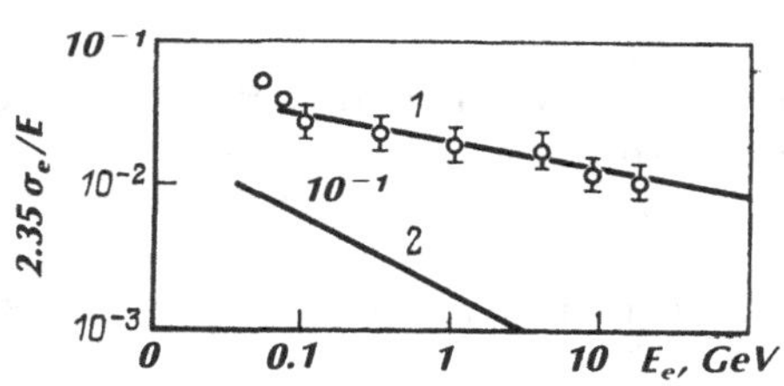


Any energy loss, however small, from the sensitive volume of the calorimeter degrades its resolution. In lead-glass calorimeters, for example, the dependence of the energy resolution on the energy loss is given by[178]

$$\sigma_E/E = [\sigma_E/E]_{t=\infty}(1+4\varepsilon+50\varepsilon^2),$$

where $\varepsilon$ is the fraction of undetected shower energy ($\varepsilon \leqslant 0.1$), and $t$ is the linear dimension of the calorimeter. In a calorimeter with fixed dimensions the fraction of lost (undetected) energy increases logarithmically with increasing energy of the primary particles. As a result, the energy resolution of such a calorimeter decreases in comparison with a calorimeter with infinite dimensions.

An experimentally measured energy resolution (Fig. 8.1) can be described by an approximate relation[97]

$$\sigma_E/E = 0.009/E^{1/4}. \tag{8.2}$$

A comparison of this expression with theoretical relation (8.1) shows that at $E \sim 1$ GeV, when the loss of shower energy from the sensitive volume of the detector is small, the experimentally measured energy resolution is approximately equal to that predicted theoretically. As $E$ is increased, the relative fraction of the energy loss also increases, resulting in a corresponding degradation of the resolution, in comparison with that predicted theoretically, for a detector with infinite dimensions, in which there are no energy losses. It should be emphasized that the difference in the laws governing the behavior of (8.1) and (8.2) cannot be explained by the fluctuation in the number of photoelectrons, which also give rise to $\sim E^{-1/2}$.

In contrast with the NaI calorimeters, the lead-glass calorimeters measure the energy with an accuracy which is determined primarily by the fluctuation in the number of photoelectrons. The measurement accuracy of these calorimeters is given by the equation[178]

$$[\sigma_E/E]_{t=\infty} = [\sigma_0^2(E)/E^2 + 1/\delta g E]^{1/2} \simeq 0.006 + 0.03\delta^{-1/2}E^{-1/2},$$

where $g$ is the average number of photoelectrons per GeV of released energy (as a rule, $g \simeq 10^3$), $\delta$ is the ratio of the photocathode area to the radiator output area (in general, $\delta \leqslant 0.5$), and $\sigma_0(E)$ is a quantity which is determined not only by the shower fluctuations but also by the Cerenkov radiation absorption in the radiator. A typical energy resolution of a lead-glass calorimeter is given by an approximate equation[96]

$$\sigma_E/E \simeq (0.04-0.05)/\sqrt{E}.$$

*Multilayer calorimeters* can be divided into discrete calorimeters and proportional calorimeters.[86] Discrete calorimeters measure the number of charged particles that cross the active layers. A typical example of such a calorimeter is a calorimeter which consists of 3.5 mm $\times$ 5 mm layers of Conversi tubes which alternate with lead plates of thickness $t = 0.9t_r$.[179] The experimentally measured energy resolution of such a detector is given by [179]

$$\sigma_E/E = 0.12/\sqrt{E},\ 0.5 < E < 3\ \text{GeV}. \tag{8.3}$$

The energy resolution of a discrete calorimeter, which is determined by the fluctuation in the number of times charged particles cross the active layers, is given by[86]

$$\sigma_E/E = 0.032\sqrt{\varepsilon_c t/[F(E_{\min}/\varepsilon_c)\cos(E_{\min}/\pi\varepsilon_c)E]}\ . \tag{8.4}$$

Here $E_{\min}$ is the minimum detectable energy of an electron (MeV), $\varepsilon_c$ is the critical energy (MeV), $E$ is the shower energy (GeV), $t$ is the thickness of the passive layer of a calorimeter (radiation length), and $F(E_{\min}/\varepsilon_c)$ is a coefficient which takes into account the decrease in the total length of charged-particle tracks as a result of the introduction of the energy detection threshold $E_{\min} \neq 0$ (see Sec. 4.5). A calculation based on Eq. (8.4) for a calorimeter with Conversi tubes[179] yields the following result: $\sigma_E/E = 0.11/\sqrt{\text{E}}$ ($E_{\min} \simeq 0$), in good agreement with the experimental result for the energy resolution of such a calorimeter [see Eq. (8.3)].

The multilayer proportional calorimeters measure the energy released in the active layers due to the ionization energy loss of charged particles. In these calorimeters the accuracy of the measurement of the energy depends not only on the fluctuation in the number of times the charged particles cross the active layers but also on the fluctuation in the loss of energy due to ionization, as well as on the fluctuation in the total length of charged-particle tracks in the active layers. The contribution of the last two sources to the error in measuring the primary-particle energy depends on the density of the material of the active layers of the calorimeter.

Let us consider the effect of the fluctuation in the ionization energy loss in the active layer of the material on the accuracy with which the proportional calorimeters determine the energy. In active dense substances (liquids, solids) the fluctuation of the ionization loss is described by the Landau function (see Sec. 2.5), which does not depend on the particle energy. The distribution width $\Delta$ of the ionization loss can therefore be characterized by

$$\sigma(\Delta)/\Delta = 2/\ln(4W/E_{\min}),$$

where $E_{\min}$ is the minimum energy of the $\delta$ electrons, which is usually assumed to be 30 eV, and $W$ is the energy at which one $\delta$ electron, on the average, with an energy higher than $W$, is produced in a layer of thickness $x$, g/cm$^2$. The numerical value of $W$ (eV) is given by the equation $W = 1.5\times 10^5 xZ/A$. If the asymmetry of the ionization loss distribution is ignored, the contribution of the ionization energy loss fluctuation to the accuracy of determining the primary-particle energy in first approximation is given

**Figure 8.2. Contribution of the fluctuations in the number of crossings (1), ionization energy loss (2), and electron path lengths to the total energy resolution (4) of a multilayer calorimeter (1.3$t_r$ lead-gas). Calculation for the primary electron was carried out by the Monte Carlo method (Ref. 182).**

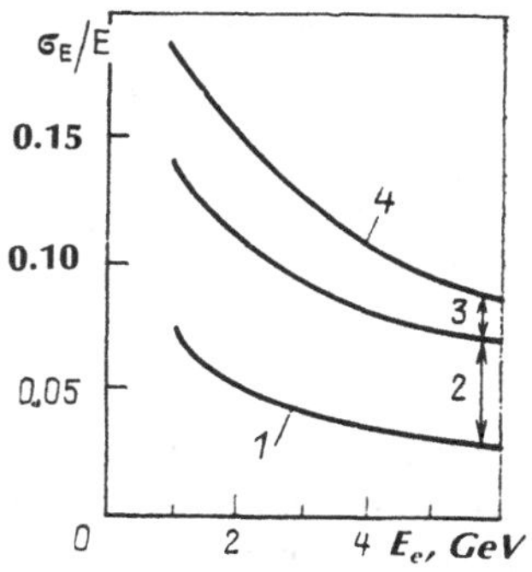


by[86] $(\sigma_E/E)\,L \simeq 2/\sqrt{N}\,\ln(10^4 x)$, where $N$ is the number of times the charged-particle tracks cross in the active layers of the calorimeter: $N \simeq (E/\varepsilon_c)t$ ($t$ is the thickness of the passive layer of the calorimeter, in radiation lengths). If the thickness of the active layer is $x = 1\ \mathrm{g/cm^2}$, the contribution of the ionization loss fluctuations is small: it is no greater than 3% of the principal contribution coming from the fluctuation in the number of crossings $N$ [see Eq. (8.4)]. If a gas is the active material, then $x \simeq 10^{-3}\ \mathrm{g/cm^2}$. The energy resolution of the calorimeter in this case decreases by approximately a factor of $\sqrt{2}$ due to the ionization loss fluctuations. The actual situation is even worse, since in a thin layer of gas the ionization loss fluctuates much more strongly than that predicted by the Landau equation (see Fig. 2.6).

The secondary low-energy shower electrons have a wide angular distribution, causing the distances they traverse in the active layers of the calorimeter to fluctuate considerably. These fluctuations are particularly strong in active layers of gas. The primary reason for this is that the minimum detectable energy in layers of gas is lower than that in solid layers, so that the electrons in gaseous layers have even a wider angular distribution. The low-energy electrons which move along the layer release in it much greater energy than the electrons which move perpendicular to the layer. Secondly, a multiple Coulomb scattering in a solid layer causes the electron to rapidly escape from this layer. As a result, the electron range and its fluctuation in the active layer decrease. The role played by the processes mentioned above is illustrated in Fig. 8.2.

In summary, we can say that the energy resolution of calorimeters with solid active multilayers is determined primarily by the fluctuation in the number of times the charged-particle tracks cross the active layers [Eq. (8.4)]. In very thin active layers, however, the fluctuations of the ionization energy loss and of the electron range are, as in the gas layers, a factor. In discrete calorimeters, in which the number of secondary electrons in the active layers is calculated, only the fluctuation in the number of crossings contributes to the error in determining the energy. The disadvantage of these calorimeters is that at high primary-particle energies, when the density of secondary shower particles is reasonably high, two or more secondary-electron tracks can

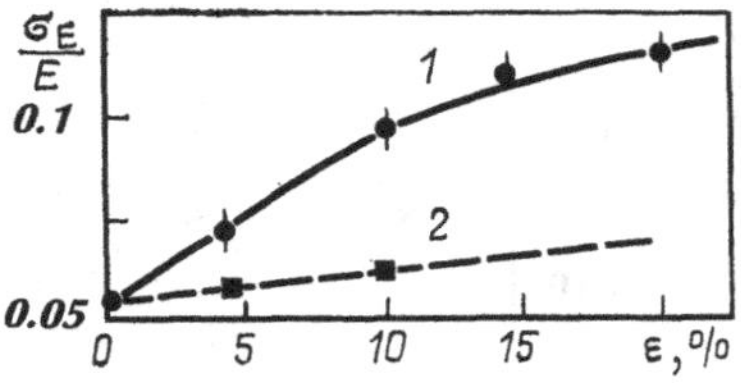


**Figure 8.3. Energy resolution of a sandwich-type calorimeter measured experimentally vs the fraction of energy ε lost in the longitudinal direction (1) and in the transverse direction (2) (Ref. 181) for a 15-GeV electron.**

simultaneously cross the detecting cell of the active layer, causing the signal to depend nonlinearly on the primary energy and the accuracy in determining the energy to be lowered.

As in the case of homogeneous calorimeters, the cascade energy loss in the sensitive region of the calorimeter affects the energy resolution of multilayer calorimeters (Fig. 8.3). The energy resolution of the calorimeter is more sensitive to the energy loss in the longitudinal direction.

## The transition effect

In multilayer detectors the active and passive layers of the material have various densities, various atomic numbers, and accordingly various critical energies. The number of secondary electrons and protons and their spectra change at the boundary of such layers. The difference in the multiple scatterings in these layers also contributes to the change in the equilibrium spectrum of secondary particles. This effect, called *transition effect*, influences the absolute value of the energy which is released in the active layers and which is detected by the calorimeter. Measurements have shown that in the multilayer detectors $E_v/E<1$, where $E_v$ is the energy detected by the detector, and $E$ is the energy of the particle that initiates a cascade. In a calorimeter consisting of alternating layers of lead ($x_p = 24$ g/cm$^2$) and a scintillator ($x_a = 0.63$ g/cm$^2$), for example, the ratio[180] $E_v/E = 0.52$. In a calorimeter with passive layers consisting of a light material—marble ($x_p = 23$ g/cm$^2$) the ratio[181] $E_v/E = 0.85$. In practice, the transition effect does not complicate the use of multilayer calorimeters, since they are calibrated in electron beams whose energy is known beforehand. As a result, a relationship between $E$ and $E_v$ is established, making it possible to determine the primary particle energy $E$ from the measured value of $E_v$. The remark concerning calorimeter calibration also applies to homogeneous detectors.

## Hadronic calorimeters

The cascades initiated by hadrons fluctuate much more strongly than the electromagnetic cascades. A substantial part of the energy released from an electromagnetic cascade cannot be detected. The undetected energy is principally the energy required to excite and break up the nuclei of the atoms of the medium and the energy carried off from the sensitive volume of the calorimeter by neutrons, muons, and neutrinos (see Fig. 8.4 and Sec. 6.4). As can be

**Figure 8.4. Relative contribution η of various processes to the energy release of a cascade initiated by a proton in a multilayer calorimeter (iron-liquid argon) (Ref. 177) (dashed curves, undetectable energy). 1, ionization losses of charged hadrons; 2, electromagnetic showers; 3, excitation and decay of nuclei (neutrinos); 4, nuclear reactions; 5, escape of particles from the sensitive region of the detector.**

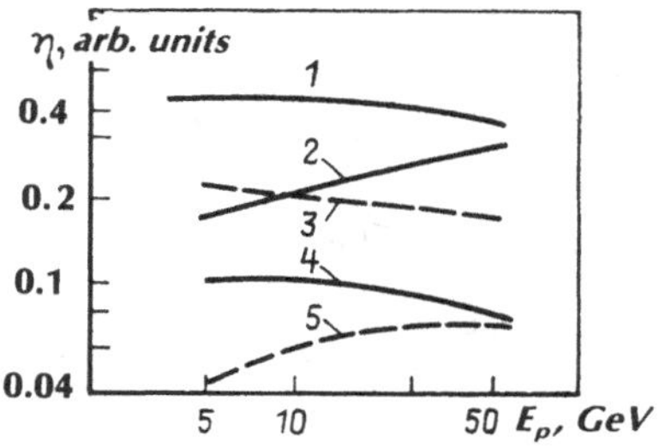


seen from the results of calculations, nearly one-half of the energy of an electromagnetic cascade initiated by a primary proton is released in the form of electromagnetic showers and about one-third of its energy is not detectable. The fraction of energy released due to the electromagnetic showers increases with increasing primary-hadron energy (Fig. 6.13). Fluctuation in the electromagnetic energy per standard deviation for a 10-GeV $\pi^+$ meson, for example, amounts to 40%.

*The energy resolution of hadronic calorimeters* is determined primarily by the hadronic-cascade fluctuations. The principal fluctuations are those of the energy released as an electromagnetic shower and those of the undetected energy. These fluctuations have a negative correlation: With increasing transfer of energy to $\pi^0$ mesons (the principal source of electron–photon showers; see Sec. 1.2), less energy is required to excite and to split the nuclei of the atoms of the medium, and vice versa. The escape of secondary particles from the limited sensitive region of the calorimeter has an appreciable effect on the accuracy of determining the primary-hadron energy.

So far, the best energy resolution was achieved using a homogeneous calorimeter with a liquid scintillator[183]: $\sigma_E/E = 0.09 + 0.11/\sqrt{E}$ $(10 < E < 150$ GeV). The presence of a constant in this equation indicates that the energy resolution is not determined exclusively by the fluctuations of the elementary processes, whose number is proportional to the primary energy $E$.

The multilayer calorimeters are used most widely for detecting hadronic cascades. For passive layers of thickness $x_p \leqslant 100\,\mathrm{g/cm^2}$, Amaldi[86] proposed an empirical equation which describes well the detectable values of the energy resolution:

$$\sigma_E/E = \sqrt{\sigma_h^2/E + \sigma_{EM}^2/E}\,, \tag{8.5a}$$

where $\sigma_h/\sqrt{E} = 0.5/\sqrt{E}$ is a term which corresponds to the fluctuation of the hadronic cascade (primarily the fluctuation of the undetectable energy);

$$\sigma_{EM}/\sqrt{E} = R\sqrt{(x_p/t_r)/KE} \tag{8.5b}$$

is a typical relation for the energy resolution of the electromagnetic calorimeter; the coefficient $K = 0.75$ corresponds to the average energy released as an electromagnetic shower; and the coefficient $R = 0.4$ is determined empirical-

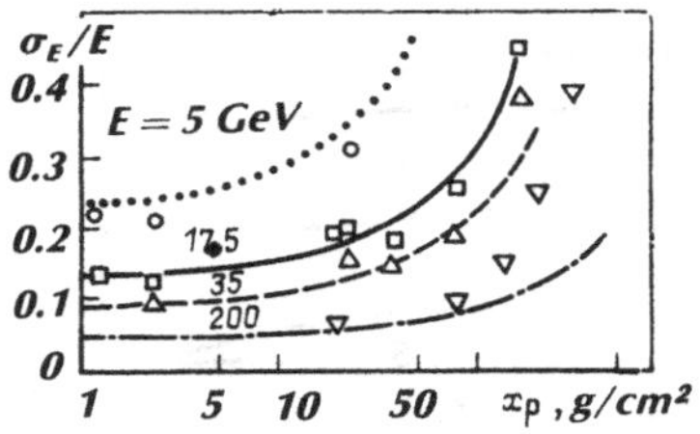


**Figure 8.5. Energy resolution of a multilayer calorimeter vs the thickness of the passive layer $x_p$. Points, experimental data (Ref. 96); curves, calculation based on Eq. (8.5).**

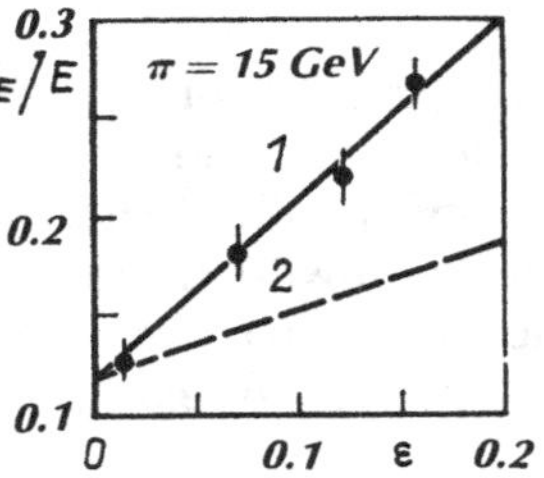


**Figure 8.6. Energy resolution of a sandwich-type hadron calorimeter measured experimentally vs the fraction of energy ε lost in the longitudinal direction (1) and in the transverse direction (3) (Ref. 181).**

ly through a best fit of the experimental data. For $x_p < 100\,\mathrm{g/cm^2}$ and $R = 0.4$ the results of the calculations based on Eq. (8.5) are in good agreement with the experiment (Fig. 8.5).

The energy resolution of hadronic calorimeters is strongly affected by the loss of cascade energy from the sensitive region of the calorimeter (Fig. 8.6). The energy resolution is particularly sensitive, as in the case of electromagnetic calorimeters, to the energy loss in the longitudinal direction. Prokoshkin[178] suggested the use of a simple semiempirical equation to determine the length of a hadronic calorimeter with passive iron layers necessary to absorb 95% of the shower energy: $L(95\%) \simeq 9 \ln E + 40$, where $L$ is the calorimeter length (in cm), and $E$ is the primary-particle energy (in GeV).

It must be noted that the energy loss from a sensitive region of the calorimeter occurs through the front wall of the calorimeter (the so-called quasi-albedo effect or backscattering; see Fig. 6.12 and Sec. 6.4). This energy loss is generally small, no greater than 2–3% of the total energy release.[97]

The principal contribution to the error in determining the particle energy by means of a hadronic calorimeter comes from the fluctuation in the energy released as an electromagnetic cascade and the fluctuation in the undetectable energy which is strongly correlated with the cascade energy and which is required to excite and split the nuclei of the atoms of the medium. Clearly, the smaller the fraction of undetectable energy, the better is the energy resolution of the calorimeter. The energy required to excite and break up the nuclei is finally removed from the sensitive volume of the calorimeter by low-energy evaporative neutrons. A unique method of reducing the fraction of undetectable internuclear cascade energy, which substantially improves the energy resolution of the calorimeter, was proposed in Ref. 184. In this method the passive layers of a sandwich-type calorimeter are made from

**Figure 8.7. Amplitude of the signal $A$ vs the energy $E_0$ of the primary particle in sandwich-type calorimeters with an iron passive layer ($x_p$ = 1.5 mm) and a uranium passive layer ($x_p$ = 1.7 mm). The thickness of liquid argon active layers is 2 mm (Ref. 184).**

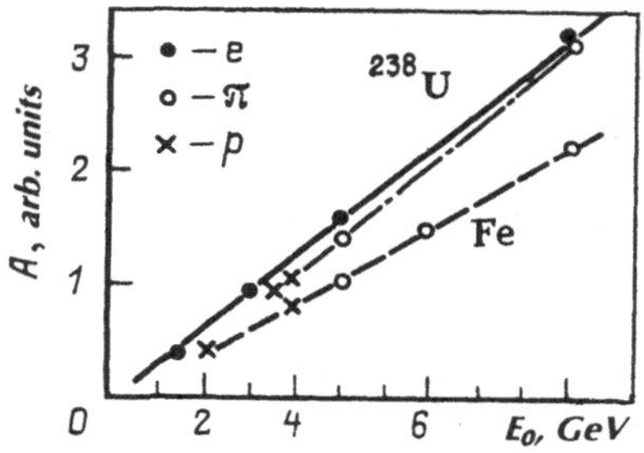


$^{238}$U. The interaction of neutrons with an energy of several MeV with $^{238}$U nuclei causes these nuclei to fission, producing 2.6 secondary neutrons on the average. The number of secondary neutrons and the number of times they interact with the nuclei of the material of the active layers of the calorimeter increase appreciably as a result of several consecutive fission reactions. Furthermore, these fission reactions and ensuing neutron capture account for the production of a large number of photons. The total energy released as photons amounts to 560 MeV per GeV of primary-particle energy, on the average (in the case of a large absorber). All these factors jointly increase the fraction of the detectable energy of the hadronic component of the internuclear cascade by approximately a factor of 1.5 in comparison with the detectable energy of a calorimeter with passive layers made from iron (Fig. 8.7). The energy resolution of the calorimeter also improves in this case. The resolution of a calorimeter with 1.5-mm-thick passive iron layers and 2-mm-thick active liquid-argon layers, for example, is $\sigma_E = 0.47\sqrt{E}$, where $E$ is the energy in GeV. The energy resolution can be improved markedly, to $\sigma_E = 0.3\sqrt{E}$, by replacing iron layers with 1.7-mm-thick uranium layers.[184]

### Determination of the coordinates and the direction of motion of the primary particle

Determination of the particle's coordinates by means of a calorimeter is usually understood to mean the determination of the coordinates of the point of its interaction (or the point at which it enters the calorimeter) in the transverse direction with respect to its momentum. To determine the coordinates of a particle, the calorimeter must be capable of measuring the development of a shower in the transverse direction. Information about the development of showers in the transverse direction also makes it possible to single out in space the cascades which were simultaneously initiated by several particles. This means that the particles responsible for these cascades can be separated spatially. Several methods of solving this problem have been developed.

### Calorimeters with a hodoscopic structure of the active layers

A typical example of these calorimeters are sandwich-type calorimeters, in which the active layers (or some of them) are an array of scintillators consisting of narrow strips or segments.[185,186] The location of the *center of mass* of the shower at several penetration depths can be determined by measuring the distribution of the energy release in the shower in the transverse direction: $y_{0j} = d\,\Sigma_{i=1}^{N}\, iA_{ij}/\,\Sigma_{i=1}^{N}\, A_{ij}$, where $A_{ij}$ is the amplitude of the signal measured by the strip with index $i$ in the active layer with index $j$, and $d$ is the width of the strip. The coordinates of the center of mass of the shower in the transverse direction $y_{0j}$ are a good approximation for determining the coordinates of the point at which the particle enters the calorimeter. A calorimeter with a hodoscopic structure[186] has a coordinate error no greater than $\sigma_y \simeq 2$ mm for energies of an electron–photon shower of several tens of gigaelectronvolts if the width of the scintillator strip is $d = 1.5$ cm. Even if the energy of a shower is $E \simeq 0.5$ GeV, when the fluctuations of the characteristics of a shower are of a particularly large magnitude, the coordinate error is $\sigma_y \simeq 6$ mm. In a hodoscopic calorimeter the cascades initiated by two photons can easily be separated spatially if the distance between the points at which they enter the calorimeter is greater than 5 cm in the transverse direction.[186] These properties and the high accuracy with which the coordinates of the particle and its energy are determined make the hodoscopic calorimeters a powerful tool for the study of events with a large multiplicity of secondary particles.

It should be noted that a precise determination of the center of mass of the transverse distribution of the shower at several depths of its development makes it possible to spatially restore the location of the shower axis, i.e., the direction of motion of the primary particle.

Hodoscopic calorimeters can be effectively used to determine the coordinates of the point at which a primary hadron enters the calorimeter. Since hadron cascades are characterized by a wider transverse distribution than the electromagnetic cascades, the hadron coordinates are determined much less accurately than the electron or photon coordinates. A typical coordinate error for hadrons is[96] $\sigma_y \simeq 2$–$5$ cm.

### Calorimeters with a cellular (segmented) structure

Typical features of calorimeters with a cellular structure may be seen in the GAMS-200/F gamma spectrometer.[187] A GAMS spectrometer is comprised of an array of cells made from a transparent lead glass. A cell is $36 \times 36 \times 420$ mm in size (the cell length is $l = 420$ mm in the direction of the particle beam, which accounts for the total absorption of an electromagnetic shower in the longitudinal direction). The energy resolution of the calorimeter is characterized by a semiempirical equation[187] $\sigma_E/E = 0.027 + 0.125/\sqrt{E}$. The coordinate error is $\sigma_y = 1.3$ mm. Its dependence on the transverse dimension $d$ of the cell (for $d > 3$ cm) can be approximated by the expression[178] $\sigma_y = \sigma_{y0} \exp(d/d_0)$, where $\sigma_{y0} = 0.8$ mm and $d_0 = 65$ mm. Such a high spatial accuracy is achieved by minimizing the difference between the measured two-dimensional spatial distribution of the shower and the distributions obtained in the electron beam calibration. Reducing the transverse dimension $d$ ($d < 3$ cm) of a cell does not materially improve the spatial resolution of the calorimeter. As the size of the cells is increased, progressively fewer cells are needed to measure the shower profile in the transverse direction, causing the fluctuations of the amplitude of the signals and the coordinate error to increase. If the width of the cell is greater than the width of the shower, so that the entire shower fits into a single cell, the coordinate is determined with accuracy of the size of the cell.

An important feature of a GAMS spectrometer is its ability to spatially separate the neighboring showers. Each of the two photons which is produced in the decay $\pi^0 \rightarrow 2\gamma$, for example, can be distinguished from the other if the distance between the points at which they enter the calorimeter is $L \geqslant 2.5$ cm. This means that one photon can be distinguished from the other even when they enter the same cell.[187]

A good spatial resolution of the particles, high coordinate accuracy, and a good energy resolution enable cell-type calorimeters to be used as devices for measuring the particle masses. A GAMS spectrometer, for example, can be used to measure the mass of a $\pi^0$ meson (from the decay $\pi \rightarrow 2\gamma$) within $\sigma_M / M = 2.4\%$ and the mass of a $\eta$ meson (from the decay $\eta \rightarrow 2\gamma$) within $\sigma_M / M = 1.3\%$.

## Identification of particles in calorimeters

Calorimeters can be used to identify particles from the specific features of their passage through matter, depending on their type. Calorimeters are widely used to identify muons, electrons, and hadrons.

In passing through matter the muons lose energy primarily as a result of ionization and excitation of the atoms of the medium. Processes such as bremsstrahlung, production of $e^-e^+$ pairs, and multiple production of hadrons by nuclei become a factor only at muon energies $E_\mu > 100$ GeV (see Sec. 2.5). Muons therefore have a greater penetrating power than hadrons. Muons also experience a multiple Coulomb scattering in their passage through matter. If, however, the muon energy is high enough, the direction of motion of a muon does not change appreciably from the original direction as a result of Coulomb scattering (see Chap. 7). The average angle through which a 10-GeV muon scatters as a result of passage through a 1.8-m-thick layer of iron is only $\langle\theta\rangle \simeq 2 \times 10^{-2}$ rad.

At $E_\mu \lesssim 100$–$300$ GeV the characteristic features of muons, which can be used to identify them during their passage through matter, are the absence of interactions which would cause them to lose an appreciable fraction of their energy, amounting to a value greater than the fluctuation scale of the ionization energy loss, and the absence of interactions, which would lead to a strong particle scattering exceeding the scale set by multiple Coulomb scattering.

The principal criterion which can be used to establish (at a certain confidence level) that the particle that passes through a calorimeter is a muon can be stated as follows: A muon is assumed to be a particle with a minimal ionizing capacity, which passes through the entire calorimeter without apparent interactions and without undergoing a large-angle scattering.

The probability that $\pi$ mesons in a wide energy range will pass through a layer of iron of thickness $l$ without apparent interactions is given by the equation

$$W(l) = \exp(-l/\lambda), \tag{8.6}$$

where $\lambda \simeq 19$ cm (Ref. 188). Since this value is approximately equal to the mean free path of a $\pi$ meson before the inelastic nuclear interaction $\lambda_{\text{in}}$, it can be concluded that the $\pi$ mesons, which have passed through a layer of iron of thickness $l$, are primarily pions which have experienced only a multiple Coulomb scattering (by analogy with muons) and small-angle elastic nuclear scattering. Accordingly, the criterion for the absence of large-angle scattering events in this case does not appreciably decrease the probability for the simulation by a $\pi$ meson of the passage of a muon through a calorimeter in comparison with the probability given by Eq. (8.6).

Upon passage through matter, the electrons and photons initiate an electromagnetic shower which differs from a hadronic cascade. The principal differences in the characteristics of electromagnetic and hadronic cascades which are used to distinguish electrons (and photons) from hadrons by means of a calorimeter are as follows.

(1) The longitudinal and transverse dimensions of an electron–photon shower are smaller by several factors than the corresponding dimensions of a hadronic cascade if the energy of the primary particle is the same.

(2) The fraction of the shower energy detected in a calorimeter in the case of an electromagnetic cascade is higher than the fraction of the shower energy detected in the case of a hadronic cascade.

(3) In contrast with a hadronic cascade, nearly all secondary charged particles in an electron–photon shower are ultrarelativistic particles: $\beta = v/c \simeq 1$.

In actual situations, all three differences in the behavior of the electromagnetic and hadronic cascades are used to distinguish electrons from hadrons. The calorimeters based on the detection of the Cerenkov radiation of secondary charged particles, for example, can be used effectively to identify electrons (photons), since the fraction of secondary relativistic particles which are capable of emitting this radiation in hadronic cascades is much smaller than that in an electron–photon shower, where nearly all secondary electrons (positrons) are ultrarelativistic particles. The relatively small longitudinal dimensions of such calorimeters, quite adequate to assure virtually total absorption of the electromagnetic shower, are totally inadequate for the development of a hadronic cascade. This factor also causes the suppression (rejection) factor of the hadron background to increase. If the momentum of the primary particle is known, the hadrons can be rejected in a more efficient manner by making use of the constraints imposed on the ratio $p/E$ ($p$ is the particle momentum measured, for example, on the basis of a magnetic analysis, and $E$ is the particle energy measured by a calorimeter*). The ratio $p/E$ for a primary electron should be close to unity. A typical value of the rejection factor† for hadrons found by using a lead-glass calorimeter is $K \sim 500$. The hadron rejection can be further improved by a factor of 2–3 by using the data on the development of a cascade in the transverse direction.[176]

If the momentum of a primary particle is not known, the information on the spatial development of a shower gives a rejection factor[189] as high as 10.

Multilayer hodoscopic calorimeters which are used to obtain information on the longitudinal and transverse development of a shower can also be used to identify electrons. A sandwich-type calorimeter with active layers consisting of liquid argon,[190] for example, yielded a rejection factor $K \simeq 400$ at an electron detection efficiency of 83%.

A calorimeter's capability to distinguish an electron from a hadron is limited by the fact that most of the energy of a primary hadron in the hadron-nucleus interaction can be transferred to a secondary $\pi^0$ meson (or several $\pi^0$ mesons). Decaying into two photons, a $\pi^0$ meson initiates an electromagnetic shower, whose characteristics are very similar to those of a shower which is initiated by an electron of the same energy as that of the primary hadron. To reduce the hadron background, substances with a large value of the ratio $\lambda_{in}/t_r \sim Z^2/A^{3/4}$ can be used as an absorber (passive layers). In the case of lead, for example, $\lambda_{in}/t_r \simeq 33$.

In addition to the use of calorimeters to separate electrons, muons, and hadrons which has now become customary, calorimeters are now widely used to identify unstable particles from the estimate of the invariant mass of their decay products[176] (for example) $\pi^0, \eta \ldots \rightarrow \gamma\gamma; \rho, J/\psi \ldots \rightarrow e^+e^-$). This technique was used to study the spectroscopy of charmonium, and it led to the discovery of a charmed $\eta_c$ meson, the process by which direct protons are produced in hadron–hadron collisions (the important point here was the simulation of single photons by photons from the meson decay), and other processes (see the review by Fabjan and Ludlam,[176] who give citations of original papers).

## 2. Acoustic detection of high-energy particles

Askar'yan and Dolgoshein[191,192] proposed a new method of detecting high-energy particles, based on the detection of acoustic radiation produced as a result of the development of an electromagnetic cascade in a dense medium. The acoustic-detection method was proposed for the detection of extremely-high-energy neutrinos[191–193] in the DUMAND (Deep Underwater Muon and

*The kinetic energy is measured in the case of electrons and nucleons, the total energy is measured in the case of mesons, and the total energy plus the rest mass, $E = E_{tot} + Mc^2$, are measured in the case of antiprotons (antineutrons).

†The rejection coefficient, $K = N$, indicates that one of $N$ hadrons, on the average, is erroneously identified as an electron (positron).

Neutrino Detection) experiment,[194] in which an enormous mass of ocean water ($\sim 10^{11}$ metric tons) was used as the active material of the detector.

A systematic theory of thermoacoustic generation of sound in water by high-energy particles was developed by Askar'yan *et al.*[140] A very simple mechanism for the generation of acoustic radiation—a thermoacoustic mechanism—involves the formation of an acoustic wave as a result of rapid expansion of a fluid in the region of effective energy release of a cascade. The expansion of a liquid due to heating depends on the energy release $Q$ of the cascade: $\Delta V = \alpha V \Delta T = \alpha VQ / VC\rho = \alpha Q / C_p$, where $\alpha$ is the thermal expansion coefficient, $C_p$ is the heat capacity of the material, $\rho$ is the density of the material, and $V$ is the volume of the region in which the cascade releases its energy.

The equation which describes the propagation of sound in water by means of the thermoacoustic-generation mechanism is[140]

$$\Delta P - \frac{1}{c_s^2} \frac{\partial^2 P}{\partial t^2} = -\frac{\alpha}{C_p} \frac{\partial^2 q(\mathbf{r}, t)}{\partial t^2}.$$

Here $P(\mathbf{r}, t)$ is the acoustic pressure, $q(\mathbf{r}, t)$ is the energy-release density, and $c_s$ is the speed of sound. The solution of this equation is given by the Kirchhoff integral

$$P(\mathbf{r}, t) = \frac{\alpha}{4\pi C_p} \int \frac{dV'}{|\mathbf{r} - \mathbf{r}'|} \frac{\partial^2}{\partial t^2} q\left(\mathbf{r}', t - \frac{|\mathbf{r} - \mathbf{r}'|}{c_s}\right), \tag{8.7}$$

which is evaluated for the volume of the region of effective energy release of the cascade.

An electromagnetic cascade develops in a time $t_h \sim 10^{-8}$ s, but the characteristic length of an acoustic signal is $t_s \sim 10^{-5}$ s. It can be assumed, therefore, that a cascade releases its energy instantaneously:

$$q(\mathbf{r}, t) = q(\mathbf{r})\delta(t). \tag{8.8}$$

Using relation (8.8), we can simplify integral (8.7) substantially

$$P(\mathbf{r}, t) = \frac{\alpha}{4\pi C_p} c_s^2 \frac{\partial}{\partial R} \int_{S^R} d\sigma \frac{q(\mathbf{r})}{R}. \tag{8.9}$$

The integration here is over a sphere of radius $R = c_s t$, where the center is at the detecting point $\mathbf{r}$. Expression (8.9), which is a Poisson equation for the solution of a homogeneous wave equation with the initial conditions $P(\mathbf{r}, 0) = (\alpha / C_p) q(\mathbf{r})$ and $\partial / \partial t\, P(\mathbf{r}, 0) = 0$, does not explicitly contain the dependence of sound on the frequency, making it impossible, for example, to take into account the absorption of sound in a medium or the frequency characteristics of an acoustic-wave detector. In calculating acoustic radiation at large distances from the source, it is necessary to take into account that absorption of sound in a medium which depends essentially on its frequency. Applying a Fourier transform to the left side and right side of relation (8.7) and introducing the function $\chi$ which takes into account the sound absorption, we obtain the following relation for the frequency component of the acoustic signal[140]:

$$P_\omega = -i\omega \frac{\alpha}{4\pi C_p} \int dV' \frac{\chi(\omega)}{|\mathbf{r}-\mathbf{r}'|} \exp\left(i\frac{\omega}{c_s}|\mathbf{r}-\mathbf{r}'|\right) q(\mathbf{r}'). \tag{8.10}$$

Since $\exp[i(\omega/c_s)|\mathbf{r}-\mathbf{r}'|]$ is a strongly oscillating function, a numerical integration of expression (8.10) over the region in which the cascade releases its energy involves serious technical difficulties. The multiplicity of the integral in Eq. (8.10) can be reduced by making use of the axial symmetry of the region in which the cascade energy is released. For practical calculations, the spatial distribution of the cascade energy release can be described approximately by the expression

$$q(z,\rho) = \frac{1}{2\pi} \sum_{i=1}^{N} A_i(z) \exp[-\rho/\lambda(z)]. \tag{8.11}$$

If the distance $R$ between the energy-release region and the point at which the signal is detected is large ($R \gg a_{\text{eff}}$, where $a_{\text{eff}}$ is the effective transverse dimension of the energy-release region), we obtain[140]

$$P_\omega = -i\omega \frac{\alpha}{4\pi C_p} \int \frac{\chi(\omega)}{R\sqrt{D}} \exp\left[i\frac{\omega}{c_s} R\sqrt{D}\left(1 - \frac{\rho}{RD}\cos\varphi\right)\right]$$
$$\times q(z,\rho)\,\rho\, d\rho\, d\varphi\, dz, \tag{8.12}$$

where $D = 1 + (Z-z)^2/R^2$, $Z$ and $R$ are the coordinates of the detecting point, and $z, \rho$, and $\varphi$ are the coordinates of the point at which the cascade energy is released, $dQ = q(z,\rho)\rho d\rho d\varphi dz$. Using relation (8.11) and integrating expression (8.12) over $\rho$ and $\varphi$, we obtain the following expression for $P_\omega$:

$$P_\omega = -i\omega \frac{\alpha}{C_p} \frac{1}{4\pi R} \int_0^\infty dz \frac{1}{\sqrt{D}} \sum_{i=1}^{N} \frac{A_i(z)\lambda_i^2(z)}{(1+\beta_i^2)^{3/2}}$$
$$\times \chi(\omega) \exp\left(i\frac{\omega}{c_s} R\sqrt{D}\right), \tag{8.13}$$

where $\beta_i^2 = (\omega/c_s)^2 \lambda_i^2(z)/D$.

Applying the inverse Fourier transform to expression (8.13), we find

$$P(\mathbf{r},t) = \frac{\alpha}{C_p} \frac{1}{4\pi^2 R} \int_0^\infty \frac{dz}{\sqrt{D}} \int_0^\infty \omega\, d\omega\, \chi(\omega)$$
$$\times \sum_{i=1}^{N} \frac{A_i(z)\lambda_i^2(z)}{(1+\beta_i^2)^{3/2}} \sin\{\omega(t_0 - t)\}, \tag{8.14}$$

where $t_0 = R\sqrt{D}/c_s$.

Expressions (8.13) and (8.14), which describe the acoustic-radiation field of an electromagnetic cascade, are quite lengthy and can therefore be used only for numerical calculations. The principal features of the acoustic-radiation field of a cascade can, however, be traced if the region of the cascade energy release is roughly represented as a narrow cylinder of length $L$ and radius $a$, where $L$ and $a$ are the effective longitudinal and transverse dimensions of the energy-release region of an electromagnetic cascade. In this approximation the frequency component of the acoustic-pressure signal in the near wave zone ($L \ll R \ll L^2/\lambda = R^*$) and for optimum emission angle ($L \cos\theta/\lambda \ll 1$) is given by the expression[191]

$$|P_\omega| = \frac{\omega}{4\pi^{3/2}} \frac{\alpha}{C_p} \frac{E}{\sqrt{RR^*}},$$

where $E$ is the cascade energy released in the volume $V = \pi a^2 L$, and $2\pi\lambda$ is the wavelength of an acoustic wave. In the far wave zone ($R \gg R^*$) we have $|P_\omega| \simeq (\omega/4\pi^2)(\alpha/C_p)(E/R)$.

The principal features of the acoustic radiation of an electromagnetic cascade can be formulated as follows (in the approximation of the cylindrical energy-release region):

(1) The radiation is coherent ($P_{\rm eff} \sim \omega|P_\omega| \sim E$) in the frequency region $f \leqslant f_c = c_s/(2\pi\lambda)$, where $2\pi\lambda \sim 2a$, and $a$ is the radius of the region in which the cascade energy is released. For $a = 2$–$3$ cm we have $f_c \simeq 25$ kHz.

(2) The radiation is quasi-cylindrical ($P_{\rm eff} \sim 1/\sqrt{R}$) in the near wave zone. This type of radiation occurs in the region of a thin disk of thickness $L$ and radius $R$ on the order of the absorption length of sound in water ($\sim 1$ km at a frequency of $\sim 25$ kHz in the sea water); the plane of the disk is perpendicular to the cascade axis.

(3) The length of the acoustic signal is $t_s \sim 2a/c_s \sim 10^{-5}$ s.

Let us estimate the acoustic pressure of the signal, $P_{\rm eff} \simeq \omega|P_\omega|$, when a cascade with an energy $E$ develops in sea water ($\alpha \simeq 2\times10^{-4}$ deg$^{-1}$, $C_p = 3.6\times10^7$ ergs/(g·deg), $c_s = 1.5\times10^5$ cm/c). Assuming $a = 3$ cm, $L = 7$ m, and $\lambda = 2a/(2\pi)$, we find $P_{\rm eff} \simeq 0.01\,(E/E_0)(1/\sqrt{R})$, where $E_0 = 10^{16}$ eV, $P_{\rm eff}$ is the pressure (Pa), and $R$ is the distance from the cascade axis to the detecting point (cm). At a distance $R = 100$ m from the axis of a shower initiated by a particle with an energy $E_0 = 10^{16}$ eV, the pressure of an acoustic signal is

$$P_{\rm eff} \simeq 10^{-4}\ \text{Pa}; \tag{8.15}$$

i.e., it is comparable with the intrinsic thermal noise of an acoustic radiation detector—a hydrophone ($P_{\rm th} \sim 10^{-4}$ Pa at a frequency of 25 kHz). Consequently, acoustic radiation can be used to detect extremely-high-energy particles ($\gtrsim 10^{16}$ eV) in water.

In the DUMAND project it was suggested that a layer of water at a depth of about 5 km in the ocean, which is "viewed" by many photomultipliers (about $10^5$ pieces), be used as a detector. Photomultipliers detect Cerenkov light flashes from showers which develop as a result of the interactions $\nu(\tilde{\nu}) + N \to \mu(\nu) +$ hadrons in the sensitive region of the detector. The sensitive volume of the installation, which is determined by the number of photomultipliers it has and by the attenuation length of light in sea water (15–20 m), is $10^9$ m$^3$ when it has approximately $10^5$ photomultipliers. The sensitive volume of the detector is limited by the difficulties of providing a reliable operation of the photomultipliers at depths of about 5 km (the pressure at such a depth is $5\times10^7$ Pa). The cost of the large, required number of photomultipliers is high.

**Figure 8.8. Acoustic pressure vs the energy release of a 200-MeV proton beam in a layer of water. Points, results of the experiment of Ref. 195; curves, calculation based on Eq. (8.9); the number of protons is ~$10^{12}$.**

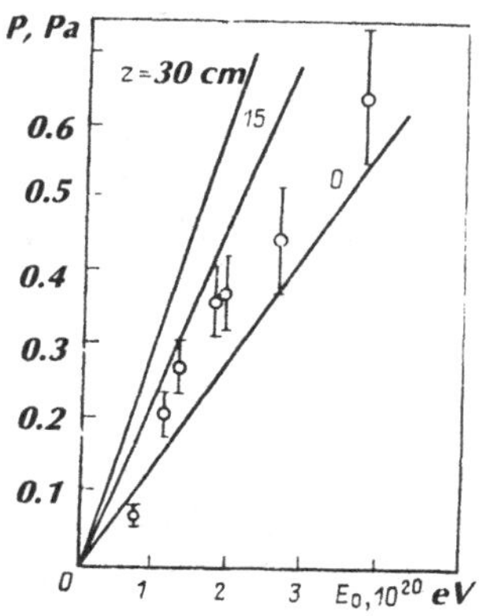


The use of acoustic radiation from high-energy cascades, on the other hand, to detect these cascades makes it possible to increase the sensitive volume of the detector by two orders of magnitude, reaching $10^{11}$ $m^3$, while keeping the number of detectors the same (in this case hydrophones, which cost much less than photomultipliers), since the attenuation length of sound in water (which is about 1 km) is much greater than the absorption length of Cerenkov radiation, 15–20 m.

The principal factors governing the formation of an acoustic signal, which are predictable on the basis of a thermoacoustic sound-generation mechanism, have been confirmed experimentally[195] (Fig. 8.8). Figure 8.8 also shows the results of a calculation, based on Eq. (8.9), of an acoustic signal situated a distance $R = 1$ m from the axis of a beam of diameter $d = 4.5$ cm for three positions of the detector along the beam axis (from $z = 0$, the point at which the beam enters the water, to $z = 30$ cm, the end point of the energy-release region). The amplitude of the acoustic pressure is found to depend linearly on the released energy, as was predicted theoretically on the basis of a thermoacoustic sound-generation mechanism. The theoretical calculations are in satisfactory agreement with the experimental results. The difference in the slopes of curves 1–3 stems from a sharp increase in the energy-release density of protons at the end of their paths (the Bragg peak). A good agreement between theory and experimental data is seen also in the plot of the acoustic pressure versus the distance from the source of the sound and versus the diameter of the energy-release region.[140,195]

The strongest argument in favor of the thermoacoustic mechanism of sound generation is the fact that the amplitude of the acoustic pressure depends on the water temperature. According to the thermoacoustic mechanism, the amplitude of the signal is proportional to the thermal expansion coefficient $\alpha$ [See Eq. (8.7)]. For water the thermal expansion coefficient $\alpha$ decreases linearly with decreasing temperature, vanishing at $T \simeq 4$ °C. A further decrease in the temperature causes the thermal expansion coefficient to become negative. If sound is generated as a result of a thermoacoustic mechanism, then the plot of the amplitude of the acoustic signal as a function

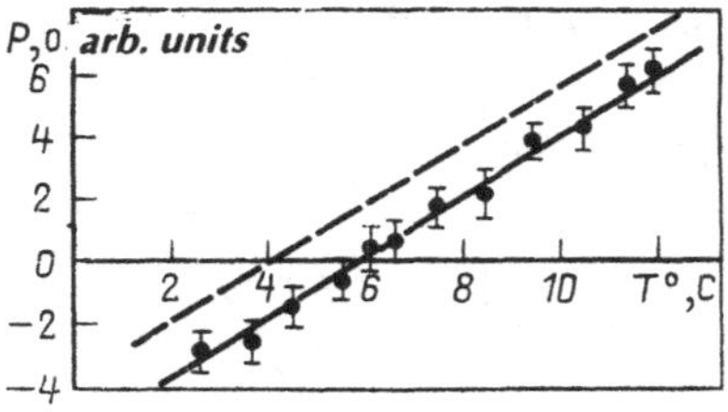


**Figure 8.9. Amplitude of the acoustic signal vs the water temperature. Points, results of the experiments of Ref. 195; dashed line, theoretical dependence without allowance for the electrostriction; solid line, with allowance for the electrostriction on the charged-particle tracks.**

of the water temperature should be similar to the behavior of the thermal expansion coefficient. In the experiment carried out by Sulak *et al.*[195] the amplitude of the signal was found to depend linearly on the water temperature, as predicted theoretically on the basis of the thermoacoustic generation mechanism (Fig. 8.9). The negative values of the amplitude of the acoustic signal correspond to a rarefaction signal, in contrast with a compression signal, which corresponds to the positive values of the amplitude. It follows from the results in Fig. 8.9 that the acoustic signal vanishes at a water temperature $T \simeq 6$ °C, whereas the thermoacoustic mechanism shows that the acoustic signal vanishes at $T = 4$ °C, with $\alpha = 0$. This difference can be explained in terms of the electrostriction on the charged-particle tracks. The ions produced as a result of passage of a charged particle through matter (water) are, because of their electrostatic field, the center of attraction of the adjacent molecules, and centers of local compression form around them. The electrostrictional compression of a material does not depend on the water temperature. Estimates obtained by Askar'yan *et al.*[140] show that allowance for the electrostriction accounts for the fact that the acoustic signal vanishes at $T \simeq 6$ °C, rather than at $T = 4$ °C, in complete agreement with the results of the experiment by Sulak *et al.*[195]

A good agreement between theory and the results of the experiment justifies the use of the theory of the generation of sound in water, based on the thermoacoustic mechanism, in the calculation of the acoustic radiation field produced as a result of the development in water of electromagnetic cascades initiated by extremely-high-energy particles. The particular shape of the acoustic field which is produced makes it possible to reproduce not only the spatial position of the axis of the electromagnetic cascade but also to determine the direction in which it develops, i.e., the direction of motion of the particle that initiates the cascade. Figure 8.10 shows the frequency spectra situation at a distance $R = 100$ m from the axis of the cascade with an energy of $10^7$ GeV for the various positions $z$ of the detector along the cascade axis. There is a softening of the frequency spectrum as the detector moves along the cascade axis in the direction of the momentum of the primary particle. This softening is attributed to the broadening of the cascade with increasing depth of its penetration. Figure 8.11 shows the time evolution of the acoustic pressure at various distances from the cascade axis near the maximum of the shower ($z = 8$ m). To allow for the absorption of sound in water, the function $\chi(\omega)$ in Eq. (8.14) was applied in the form[196] $\chi(\omega) = \exp(-0.115\eta S)$;

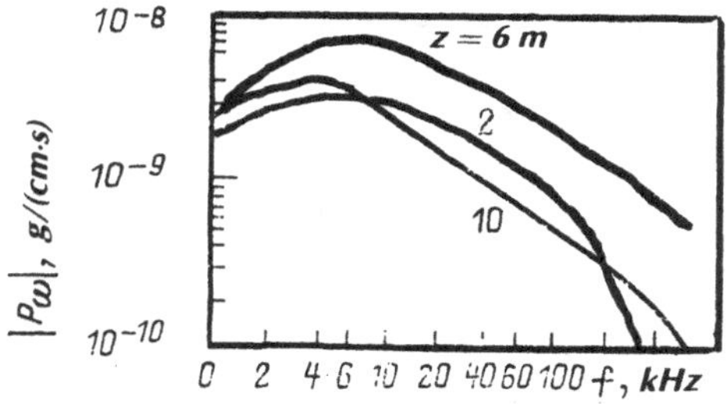


**Figure 8.10.** Frequency spectra of the acoustic signal.

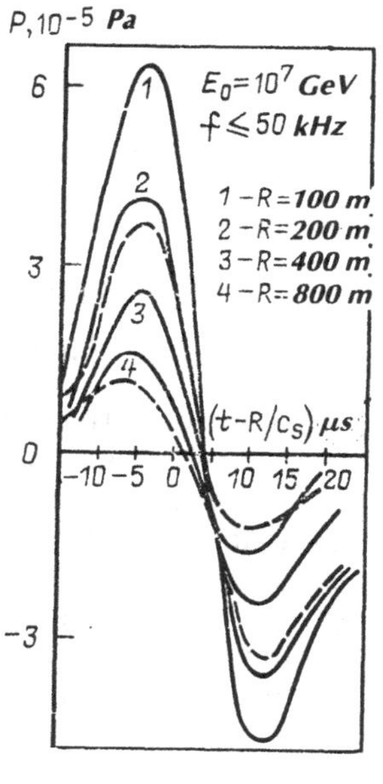


**Figure 8.11.** The pressure signal (time evolution) with allowance for the absorption of sound in sea water (- - - ) and without allowance for it (—).

$\eta = 0.036\,(\omega/2\pi)^{3/2}\ \mathrm{km}^{-1}$, where $S = |\mathbf{r} - \mathbf{r}'|$ is the distance from the detecting point to the volume element $dV'$ in which the energy is released $q(\mathbf{r}')\,dV'$ [see Eq. (8.7)]. The pressure signal is comprised of compression and rarefaction signals, each with an effective length $\tau \sim 10$–$20\ \mu$s, which follow each other. The position of the pressure signals on the time scale $(t - R/c_s)$ and the length of these signals show that the effective source of the sound is the region, approximately 2 cm in radius, of the electromagnetic cascade near its axis. The amplitude of the compression signal, which is $6.2 \times 10^{-5}$ Pa at a distance of 100 m from the cascade, is close to the value $P_{\mathrm{eff}} \simeq 10^{-4}$ Pa found in the approximation of the cylindrical energy-release region [see Eq. (8.15)]. The acoustic pressure increases almost linearly ($\sim E_0^{1.07}$) with an increase in the total energy of the electromagnetic cascade (Fig. 8.12).

In summary, the acoustic radiation of an electromagnetic cascade gives the spatial and energy characteristics of the cascade, actually reconstructing

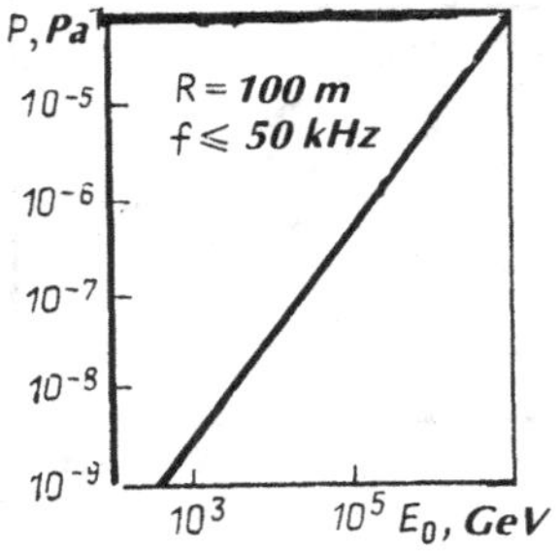


**Figure 8.12. Amplitude of the acoustic signal vs the energy of an electromagnetic cascade $E_0$.**

its acoustic image, and it can be used to detect extremely-high-energy showers ($\gtrsim 10^{16}$ eV) in detectors of enormous mass (on the order of $10^{11}$ metric tons).

## 3. Simulation of physical experiments and radiation problems

Let us consider other applications of the methods, described in the preceding chapters, which are used to calculate the electromagnetic cascades. With increasing particle energy and complexity of the experiments and with increasing capabilities of the computer programs, the number of applications of these calculation methods in high-energy physics and cosmic-ray physics increases dramatically. We shall therefore briefly summarize the key situations in which these methods are used to calculate electromagnetic cascades, principally involving high-energy accelerators.

### Simulation of physical experiments

In Secs. 8.1 and 8.2 this approach was partially illustrated. Calculation of electromagnetic cascades is the foundation upon which the planning stage of the present-day experiments with high-energy-particle beams is based. The main objectives here are the estimate of the expected result, upgrading the individual elements of the experimental installation, and the determination of the background conditions of the detectors. Some examples of the solution of these problems using the Monte Carlo programs described in Chap. 6 are calculation of the background hadron flux in the region of the detecting apparatus and upgrading of a uranium-tungsten collimator in the hyperon experiment[197]; improvement of the thick target and computational reconstruction of the neutrino spectrum in neutrino experiments at the Institute of High Energy Physics, Serpukhov[46,198]; optimization of the yield of the antiprotons from targets in a magnetic field[80] (the essential-sampling methods were used effectively in this study to calculate the improbable reaction channels; see Chap. 6); simulation of the characteristic radiation from hadronic atoms

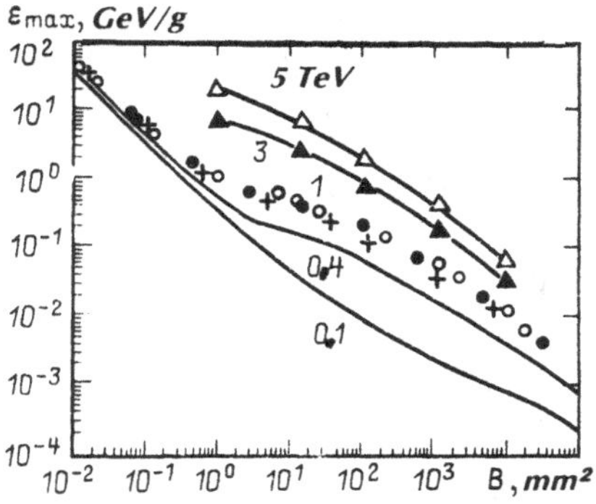


**Figure 8.13.** $\varepsilon_{max}$ vs the surface area of the proton beam. The standard deviation ratios for both directions are given for the energy of 1 TeV. ●, $\sigma_h = \sigma_v$; ○, $\sigma_h = 2\sigma_v$; +, $\sigma_h = 10\sigma_v$.

caused by the stopping of $\pi^-$ and $K^-$ mesons and $\Sigma^-$ hyperons[139]; optimization of the neutrino channel of the 3000-GeV proton accelerator-storage ring (Chap. 7).

**Radiation heating of targets and absorbers**

At high energies and high particle beam intensities the energy release in matter caused by the development of electromagnetic cascades may lead to several macroscopic effects: fusion, evaporation, cracking, and buildup of radiation damage.[135,142,143,199,200]

Instantaneous radiation heating can be determined from the calculated energy-release density $\varepsilon(\mathbf{r})$ and from the enthalpy reserve of the material, $\Delta H(T)$. The quantity $\varepsilon(\mathbf{r})$ depends on many parameters (see Sec. 6.4). The maximum values of the energy release density, $\varepsilon_{max}$, in a graphite target, which determine the maximum temperature, are plotted in Fig. 8.13 as a function of the cross-sectional area of the beam, $B = 4\pi\sigma_v\sigma_h$, where $\sigma_v$ and $\sigma_h$ are the standard deviations in the vertical and horizontal directions of the normal density profile of protons in the beam. The calculations were carried out on the basis of the MARS-8 program[135] for protons with energies $E_0$ in the range 100–5000 GeV. The results for $E_0 = 1000$ GeV, given for beams of three different shapes, show that $\varepsilon_{max}$ does not depend, as was reported in Ref. 142 and discussed in Sec. 6.4, on the beam shape.

The enthalpy (Fig. 8.14) is given by

$$\Delta H(T) = \int_{T_0}^{T} C_p(T')\,dT', \tag{8.16}$$

where $T_0$ and $T$ are the initial and final temperatures of the target, respectively, and $C_p(T)$ is the specific heat.

The spatial distribution of the temperature in the target immediately after a short (in comparison with the characteristic thermal conductivity time) proton pulse can be determined from the equation $\Delta H[T(\mathbf{r})] = 1.6\times10^{-10} I\varepsilon(\mathbf{r})$, where $\Delta H$ is the enthalpy reserve ($J/g$), $\varepsilon$ is the energy release density (GeV/g), and $I$ is the number of protons per pulse.

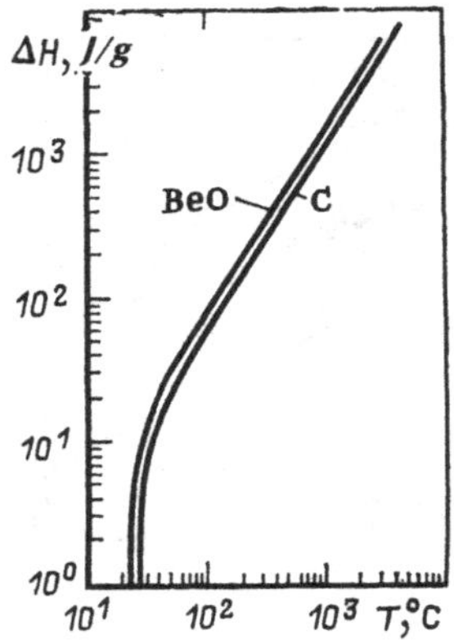


**Figure 8.14. Enthalpy (8.16) for the ceramic compound BeO and graphite plotted as a function of the temperature at the initial temperature $T_0 = 20$ °C (Ref. 135).**

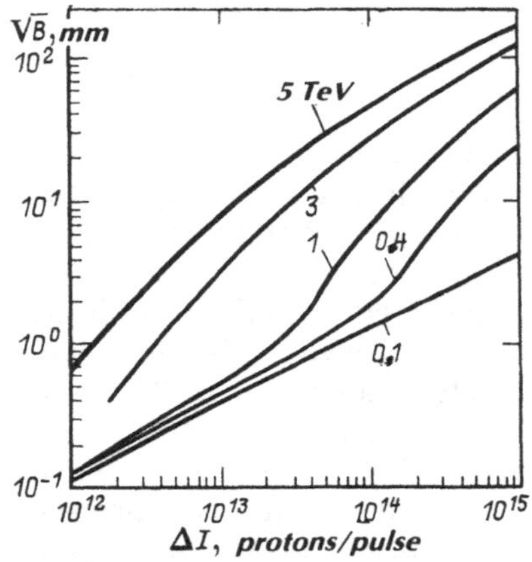


**Figure 8.15. Minimum size of the beam with a graphite absorber ($T_{max} = 2300$ °C) vs the number of protons of various energies in the beam with an instantaneous momentum.**

If the goal is to achieve maximum heating $T_{max}$, then the minimum acceptable beam size at the target ($\sqrt{B}$) can be calculated. Figure 8.15 shows the value of $\sqrt{B}$ calculated in Ref. 135. The functional dependences shown here make it possible to optimize the design of the targets and absorbers and to improve their operating conditions for accelerators that are being designed. In particular, these data were used to build an absorber for a 1000-GeV beam dump.[200]

Worth noting also is the extreme behavior of matter in special (usually mercury) targets for antiproton storage rings[142,199] (Fig. 8.16), when the energy release reaches values at which there is a marked spatial density distribution of the material.[199] Figure 8.16 also shows the effect of low-energy proton transfer initially observed by Mokhov and van Ginneken[142] (see Chap. 6).

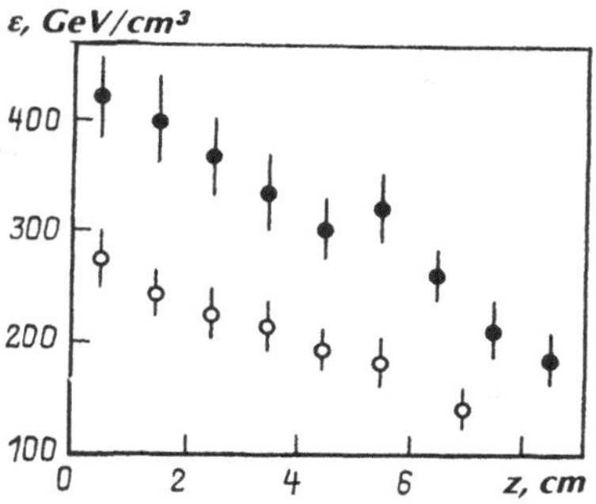


**Figure 8.16. Longitudinal density profile of the energy release in a mercury target bombarded by 80-GeV protons. Beam, Gaussian $\sigma_v = 30$ μm, $\sigma_h = 60$ μm; $0<r<15$ μm; ●, calculation in the approximation of the local absorption of nucleons with energies $E<20$ MeV; ○, calculation with allowance for the transfer of these particles.**

## Radiation heating of superconducting magnets

Deflecting and focusing superconducting magnet systems are the basis of the projects dealing with the latest generation of accelerators. Since protons are unavoidably lost at various stages of the working cycle of accelerators, the superconducting magnets are subjected to an almost constant exposure to ionizing radiation. The energy released in magnet elements due to the development of electromagnetic cascades leads to radiation damage of materials, to transition of superconducting windings to the normal state, and to additional thermal load on the cryogenic system. The interaction of high-energy particles with superconducting magnets has recently attracted considerable research interest.[134,197,201–205] If special measures are not taken to protect superconducting magnets against radiation heating, this problem may prove to be the principal factor that would limit the design intensity of the beams of current accelerators.

The permissible proton loss, in terms of radiation heating, per unit length of the accelerator magnet can be determined in the following way:

$$\Delta I(\mathbf{r}, \xi, \tau, E) = \varepsilon_q (\mathbf{r}, \xi, \tau)/\varepsilon_{\max} (\mathbf{r}, E),$$

where $\mathbf{r}$ is the radius vector of the part of the superconducting windings under study; $\xi = j/j_{\max}$; $j$ is the current density at which the transition of the magnet is studied; $j_{\max}$ is the maximum current density of the magnet at the initial temperature $T_0$, which is equal to the critical current density of a cable in a maximum magnetic field; $\tau$ is the radiation pulse length; $\varepsilon_q$ is the energy-release density in a winding, which corresponds to the transition of a magnet to the normal state; and $\varepsilon_{\max}$ is the maximum density of the energy release in a magnet due to instantaneous loss in it of a single proton with an energy $E$.

The spatial density function of the energy release in superconducting magnets of the 3000-GeV proton accelerator-storage ring (and, correspondingly, the quantity $\varepsilon_{\max}$) were calculated in Refs. 134, 201, and 202 using the MARS package of programs (see Sec. 6.3). Various conditions under which proton beams with energies 200–3000 GeV can be used for bombardment were studied. Typical results of a calculation based on the MARS-9M program (see Sec. 6.3) are shown in Fig. 8.17. The calculation was carried out for a case in which the magnetic field was present and without it. In the studies cited

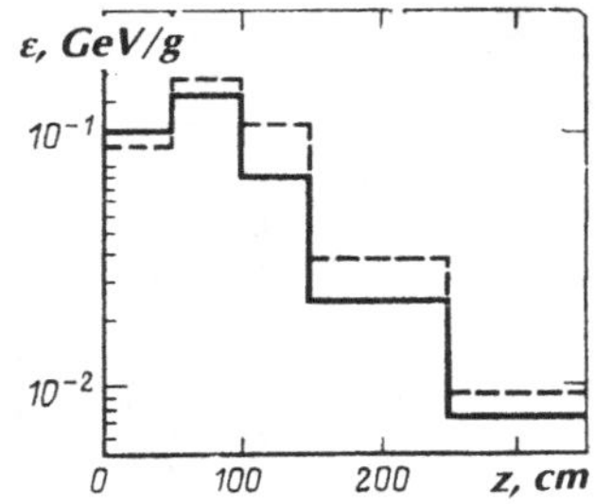


**Figure 8.17. Longitudinal distribution of the energy release in the superconducting coil of a dipole magnet of the proton accelerator-storage ring bombarded by a narrow 3000-GeV proton beam at an angle of 1 mrad with respect to the vacuum chamber (the inside radius of the chamber is 3.5 cm and the inside radius of the coil is 4.5 cm; the results are given for the maximum values of the radial and azimuthal distributions; solid line, $B = 5$ T; dashed line, $B = 0$).**

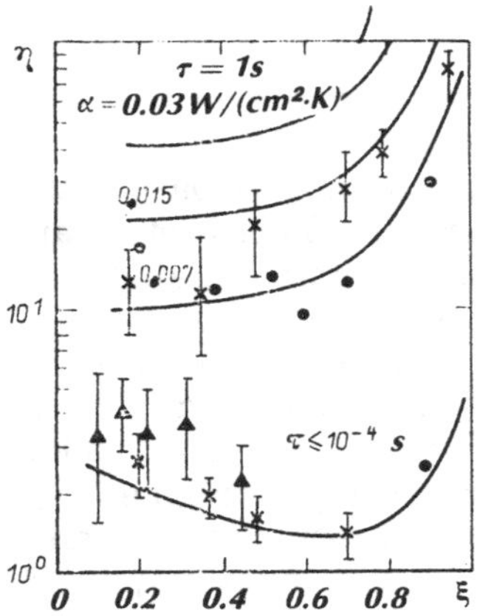


**Figure 8.18. The function $\eta(r, \xi, \tau)$ obtained when the dipoles of the proton accelerator-storage ring are bombarded during the acceleration. $r$ is the median plane, with the magnetic field induction $B = 0.8B_{max}$; thermal conductivity $\lambda_\perp = 0.25$ W/(cm·K); curves, calculation (Ref. 205); points, experiment (Refs. 203 and 205).**

above, $\varepsilon_{max}$ in the winding increased almost linearly with increasing proton energy (at $E > 200$ GeV): $\varepsilon_{max} = CE^n$, where $n \approx 1$, and the parameter $C$ depends on the bombardment conditions.

Before the publication of a paper by Maslov and Mokhov,[205] it was generally assumed that in the case of adiabatic heating $\varepsilon_q$ is equal to the enthalpy (8.16) of a superconducting cable. Maslov and Mokhov[205] showed that in actual superconducting magnet systems the maximum permissible energy-release density can in fact be estimated by means of the expression

$$\varepsilon_q(\mathbf{r}, \xi, \tau) = \Delta H[T(\xi)]\eta(\mathbf{r}, \xi, \tau),$$

where the function $\eta$ is rigorously greater than unity in all cases.

The function $\eta$ calculated in Ref. 205 was found to be dependent on the radial gradient of the energy release in the winding, on the magnetic field induction at the point $\mathbf{r}$, on the effective thermophysical parameters of the winding (primarily on the radial thermal conductivity $\lambda_\perp$), and on the heat-removal coefficient $\alpha$. Figure 8.18 shows the results of the calculation, along with the experimental data, for the two limiting cases of the irradiation time: "instantaneous" irradiation with no time dependence and "stationary" irradiation in which the ratio $\eta/\tau$ does not depend on $\tau$.

Since the constraints imposed on the permissible loss $\Delta I$, which are calculated from the functions $\varepsilon_{max}$ and $\varepsilon_q$, are extremely rigorous, Maslov and Mokhov[134] and Balbekov *et al.*[206] on the basis of an analysis of the characteristics of electromagnetic cascades and special cascade calculations have found that some measures can be taken to substantially reduce $\varepsilon_{max}$ in actual superconducting systems.

**Radiation shielding**

The methods of calculation of the transport of radiation through matter are used most frequently for radiation shielding. Although the number of problems to be solved here is exceptionally large, the task reduces in the final analysis to the determination of the configuration and thickness of the shielding which would hold the radiation level down to the established norms of radiation safety.[207] Since the factors by which the particle fluxes must be reduced frequently are very large ($10^5$–$10^{10}$), and since the shielding configurations are complex, various synthetic methods and modified Monte Carlo methods can be used very effectively in these problems (see Secs. 1.5, 5.5, and 6.1).

The results of calculations of electromagnetic cascades, which can be applied directly to the study of radiation shielding, were presented in Sec. 6.4. We shall consider here several typical problems which can be solved by using methods described in Chaps. 5–7.

Mokhov and Frolov[131] calculated the energy spectra of protons, neutrons, $\pi^{\pm}$ and $K^{\pm}$ mesons, muons, and photons behind thick composite shields bombarded by protons with energies from 1 to 1000 GeV. The calculations were based on the SYNHET program, which synthesized the numerical algorithm for the solution of the kinetic equations of the HAMLET program (see Sec. 5.4), and the Monte Carlo method of the MARS program (see Sec. 6.3). The dose composition of the radiation behind the shielding of the high-energy accelerators can be estimated in difficult situations by using this approach.

Example of an estimate of a radiation dose behind concrete shielding is shown in Fig. 8.19. The energy of protons which interact with a 40-cm-long aluminum target is 70 Gev. The calculation carried out by Baĭshev *et al.*[133] is in good agreement with the results of measurements with an ionization chamber.

A set of experiments and calculations in a similar arrangement was carried out at the FNAL proton synchrotron with energies between 200 and 400 GeV (Ref. 208). The lateral ground shielding that was studied ranged in thickness from 2 to 17 mean free paths of hadrons before the absorption (260–1610 g/cm$^2$). The results of the Monte Carlo calculations of the absorbed dose differed from the experimental data obtained with use of the ionization chamber by no more than a factor of 2 over the entire range.

Kidd *et al.*[200] solved many radiation problems using the results of calculations of electromagnetic cascades upon construction of a beam dump system for the extraction of a 1000-GeV proton beam from an accelerator.

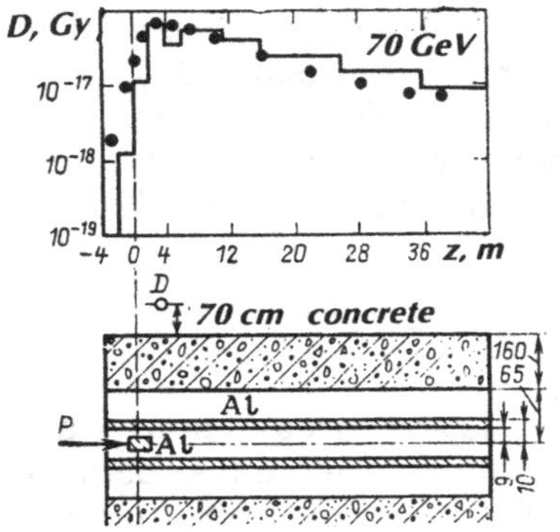


**Figure 8.19. Distribution of the absorbed dose behind the side concrete shield of channel No. 8 of the IHEP proton synchrotron (Ref. 133). Histogram, calculation based on MARS-4 program; points, results of the measurements using the ionization chamber.**

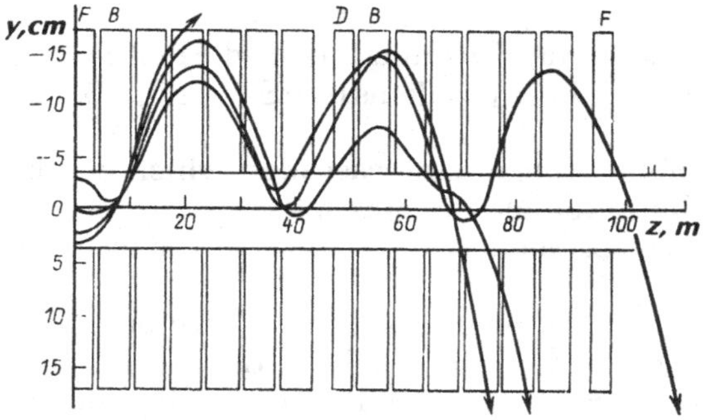


**Figure 8.20. Projections onto the *YZ* plane of the muon trajectories in the magnetic structure of the proton accelerator-storage ring. *F*, *D*, focusing and defocusing quadrupole lenses, respectively; B, dipole magnet with an induction $B = 5$ T; the initial muon energy is 500 GeV.**

In Chap. 7 we described methods of calculation of the production and transport of high-energy muons in matter. In the latest generation of accelerators such as the 3000-GeV proton accelerator-storage ring complex of the Institute of High Energy Physics the muons determine the global shielding completely in many cases. Maslov *et al.*[174] calculated the muon fields around the proton synchrotron ring tunnels. Figure 8.20 shows the calculated muon trajectories. All the processes described in Chap. 7 were taken into account in the calculation of the passage of muons through the magnet elements and the ground shielding. Figure 8.21 shows the calculated muon isoflux in the ground shielding of the accelerator-storage ring complex for a uniform loss of protons at an angle $\theta \sim 1$ mrad to the vacuum chamber in the 200 to 300-meter-long sections of the circular accelerator. The results were obtained for a 3000-GeV proton energy inside and outside of the accelerator ring. Worth noting here is the heavy shielding which in some cases (a neutrino channel, for example) may be even heavier. This is one of the reasons high-precision methods of calculating the passage of muons through matter must be developed.

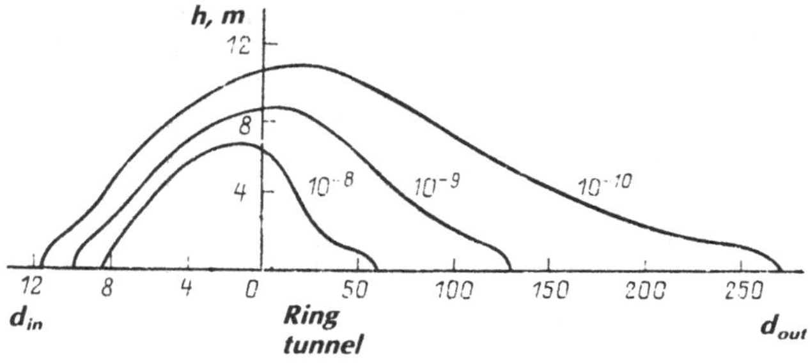


**Figure 8.21.** Muon isoflux (muons/cm² per proton/m).

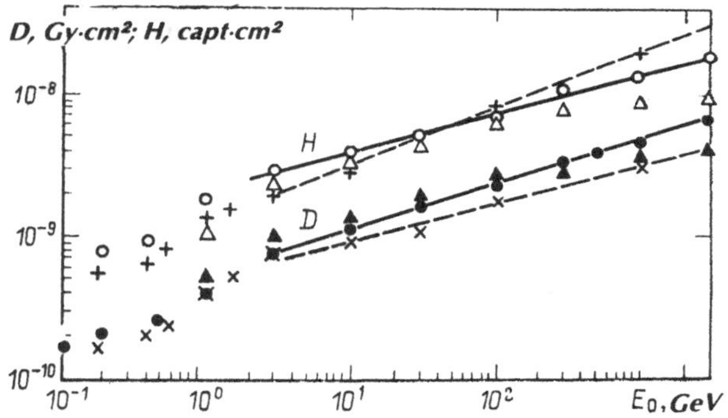


**Figure 8.22.** The maximum absorbed dose (the lower curves) and the equivalent dose (the upper curves) in a flat tissue-equivalent phantom bombarded by a wide neutron beam with an energy $E_0$. ●, ○, calculation based on MARS-4 program; ▲, △, calculation (Ref. 210); ×, +, calculation (Ref. 211).

## Dosimetry and detector response

Methods for the calculation of the passage of high-energy particles through matter are also used extensively in dosimetry. The following are some typical examples:

- Study of dose profiles of biological systems in the radiation fields behind the shielding of high-energy accelerators and space vehicles.[209]
- Comprehensive study of the dose profiles in tissue-equivalent phantoms which are irradiated with beams of high-energy particles.[210–212] The principal purpose of such studies is to establish radiation and cancer therapy norms. Figure 8.22 shows calculated maximum doses in a 30-cm phantom irradiated with neutrons. We see that the danger posed by radiation increases markedly with increasing energy.
- Study of the processes which accompany the stopping of $\pi^-$ mesons in matter. Also studied here is the feasibility of using $\pi^-$ meson beams for radiation therapy and for synthesizing new isotopes; radiation damage of semiconductor detector materials and superconducting magnet materials is analyzed as well. Typical curves for the capture of $\pi^-$ mesons in silicon and in biological tissue are shown in Figs. 8.23 and 8.24 (Ref. 213).

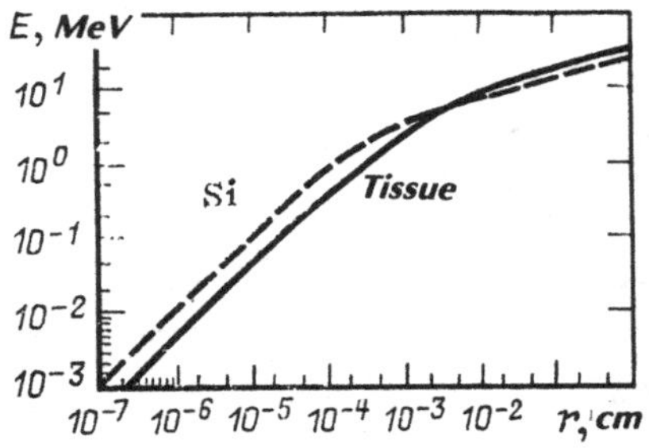


**Figure 8.23. Energy absorbed inside a sphere of radius *r* as a result of stopping of a $\pi^-$ meson in biological tissue and in silicon at the point $r = 0$.**

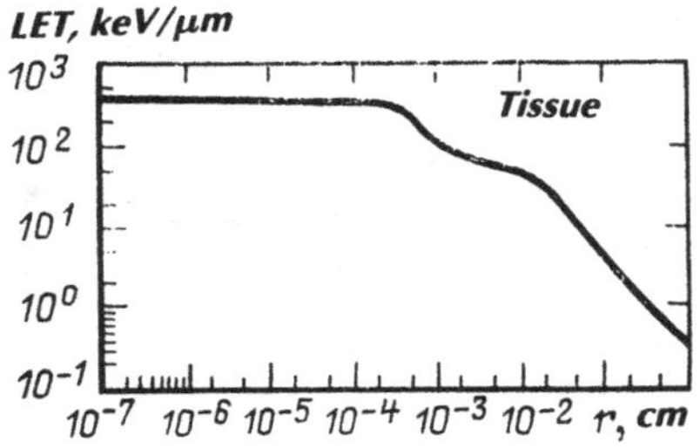


**Figure 8.24. Average linear energy transfer vs the distance to the point at which a $\pi^-$ meson is captured in biological tissue.**

- Study of the response of various detecting systems: study of the response-versus-energy curves for detectors which are bombarded by high-energy particle beams. Electromagnetic cascade calculations have made it possible to determine these functions for LiI and $BF_3$ detectors in polyethylene moderators[161] and for nuclear-emulsion dosimeters[214] over a broad energy range.

Even a short review of the application of the computational methods and programs shows, on the one hand, their capabilities and, on the other, the ever-increasing demands of the high-energy physics community for effective, informative, and reliable methods of calculating the passage of high-energy particles through matter.

# Constants and physical quantities

1. Classical electron radius

   $r_e = e^2/m_e c^2 \approx 2.8 \times 10^{-13}$ cm.

2. Compton wavelength of the electron

   $\bar{\lambda}_e^C = \hbar/m_e c \approx 3.85 \times 10^{-11}$ cm.

3. Compton wavelength of the muon

   $\bar{\lambda}_\mu^C = \hbar/m_\mu c \approx 2 \times 10^{-13}$ cm.

4. Compton wavelength of the nucleon

   $\bar{\lambda}_N^C = \hbar/m_N c \approx 0.2 \times 10^{-13}$ cm.

5. Nuclear radius ($A$ is the number of nucleons in the nucleus)

   $R = R_0 A^{1/3}, \quad R_0 \approx (1.1\text{–}1.2) \times 10^{-13}$ cm.

6. Bohr radius of the hydrogen atom

   $a_0 = 0.53 \times 10^{-8}$ cm.

7. Thomson cross section

   $\tau_T = (8.3)\pi v_e^2 \approx 0.67 \times 10^{-24}$ cm.

8. Electron mass

   $m_e \approx 0.511$ MeV $\approx 9.1 \times 10^{-28}$ g.

9. Muon mass

   $m_\mu \approx 105.6$ MeV.

10. Pion mass

    $m_{\pi^0} \approx 135$ MeV,

    $m_{\pi^+} \approx 139.6$ MeV.

11. Kaon mass

    $m_{K^0} \approx 498$ MeV,

    $m_{K^+} \approx 494$ MeV.

12. Proton mass

    $m_P \approx 938.3$ MeV $\approx 1.66 \times 10^{-24}$ g.

13. Muon lifetime

    $\tau_\mu \approx 2.2 \times 10^{-6}$ s.

14. Pion lifetime

$\tau_{\pi^0} \approx 0.83 \times 10^{-16}$ s,

$\tau_{\pi^\pm} \approx 2.6 \times 10^{-8}$ s.

15. Kaon lifetime

$\tau_K \approx 1.24 \times 10^{-8}$ s,

$\tau_{K^0_s} \approx 0.89 \times 10^{-10}$ s,

$\tau_{K^0_L} \approx 5.2 \times 10^{-8}$ s.

16. Velocity of light in a vacuum

$c \approx 3 \times 10^{10}$ cm/s.

17. Planck's constant

$\hbar \approx 6.6 \times 10^{-22}$ MeV·s $\approx 1.05 \times 10^{-27}$ erg·s.

18. Avogadro's number

$N_A \approx 6.02 \times 10^{23}$ (g·mole)$^{-1}$.

19. Fine-structure constant

$\alpha = e^2/\hbar c \approx 1/137$.

20. Electron charge

$e \approx 4.8 \times 10^{-10}$ esu.

21. 1 TeV = $10^3$ GeV = $10^6$ MeV = $10^9$ keV = $10^{12}$ eV,

1 eV $\approx 1.6 \times 10^{-12}$ erg $\approx$ 11 600 K.

# References

1. É. V. Bugaev, Yu. D. Kotov, and I. L. Rozental', *Cosmic Muons and Neutrinos* (Atomizdat, Moscow, 1970).
2. N. V. Mokhov, S. I. Striganov, and A. V. Uzunyan, "Fluctuation in the Energy Loss of Ultrarelativistic Muons," Institute of High-Energy Physics, Serpukhov, Preprint No. 80–56, 1980.
3. N. V. Mokhov, G. I. Semenova, and A. V. Uzunian, Nucl. Instrum. Methods **180**, 469 (1981); Institute of High-Energy Physics, Serpukhov, Preprint No. 79–101, 1979.
4. Yu. A. Budagov *et al.*, Fiz. Elem. Chastits At. Yadra **11**, 687 (1980) [Sov. J. Part. Nucl. **11**, 273 (1980)].
5. N. S. Graigie, Phys. Rep. **47**, 1 (1978).
6. J. L. Ritchie *et al.*, Phys. Rev. Lett. **44**, 230 (1980).
7. Yu. P. Nikitin and I. L. Rozental', *Theory of Multiple Processes* (Atomizdat, Moscow, 1976).
8. Yu. P. Nikitin and I. L. Rozental', *High-Energy Nuclear Physics* (Atomizdat, Moscow, 1980).
9. R. P. Feynman, *Photon–Hadron Interactions* (Benjamin, New York, 1972).
10. U. Fano, L. Spenser, and M. Berger, *Gamma Radiation Transfer,* Russian translation (Mir, Moscow, 1963).
11. K. Keiz and P. F. Zweifel, *Transport Theory*, NATO Advanced Study Institute (Middle East Tech. Univ., 1965).
12. F. S. Alsmiller, "A general category of soluble nucleon–meson cascade equations," ORNL Report No. ORNL-3746 (1965).
13. A. M. Kol'chuzhkin and V. V. Uchaĭkin, *Introduction to the Theory of the Passage of Particles Through Matter* (Atomizdat, Moscow, 1978).
14. J. Spanier and E. M. Gelbard, *Monte Carlo Principles and Neutron Transport Problems* (Addison-Wesley, Reading, 1969).
15. J. Lewins, *Importance, the Adjoint Function*, 1st ed. (Pergamon, Oxford, New York, 1965).
16. I. M. Sobol', *Numerical Monte Carlo Methods* (Nauka, Moscow, 1973).
17. A. D. Frank–Kamenetskiĭ, *Simulation of the Neutrino Trajectories in the Monte Carlo Reactor Calculations* (Atomizdat, Moscow, 1978).
18. G. I. Marchuk *et al.*, *Monte Carlo Method in Atmospheric Optics* (Nauka, Novosibirsk, 1976).
19. N. V. Mokhov, "Monte Carlo techniques applied to the transport of high-energy particles through matter," Institute of High-Energy Physics, Serpukhov, Preprint No. 76–64, 1976.
20. A. I. Akhiezer and V. B. Berestetskiĭ, *Quantum Electrodynamics* (Nauka, Moscow, 1981).
21. H. Messel and D. F. Crawford, *Electron–Photon Shower Distribution Function* (Pergamon, New York, 1970).
22. H. Davies, H. A. Bethe, and L. C. Maximon, Phys. Rev. **93**, 788 (1954).
23. L. D. Landau and E. M. Lifshitz, *Quantum Mechanics: Non-Relativistic Theory*, 3rd ed. (Pergamon, Oxford, 1977).
24. V. B. Berestetskiĭ, E. M. Lifshitz, and L. P. Pitaevskiĭ, *Relativistic Quantum Theory, Part 1* (Pergamon, Oxford, 1971).
25. J. A. Wheeler and W. A. Lamb, Phys. Rev. **55**, 858 (1939).

26. E. M. Lifshitz and L. P. Pitaevskiĭ, *Relativistic Quantum Theory, Part 2* (Pergamon, Oxofrd, 1974).
27. A. B. Migdal, Phys. Rev. **103**, 1811 (1956).
28. H. W. Koch and J. W. Motz, Rev. Mod. Phys. **31**, 920 (1959).
29. E. Storm and H. I. Israel, Nucl. Data Tables **A7**, 565 (1970).
30. *Muon Physics. Electromagnetic Interactions, Vol. 1*, edited by V. W. Hughes and C. S. Wu (Academic, New York, 1977).
31. D. M. Ritson, "Instrumentation. Proceedings of Summer Institute on Particle Physics," SLAC Report No. 239 (1981).
32. Review of Particle Properties/Data particle group, Rev. Mod. Phys. **52**, No. 2 (1980).
33. E. Fermi, Phys. Rev. **57**, 475 (1940).
34. S. Hayakawa, *Cosmic Ray Physics: Nuclear and Astrophysical Aspects* (Wiley-Interscience, New York, 1969).
35. R. M. Sternheimer, Phys. Rev. **103**, 511 (1956).
36. L. D. Landau, J. Phys. USSR **8**, 201 (1944); *A collection of papers, Vol. 1* (Nauka, Moscow, 1969), p. 482.
37. P. V. Vavilov, Zh. Eksp. Teor. Fiz. **32**, 920 (1957) [Sov. Phys. JETP **5**, 749 (1957)].
38. V. K. Ermilova and V. A. Chechin, P. N. Lebedev Physics Institute, Preprint No. 10, 1976.
39. Z. Moliere, Z. Naturforsch. **2a**, 133 (1947).
40. H. A. Bethe, Phys. Rev. **89**, 1256 (1953).
41. W. T. Scott, Rev. Mod. Phys. **35**, 231 (1963).
42. R. L. Ford and W. R.Nelson, "The EGS code system: Computer program for the Monte Carlo simulation of electromagnetic cascade showers (Version 3)," SLAC Report No. 210 (1978).
43. W. R. Nelson, Nucl. Instrum. Methods **66**, 293 (1968).
44. L. B. Okun', *Leptons and Quarks* (Nauka, Moscow, 1981).
45. G. I. Kopylov, *Principles of the Resonance Kinematics* (Nauka, Moscow, 1970).
46. D. S. Baranov *et al.*, "Computational method of reconstructing the neutrino spectrum in neutrino experiments at proton accelerators," Institute of High-Energy Physics, Serpukhov, Preprint 77-22 (1977).
47. S. Hayakawa, Phys. Rev. **108**, 1533 (1957).
48. V. S. Barashenkov and V. D. Toneev, *Interaction of High-Energy Particles and Atomic Nuclei with Nuclei* (Atomizdat, Moscow, 1972).
49. Yu. P. Nikitin, I. L. Rozental', and F. M. Sergeev, Usp. Fiz. Nauk **121**, 3 (1977) [Sov. Phys. Usp. **20**, 1 (1977)].
50. I. V. Andreev and I. M. Dremin, Usp. Fiz. Nauk **122**, 37 (1977) [Sov. Phys. Usp. **20**, 381 (1977)].
51. N. N. Nikolaev, Usp. Fiz. Nauk **134**, 369 (1981) [Sov. Fiz. Usp. **24**, 531 (1981)].
52. A. K. Likhoded and P. V. Shlyapnikov, in Proceedings of the International Conference on Multiple Production Processes and Inclusive Reactions at High Energies (Serpukhov, 1977), pp. 5–79.
53. L. D. Landau, Izv. Akad. Nauk SSSR, Ser. Fiz. **17**, 51 (1953).
54. P. D. B. Collins, *An Introduction to Regge Theory and High-Energy Physics* (Cambridge Univ. Press, Cambridge, 1977).
55. K. A. Ter-Martirosyan and Yu. M. Shabel'skiĭ, Yad. Fiz. **25**, 670 (1977) [Sov. J. Nucl. Phys. **25**, 356 (1977)].

56. V. A. Abramovskiĭ, V. N. Gribov, and O. V.Kancheli, Yad. Fiz. **18**, 595 (1973) [Sov. J. Nucl. Phys. **18**, 308 (1974)].
57. K. A. Ter-Martirosyan, Phys. Lett. **44B**, 179 (1973).
58. A. B. Kaĭdalov, in *Elementary Particles. Second School of Physics, Institute of High-Energy Physics* (Atomizdat, Moscow, 1975), issue 3, p. 5.
59. Yu. M. Shabel'skiĭ, Yad. Fiz. **27**, 1084 (1977) [Sov. J. Nucl. Phys. **27**, 574 (1978)].
60. V. N. Gribov, in *Elementary Particles. First School of Physics, Institute of High-Energy Physics* (Atomizdat, Moscow, 1973), issue 1, p. 65.
61. A. A. Ansel'm, in *Elementary Particles. First School of Physics, Institute of High-Energy Physics* (Atomizdat, Moscow, 1973), issue 2, p. 3.
62. V. V. Anisovich, in *Proceedings of the Fourteenth Winter School on Nuclear Physics and Elementary Particles Physics,* (Leningrad Institute of Nuclear Physics, Leningrad, 1979), Vol. 1, p. 3.
63. S. A. Voloshin, Yu. P. Nikitin, and P. I. Porfirov, Yad. Fiz. **31**, 762 (1980) [Sov. J. Nucl. Phys. **31**, 395 (1980)].
64. S. A. Voloshin, Yu. P. Nikitin, and V. M. Emel'yanov, Yad. Fiz. **26**, 1104 (1977) [Sov. J. Nucl. Phys. **36**, 584 (1977)].
65. S. A. Voloshin and Yu. P. Nikitin, Yad. Fiz. **27**, 223 (1978) [Sov. J. Nucl. Phys. **27**, 119 (1978)].
66. K. P. Dar and R. C. Hwa, Phys. Lett. **68B**, 459 (1977).
67. J. J. Sakurai, *Currents and Mesons* (Chicago Univer. Press, Chicago, 1969).
68. V. I. Zakharov and N. N. Nikolaev, Yad. Fiz. **21**, 434 (1975) [Sov. J. Nucl. Phys. **21**, 227 (1975)].
69. Y. L. Dokshitzer, D. I. Dyakonov, and S. I. Troyan, Phys. Rep. **58C**, 269 (1980).
70. W. Marchiano and H. Pagels, Phys. Rep. **36C**, 137 (1978).
71. V. Yu. Glebov, in *High-Energy Physics. Proceedings of the Sixteenth Winter School* (Leningrad Institute of Nuclear Physics, Leningrad, 1981), p. 54.
72. Yu. Shabelsky, "On the multiplicity of the secondaries produced in collisions of relativistic nuclei," LNPI Preprint No. 464, 1979.
73. G. W. Brandenburg, *Quarks, Gluons, and Jets, Proceedings of the XIV Rencontre de Moriond* (Les Arcs-Savoie, France, 1979), Vol. 1, p. 507.
74. Yu. M. Shabel'skiĭ, Fiz. Elem. Chastits At. Yadra **12**, 1070 (1981) [Sov. J. Part. Nucl. **12**, 430 (1981)].
75. N. V. Mokhov and Yu. P.Nikitin, in *Nuclear Physics and Cosmic Rays*, 6th ed. (State University, Khar'kov, 1977), p. 19.
76. J. Ranft and J. T. Routti, Part. Accel. **4**, 101 (1972).
77. V. N. Folomeshkin, "Empirical formula for pion and kaon spectra in *pp* collisions," Institute of High-Energy Physics, Serpukhov, Preprint 71-22, 1971.
78. L. P. Kimel' and N. V. Mokhov, Izv. vuzov SSSR, Ser. Fiz., No. 10, p. 18 (1974).
79. L. P. Kimel' and N. V. Mokhov, in *Dosimetry and Radiation Shielding,* edited by L. R. Kimel' and V. K. Sakharov (Atomizdat, Moscow, 1975), issue 14, p. 37.
80. B. V. Chirikov, V. A. Tayurski, H. J. Mörhing, and J. Ranft, Nucl. Instrum. Methods **144**, 129 (1977).
81. B. S. Sychev, A. Ya. Serov, and B. V. Man'ko, MRTI, Preprint No. 799, 1979.
82. V. S. Barashenkov and N. V. Slavin, Acta Phys. Polo. **B12**, 959 (1981).
83. M. V. Kazarnovskiĭ, G. K. Matushko, V. L. Matushko, É. Ya. Par'ev, and S. V. Serezhnikov, At. Energ. **50**, 190 (1981).
84. A. E. Brenner *et al.*, Fermilab Conf. 80/47-Exp. 1980.

85. J. Ranft. Part. Accel. **3**, 129 (1972).
86. U. Amaldi, Phys. Scr. **23**, 409 (1981).
87. O. F. Nemets and Yu. V. Gofman, *Handbook on Nuclear Physics* (Naukova dumka, Kiev, 1975).
88. S. Z. Belen'kiĭ, *Avalanche Processes in Cosmic Rays* (Gosatomizdat, Moscow, 1948).
89. S. Z. Belen'kiĭ and I. P. Ivanenko, Ukr. J. Fiz. **69**, 491 (1959).
90. B. B. Rossi, *High-Energy Particles* (Prentice-Hall, New York, 1952).
91. L. D. Landau and Yu. B. Rumer, *Cascade Theory of Electromagnetic Showers. A collection of papers*, (Nauka, Moscow, 1969), Vol. 1, p. 302.
92. L. D. Landau, *Angular Distribution of Particles in Showers. A collection of papers* (Nauka, Moscow, 1969), Vol. 1, p. 328.
93. B. E. Shtern, "Library of subprograms for simulating electromagnetic processes in matter at high energies," Institute for Nuclear Research, Preprint No. P-0081, 1978; "SIMEX-1. Program for generating electromagnetic cascades in matter," Institute or Nuclear Research, Preprint No. P-0082, 1978.
94. I. S. Baĭshev and N. V. Mokhov, "Spatial distributions of the energy release from an electron–photon shower," Institute of High-Energy Physics, Serpukhov, Preprint No. 79-124, 1979.
95. V. V. Akimov *et al.*, "A method of calculating the physical characteristics of 'Gamma-1' gamma telescope," IKI, Preprint No. P-684, 1981.
96. D. Müller, Phys. Rev. D **5**, 2677 (1972).
97. S. Iwata, "Calorimeter," Preprint DPNU, No. 13, 1980.
98. H. A. Gordon, "Sampling calorimeters in high-energy physics, Proceedings of Summer Institute on Particle Physics, 1980," SLAC Report No. 239, p. 241.
99. I. L. Rozental', Dokl. Akad. Nauk **80**, 731 (1951).
100. V. S. Murzin and L. I. Sarycheva, *Cosmic Rays and Their Interaction* (Atomizdat, Moscow, 1968).
101. A. Liland, Fortschr. Phys. **23**, 571 (1975).
102. A. I. Dem'yamov, V. S. Murzin, and L. I. Sarycheva, *Nuclear Cascade Process in Solids* (Nauka, Moscow, 1977).
103. F. S. Alsmiller, "A general category of soluble nucleon–meson cascade equations," ORNL Report No. ORNL-3746, 1965.
104. K. O'Brien, Nucl. Instrum. Methods **72**, 93 (1969).
105. R. G., Alsmiller and F. S. Alsmiller, "A perturbation method for solving the angle-dependent nucleon–meson cascade equations," ORNL Report No. ORNL-3467, 1963.
106. V. S. Endovitskiĭ, L. P. Kimel', and N. V. Mokhov, in *Dosimetry and Radiation Shielding,* edited by L. P. Kimel' (Atomizdat, Moscow, 1971), issue 12, p. 15.
107. V. S. Endovitskiĭ, L. P. Kimel', N. V. Mokhov, and A. Kh. Rakhmatulina, in Ref. 106, p. 24.
108. V. S. Endovitskiĭ, L. P. Kimel', N. V. Mokhov, V. N. Britvich, and V. N. Lebedev, Institute of High-Energy Physics, Serpukhov, Preprint No. 71-96, 1971.
109. N. Brikov *et al.*, in *Dosimetry and Radiation Shielding,* edited by L. P. Kimel' (Atomizdat, Moscow, 1971), issue 12, p. 3.
110. E. K. Gel'fand, A. Ya. Serov, and B. S. Sychev, "Use of the method of successive collisions to calculate accelerator radiation shielding" in Proceedings of RTI, Moscow, 1974, No. 20, p. 136.
111. T. A. Germogenova *et al.*, *Fast Neutron Transport in Plane Shielding* (Atomizdat, Moscow, 1971).

112. G. I. Britvich, L. P. Kimel', V. N. Lebedev, and N. V. Mokhov, "Theoretical and experimental studies of the passage of high-energy hadrons through shielding materials," Institute of High-Energy Physics, Serpukhov, Preprint No. 74-86, 1974.

113. L. P. Kimel' and N. V. Mokhov, in *Dosimetry and Radiation Shielding*, edited by L. P. Kimel' and V. K. Sakharov (Atomizdat, Moscow, 1975), issue 14, p. 37.

114. K. O'Brien and J. E. McLaughlin, Nucl. Instrum. Methods **60**,. 129 (1968).

115. L. N. Zaĭtsev, M. M. Komochkov, and B. S. Sychev, *Principles of Accelerator Shielding* (Atomizdat, Moscow, 1973).

116. R. G. Alsmiller, F. R. Mynatt, J. Barish, and W. Engle, "Shielding against neutrons in the energy range 50 to 400 MeV," ORNL Report No. ORNL-TM-2554, 1969.

117. E. A. Belogorlov *et al.*, "Optimization of the concrete thickness in heterogeneous side shielding of high-energy accelerators," in Proceedings of the Seventh All-Union Conference on Charged Particle Accelerators, Dubna, 1981.

118. B. V. Man'ko and B. S. Sychev, in Proceedings of RTI, Moscow, RIAN SSSR, 1974, No. 20, p. 147.

119. N. V. Mokhov, "Investigation of internuclear cascade initiated by hadrons with energies 1–5000 GeV in matter." Author's abstract of candidate's dissertation in physicomathematical sciences, Moscow Engineering-Physics Institute, Moscow, 1975.

120. B. R. Bergel'son, A. P. Suvorov, and B. Z. Torlin, *Multigroup Methods of Calculating Neutron Shielding* (Atomizdat, Moscow, 1970).

121. D. L. Broder, S. A. Kozlovskiĭ, V. S. Kyz'yurov, K. K. Popkov, and S. M. Rubanov (Atomizdat, Moscow, 1969).

122. N. V. Mokhov and V. N. Grebenev, in *Dosimetry and Radiation Shielding*, edited by V. K. Sakharov (Atomizdat., Moscow, 1977), issue 16, p. 7.

123. I. Kataoka and K. Takeuchi, "Accord and some results of a numerical integration of the photon transport equation in slab geometry." Papers of Ship Res. Inst., 1965, No. 6.

124. F. James, "Monte Carlo method for particle physicist," (Strasburg, 1968).

125. V. V. Uchaĭkin, Zhurn. Vychisl. Mat. Mat. Fiz. **16**, 758 (1976).

126. T. W. Armstrong and K. C. Chandler, "HETC—a high-energy transport code," Nucl. Sci. Eng. **49**, 110 (1972).

127. W. A. Coleman and T. W. Armstrong, NMTC—a nucleon–meson transport code, Nucl. Sci. Eng. **43**, 353 (1971).

128. V. S. Barashenkov, N. M. Sobolevskiĭ, and V. D. Toneev, At. Energ. **32**, 217 (1972).

129. J. Ranft and J. T. Routti, Comput. Phys. Commun. **7**, 327 (1974).

130. N. V. Mokhov, in *Proceedings of the Fourth All-Union Conference on Charged Particle Accelerators, Moscow, 1974* (Nauka, Moscow, 1975), Vol. 2, p. 222.

131. N. V. Mokhov and V. V. Frolov, At. Energ. **38**, 42 (1975).

132. S. L. Kuchinin, N. V. Mokhov, and Ya. N. Rastsvetalov, Institute of High-Energy Physics, Serpukhov, Preprint No. 75-74, 1975.

133. I. S. Baĭshev, S. L. Kuchinin, and N. V. Mokhov, Institute of High-Energy Physics, Serpukhov, Preprint No. 78-2, 1977.

134. M. A. Maslov and N. V. Mokhov, Part. Accel. **11**, 91 (1980); Institute of High-Energy Physics, Serpukhov, Preprint No. 79-135, 1979.

135. N. V. Mokhov, "Energy deposition in targets and beam dumps at 0.1-5 TeV proton energy," Fermilab report No. FN-328, 1980.

136. A. Van Ginneken, "CASIM—program to simulate transport of hadronic cascades in bulk matter, " Fermilab Report No. FN-272, 1975.

137. J. Ranft, CERN Report No. LAB II-RA/75-1, 1975.

138. H. W. Bertini, Phys. Rev. **188**, 1711 (1969).

139. D. S. Denisov *et al.*, Leningrad Institute of Nuclear Physics, Leningrad, Preprint No. 459, 1979.

140. O. A. Askar'yan, B. A. Dolgoshein, A. N. Kalinovskiĭ, and N. V. Mokhov, Nucl. Instrum. Methods **164**, 267 (1979).

141. H. Edwards, S. Mori, and A. Van Ginneken, "Studies on radiation shielding of energy doubler magnets," Fermilab Report No. UPC-309, 1979.

142. N. V. Mokhov and A. Van Ginneken, in *Proceedings of High-Intensity Targeting Workshop* (Fermilab, 1980), p. 64; Fermilab Report No. TM-977, 1980.

143. N. V. Mokhov, Institute of High-Energy Physics, Serpukhov, Preprint No. 82-168, 1982.

144. J. Ranft and W. R. Nelson, CERN Report No. HS-RP/031/PP, 1979.

145. A. Van Ginneken, "AEGIS—a program to calculate the average behavior of electromagnetic showers," Fermilab Report No. FN-309, 1979.

146. H. Grote, R. Hagedorn, and J. Ranft, "Atlas of particle spectra," CERN Report, 1970.

147. V. P. Kryuchkov, V. N. Lebedev, and N. V. Mokhov, Institute of High-Energy Physics, Serpukhov, Preprint No. 76-132, 1976.

148. K. L. Brown and Ch. Iselin, Decay TURTLE—a program for simulating charged particle beam transport systems, including decay calculations," CERN Report No. 74-2, 1974.

149. K. Chen *et al.*, Phys. Rev. **166**, 949 (1968).

150. E. Brassi *et al.*, CERN/HERA Report No. 73-1, 1973.

151. N. Metropolis *et al.*, Phys. Rev. **110**, 185 (1958).

152. T. A. Gabriel, R. T. Santoro, and J. Barish, "A calculational method for predicting particle spectra from high-energy nucleon and pion collisions ($E \geqslant 3$ GeV) with protons," ORNL Report No. ORNL-TM-3615, 1971.

153. I. S. Baĭshev, M. A. Maslov, and N. V. Mokhov, in Abstracts of papers presented at the Eighth All-Union Conference on Charged Particle Accelerators, Serpukhov, 1982, p. 92.

154. J. Wachter *et al.*, Phys. Rev. **161**, 971 (1967).

155. R. G. Alsmiller *et al.*, Nucl. Sci. Eng. **36**, 291 (1969).

156. G. S. Levine *et al.*, Part. Accel. **3**, 91 (1972).

157. F. D. Vlatskiĭ *et al.*, Institute of High-Energy Physics, Serpukhov, Preprint No. 74-142, 1974.

158. R. L. Childers, C. D. Zerby, C. M. Fisher, and R. H. Thomas, Nucl. Instrum. Methods **32**, 53 (1965).

159. A. Citron *et al.*, Nucl. Instrum. Methods **32**, 48 (1965).

160. K. Anderson, J. Curry, J. Pilcher, and J. Sidles, "Measurements of cascade of 150 GeV pions in iron,"FNAL Prop. 331, 1974.

161. I. S. Baĭshev, A. S. Makhon'kov, and N. V. Mokhov, "Low-energy radiation component of high-energy accelerators," Institute of High-Energy Physics, Serpukhov, Preprint No. 77-134, 1977.

162. T. W. Armstrong and R. G. Alsmiller, Nucl. Sci. Eng. **38**, 53 (1969).

163. K. O'Brien, "Transverse shielding calculations for the components of a $\frac{1}{2}$ TeV proton synchrotron," USAEC Report No. HASL-199, 1968.

164. R. T. Santoro and T. A. Gabriel, "Nucleon and pion star production from high-energy (> 40-GeV) protons incident on iron," ORNL Report No. ORNL-TM-3355, 1971.

165. L. R. Kimel', M. M. Komochkov, V. P. Sidorin, B. S. Sychev, and A. P. Cherevatenko, Dubna Preprint No. P16-3514, 1967.

166. V. A. Grigor'ev *et al.*, JINR, Dubna, Preprint No. P15-6729, 1972.

167. M. Awschalom, P. J. Gollon, C. Moore, and A. Van Ginneken, Nucl. Istrum. Methods **131**, 235 (1975).

168. P. Sievers, "Measurements of the energy deposition of 200 and 400 GeV/c protons in aluminum and copper," CERN Report No. CERN-TM SPS/ABT/77-1, 1977.

169. T. P. Amineva *et al.*, *Study of Superhigh-Energy Muons* (Nauka, Moscow, 1975).

170. V. I. Balbekov, Yu. P. Dmitrevskiĭ, and O. V. Kurnaev, in Proceedings of the Sixth All-Union Conference on Charged Particles Accelerators, Dubna, 1981, Vol. 1, p. 1.

171. L. Eyges, Phys. Rev. **74**, 1534 (1948).

172. R. G. Alsmiller, F. S. Alsmiller, J. Barish, and Y. Shima, "Muon transport and the shielding of high-energy (⩽500-GeV) proton accelerators," CERN Report No. CERN-71-16, 1971, Vol. 2, p. 601.

173. G. I. Britvich, N. V.Mokhov, and A. V. Uzunyan, "Production and transport of high-energy muons in matter," Institute of High-Energy Physics, Serpukhov, Preprint No. 76-66, 1976.

174. M. A. Maslov, N. V. Mokhov, and A. V. Uzunyan, "Calculation of muon fields at high-energy proton accelerators," Institute of High-Energy Physics, Serpukhov, Preprint No. 82-75, 1982.

175. M. A. Maslov, N. V. Mokhov, and A. V. Uzunyan, "Simulation of high-energy particle trajectories in a medium with an arbitrary magnetic field," Institute of High-Energy Physics, Serpukhov, Preprint No. 78-153, 1978.

176. C. W. Fabjan and T. Ludlam, "Calorimetry in high-energy physics," CERN Preprint No. CERN EP/82-37, 1982.

177. E. Longo and I. Sestili, Nucl. Instrum. Methods **128**, 283 (1975).

178. Yu. D. Prokoshkin, in Proceedings of the Second ICFA Workshop on Possibilities and Limitations of Accelerators and Detectors, 1980, p. 405.

179. L. Federici, N. Nardi, F. Ceradini, and M. Conversi, Nucl. Instrum. Methods **151**, 103 (1978).

180. S. L. Stone *et al.*, Nucl. Instrum. Methods **151**, 387 (1978).

181. A. N. Diddens *et al.*, Nucl. Instrum. Methods **178**, 27 (1980).

182. H. G. Fisher, Nucl. Instrum. Methods **156**, 81 (1978).

183. A. Benvenuti *et al.*, Nucl. Instrum. Methods **125**, 447 (1975).

184. C. W. Fabjan *et al.*, Nucl. Instrum. Methods **141**, 61 (1977).

185. Yu. B. Bushnin *et al.*, Nucl. Instrum. Methods **106**, 493 (1973).

186. Yu. B. Bushnin *et al.*, Nucl. Instrum. Methods **120**, 391 (1974).

187. F. Binon *et al.*, Nucl. Instrum. Methods **188**, 507 (1981).

188. M. Holder *et al.*, Nucl. Instrum. Methods **151**, 69 (1978).

189. V. A. Davydov *et al.*, Nucl. Instrum. Methods **145**, 267 (1977).

190. C. W. Fabjan *et al.*, Nucl. Instrum. Methods **158**, 93 (1979).

191. G. A. Askar'yan and B. A. Dolgoshein, FIAN Preprint No. 160, 1976.

192. G. A. Askar'yan and B. A. Dolgoshein, Pis'ma Zh. Eksp. Teor. Fiz. **25**, 232 (1977) [JETP Lett. **25**, 213 (1977)].

193. B. A. Dolgoshein, in Proceedings of the 1976 DUMAND Summer Workshop, Hawaii, 1976.

194. V. S. Berezinskiĭ and G. T. Zatsepin, Usp. Fiz. Nauk **122**, 3 (1977) [Sov. Phys. Usp. **20**, 1 (1977)].

195. L. Sulak *et al.*, Nucl. Instrum. Methods **161**, 203 (1979).

196. G. N. Sverdlin, *Applied Hydroacoustics* (Shipbuilding, Leningrad, 1976).

197. N. V. Mokhov, O. I. Pogorelko, and V. V. Frolov, "Calculation of the background conditions and radiation heating of superconducting windings in a hyperon experiment at IHEP," Institute of High-Energy Physics, Serpukhov, Preprint No. 114, 1975.

198. D. S. Baranov, A. N. Kalinovskiĭ, Yu. P. Nikitin, R. A. Rzaev, and A. V. Samoĭlov, "Spectra of secondary particles produced as a result of the interaction of 70-GeV/c protons with nuclei of an extended target," Institute of High-Energy Physics, Serpukhov, Preprint No. 74-132, 1974.

199. G. Bohannon, in Proceedings of the High-Intensity Targeting Workshop, Fermilab, 1980, p. 85.

200. J. Kidd *et al.*, in Proceedings of the National Accelerator Conference, Washington, D.C., 1981.

201. A. G. Daĭkovskiĭ, M. A. Maslov, N. V. Mokhov, and A. I. Fedoseev, "Nonstationary thermal fields in superconducting magnet systems produced as a result of the loss of high-energy particles," Institute of High-Energy Physics, Serpukhov, Preprint No. 77-139, 1977.

202. N. V. Mokhov, Zh. Tekh. Fiz. **49**, 1254 (1979) [Sov. J. Tech. Fiz. **24**, 694 (1979)].

203. R. Dixon, N. V. Mokhov, and A. Van Ginneken, "Beam-induced quench study of tevatron dipoles," Fermilab Report No. FN-327, 1980.

204. L. N. Zaĭtsev, Fiz. Elem. Chastits At. Yadra **11**, 525 (1980) [Sov. J. Part. Nucl. **11**, 201 (1980)].

205. M. A. Maslov and N. V. Mokhov, Institute of High-Energy Physics, Serpukhov, Preprint No. 81-128, 1981.

206. V. I. Balbekov *et al.*, in *Proceedings of the Workshop on Possibilities and Limitations of Accelerators and Detectors* (Fermilab, Batavia, 1979).

207. V. N. Lebedev, N. V. Mokhov, and B. S. Sychev, in *Dosimetry and Radiation Shielding*, edited by V. K. Sakharov (Atomizdat, Moscow, 1979), issue 18, p. 152.

208. J. D. Cossairt, N. V. Mokhov, and C. T. Murphy, "Absorbed dose measurements external to thick shielding at a high energy proton accelerator: comparison with Monte Carlo calculations," Fermilab Publ. No. 81/159, 1981; Nucl. Instrum. Methods 197, 465 (1982).

209. N. V. Mokhov, E. L. Potemkin, and E. L. Frolov, At. Energ. **38**, 42 (1975).

210. V. T. Golovachik, E. L. Potemkin, V. N. Lebedev, and V. V. Frolov, Institute of High-Energy Physics, Serpukhov, Preprint No. 74-58, 1974.

211. R. G. Alsmiller, T. W. Armstrong, and W. A. Coleman, "Absorbed and equivalent dose for neutrons in the energy region from 60 to 2000 MeV and for protons in the energy region from 400 to 3000 MeV," ORNL Preprint No. ORNL-2924, Oak Rdige, 1974.

212. G. I. Britvich and N. V. Mokhov, "Energy release from high-energy leptons in matter," Institute of High-Energy Physics, Serpukhov, Preprint No. 75-120, 1975.

213. N. V. Mokhov and E. L. Potemkin, "Distribution of absorbed energy in the $\pi^-$ capture, in *Proceedings of the Third All-Union Conference on Microdosimetry* (MIFI, Moscow, 1979), p. 62.

214. I. S. Baĭshev, V. N. Lebedev, and N. V. Mokhov, At. Energ. **44**, 247 (1978).

# Index

Printed in the United States
By Bookmasters